CSR, Sustainability, Ethics & Governance

In recent years the discussion concerning the relation between business and society has made immense strides. This has in turn led to a broad academic and practical discussion on innovative management concepts, such as Corporate Social Responsibility, Corporate Governance and Sustainability Management. This series offers a comprehensive overview of the latest theoretical and empirical research and provides sound concepts for sustainable business strategies. In order to do so, it combines the insights of leading researchers and thinkers in the fields of management theory and the social sciences – and from all over the world, thus contributing to the interdisciplinary and intercultural discussion on the role of business in society. The underlying intention of this series is to help solve the world's most challenging problems by developing new management concepts that create value for business and society alike. In order to support those managers, researchers and students who are pursuing sustainable business approaches for our common future, the series offers them access to cutting-edge management approaches.

CSR, Sustainability, Ethics & Governance is accepted by the Norwegian Register for Scientific Journals, Series and Publishers, maintained and operated by the Norwegian Social Science Data Services (NSD)

Samuel Chukwujindu Nwokolo • Rubee Singh •
Shahbaz Khan • Anil Kumar • Sunil Luthra

Africa's Path to Net-Zero

Exploring Scenarios for a Sustainable Energy Transition

Springer

Samuel Chukwujindu Nwokolo
University of Calabar
Calabar, Nigeria

Rubee Singh
Institute of Business Management
GLA University
Mathura, India

Shahbaz Khan
Institute of Business Management
GLA University
Mathura, India

Anil Kumar
Guildhall Faculty of Business and Law
London Metropolitan University
London, UK

Sunil Luthra
All India Council for Technical Education (AICTE)
New Delhi, India

ISSN 2196-7075 ISSN 2196-7083 (electronic)
CSR, Sustainability, Ethics & Governance
ISBN 978-3-031-44513-2 ISBN 978-3-031-44514-9 (eBook)
https://doi.org/10.1007/978-3-031-44514-9

This Springer imprint is published by the registered company Springer Nature Switzerland AG
The registered company address is: Gewerbestrasse 11, 6330 Cham, Switzerland

Paper in this product is recyclable.

Foreword

In this insightful book, "Africa's Path to Net-Zero: Exploring Scenarios for a Sustainable Energy Transition", the authors delve into the intricate web of challenges and opportunities that lie ahead for the continent in terms of moving forward towards the Net-Zero goal. Through meticulous research and analysis, the book presents a comprehensive exploration of various scenarios that could pave the way towards achieving a sustainable energy transition in Africa. By examining the unique socioeconomic and environmental factors at play, this book offers valuable insights and practical recommendations for policymakers, businesses, and individuals seeking to contribute to Africa's energy transition.

The book also highlights the need for policy reforms and regulatory frameworks that support the adoption of renewable energy technologies and encourage private sector participation. Additionally, it explores the role of international partnerships and cooperation in facilitating the transfer of knowledge, technology, and financial resources to African countries. It emphasizes the role of capacity building and knowledge sharing to empower local communities and enable them to actively participate in the energy transition process. By addressing these challenges and opportunities, the book offers a comprehensive approach to achieving a sustainable energy transition in Africa.

By understanding the potential strategies proposed in numerous African countries, future leaders and students can gain valuable insights into how to overcome challenges and replicate these achievements in their regions. Moreover, this book provides a comprehensive analysis of the economic benefits that come with transitioning to clean energy sources, empowering future leaders to make informed decisions that prioritize both sustainability and economic growth.

With the future in mind, I recommend this book not only to professionals and future leaders but also to students to help them update their ideas and knowledge as well as understand the challenges and opportunities associated with the transition to clean energy sources. This book offers valuable insights into the potential job creation and investment opportunities that arise from the clean energy sector. By equipping students with this knowledge, they can actively contribute to shaping a

Foreword

In this insightful book, "Africa's Path to Net-Zero: Exploring Scenarios for a Sustainable Energy Transition", the authors delve into the intricate web of challenges and opportunities that lie ahead for the continent in terms of moving forward towards the Net-Zero goal. Through meticulous research and analysis, the book presents a comprehensive exploration of various scenarios that could pave the way towards achieving a sustainable energy transition in Africa. By examining the unique socioeconomic and environmental factors at play, this book offers valuable insights and practical recommendations for policymakers, businesses, and individuals seeking to contribute to Africa's energy transition.

The book also highlights the need for policy reforms and regulatory frameworks that support the adoption of renewable energy technologies and encourage private sector participation. Additionally, it explores the role of international partnerships and cooperation in facilitating the transfer of knowledge, technology, and financial resources to African countries. It emphasizes the role of capacity building and knowledge sharing to empower local communities and enable them to actively participate in the energy transition process. By addressing these challenges and opportunities, the book offers a comprehensive approach to achieving a sustainable energy transition in Africa.

By understanding the potential strategies proposed in numerous African countries, future leaders and students can gain valuable insights into how to overcome challenges and replicate these achievements in their regions. Moreover, this book provides a comprehensive analysis of the economic benefits that come with transitioning to clean energy sources, empowering future leaders to make informed decisions that prioritize both sustainability and economic growth.

With the future in mind, I recommend this book not only to professionals and future leaders but also to students to help them update their ideas and knowledge as well as understand the challenges and opportunities associated with the transition to clean energy sources. This book offers valuable insights into the potential job creation and investment opportunities that arise from the clean energy sector. By equipping students with this knowledge, they can actively contribute to shaping a sustainable future and seize the numerous career prospects that emerge in this rapidly growing industry.

I congratulate the authors for their efforts and contributions.

Operations Management, Head of the Centre for Supply Chain Improvement, College of Business, Law and Social Sciences, The University of Derby, Kedleston Road Campus, Derby, UK

Jose Arturo Garza-Reyes

Preface

Africa’s Path to Net-Zero is a comprehensive analysis of the challenges and opportunities facing the continent as it seeks to transition to a sustainable energy future. The book explores a range of scenarios for achieving net-zero emissions, taking into account the unique social, economic, and political contexts of African countries. This highlights the importance of renewable energy sources such as solar, wind, and hydropower in reducing greenhouse gas emissions and improving access to electricity. The authors also emphasize the need for innovative financing mechanisms and policy frameworks that can support the development of clean energy infrastructure. Throughout the book, case studies from across Africa illustrate both the successes and challenges faced by countries in their efforts to transition to a low-carbon economy. Ultimately, Africa's Path to Net-Zero presents four possible scenarios that are required for Africa's propensity for a net-zero energy transition in this era of decarbonization. This is triggered by the fact that the window is closing and the pathway to a net-zero future for Africa is narrowing to reach a decarbonized economy by 2050 as the transition race is gaining momentum globally. Unfortunately, it is unclear how much of the burden African countries will bear and how they can benefit from an increasingly decarbonized world. There is a risk of African countries being left behind in the global energy transition due to a lack of resources, limited access to technology, poor governance, political instability, and the need for significant capital investment. To achieve this, global partners must provide technical assistance and capacity building to African countries, as well as access to finance and other resources to support the development and deployment of eco-friendly energy sources. The scenarios presented demonstrate that a net-zero energy transition in Africa is achievable but will require the right mix of policy, technology, local financing, foreign investment, expert knowledge, local political backing, policy measures, environmental, sociopolitical, and sociocultural complexities in Africa, as well as the rapidly evolving energy sector. The four potential scenarios that could result in an African net-zero energy transition are aided and harsher scenarios; collaboration and a calculated transition; unaided and self-sufficient in exports; and a tangled and trapped transition. In scenario 1, significant foreign funding for renewable energy will be provided, while in scenario 2, the money earned from the export of fossil fuels will be limited and therefore will not be able to sustain Africa's economy as much as it would in scenario 1. In scenario 4, there would be significant harm to Africa by foreign investments in renewable energy, and exports of fossil fuels would be severely constrained. Generally, financing foreign renewable energy sources in Africa will not be as beneficial to the continent in the long run as it would be if those limitations were present. Scenario 1, which includes strict limitations on the money earned from exporting fossil fuels, should be the priority in Africa to promote long-term economic stability and sustainability. Scenario 2, which does not include any such limitations, could have a negative impact if it is not monitored or regulated properly. This could lead to a situation where African countries use foreign investments to fund unsustainable practices while earning large profits from the export of fossil fuels. This could have a long-term negative impact on the environment and lead to Africa not utilizing its resources sustainably. African countries must ensure that the money earned from the export of fossil fuels is properly monitored and regulated to prevent this kind of situation from arising and to ensure that it is being used to fund renewable energy sources and sustainable practices that are in line with the environmental goals of African countries. In scenario 3, Africa will be self-sufficient and export-free, but remaining export-free would lead to greater reliance on fossil fuels and a vicious cycle of poverty, pollution, and dependency on foreign aid. This approach would not only promote economic growth and development but also ensure a more sustainable energy future for the continent. To achieve self-sufficiency and prevent potential exploitation of the continent's resources, Africa must set limitations and regulations on foreign investments and ensure that those investing in renewable energy are held accountable for their actions. Furthermore, it is crucial for African countries to invest in education and training programs that focus on renewable energy technologies. By equipping its citizens with the necessary skills and knowledge, Africa can create a workforce that is capable of developing and implementing sustainable energy solutions. In addition, governments should provide incentives for private sector investment in renewable energy projects, such as tax breaks and subsidies. This would encourage businesses to shift away from fossil fuels and towards cleaner energy sources. Finally, Africa must prioritize the protection of its natural resources by implementing strict environmental regulations and enforcing them rigorously. By taking these steps, Africa can break free from its reliance on fossil fuels and pave the way for a more sustainable future.

Calabar, Nigeria — Samuel Nwokolo
Mathura, India — Rubee Singh
London, UK — Shahbaz Khan
New Delhi, India — Anil Kumar
Sunil Luthra

Acknowledgement

We would like to express our gratitude to all those who have contributed to the creation of this book. First, we thank the researchers and experts who have shared their knowledge and insights with us, providing valuable information on the current state of energy production and consumption in Africa. We also extend our appreciation to the organizations and institutions that have supported our work, including government agencies, nongovernmental organizations, and academic institutions. Their funding and resources have enabled us to conduct extensive research and analysis on the potential pathways towards a sustainable energy transition in Africa.

Finally, we would like to acknowledge the invaluable contributions of our colleagues and editors, who have worked tirelessly to ensure that this book is informative, engaging, and accessible to a wide range of readers. We hope that this book will inspire further research and action towards a net-zero future for Africa. We would like to express our sincere gratitude to the experts and stakeholders who contributed their valuable insights and perspectives to this book. Their contributions have been instrumental in shaping our understanding of the challenges and opportunities involved in achieving a sustainable energy transition in Africa. We hope that this book will serve as a useful resource for policymakers, practitioners, and researchers working towards a net-zero future for the continent.

We would like to thank the staff at Springer, in particular Prof. Samuel O Idowu (Series Editor), Dr. Prashanth Mahagaonkar (Executive Editor), and the entire team of Springer for their help and support.

About the Book

Africa's Path to Net-Zero is a thought-provoking book that explores the complex issue of sustainable energy transition in Africa. It provides a comprehensive overview of the current energy landscape and explores various scenarios for achieving net-zero emissions. The book highlights the challenges faced by African countries in transitioning to clean energy and provides practical solutions to overcome these challenges. It emphasizes the need for public–private partnerships to finance large-scale renewable energy projects and create a sustainable future for Africa. Prioritizing clean energy and environmental stewardship can achieve net-zero goals while promoting economic growth and social development. The authors also emphasize the importance of international cooperation and collaboration in achieving a sustainable energy future for Africa. The book is a must-read for policymakers, researchers, and anyone interested in understanding the critical role Africa plays in global efforts to combat climate change. It highlights four potential scenarios for Africa to achieve net-zero emissions by 2050: a rapid transition to renewable energy, increased investment in energy efficiency, electrification of transportation, and the adoption of circular economy principles. The book also emphasizes the importance of political will and international cooperation in achieving these goals. The book presents a detailed roadmap for transitioning Africa's energy systems to renewable sources, reducing emissions from transportation and industry, and protecting natural ecosystems. The authors emphasize the importance of collaboration between governments, businesses, and communities in achieving these goals. They showcase successful case studies from South Africa and Kenya, demonstrating the potential for renewable energy projects to create jobs and stimulate economic growth. However, the authors acknowledge challenges such as securing financing for large-scale projects and developing new technologies. To address these challenges, they offer recommendations for policy changes and investment strategies that can accelerate progress towards net-zero emissions. Overall, Africa's Path to Net-Zero provides a compelling vision for a sustainable future that benefits both people and the planet.

Contents

About the Authors

Samuel Chukwujindu Nwokolo is a distinguished figure in the field of renewable energy and climate science activism. With a strong educational background, he holds a bachelor's degree and a postgraduate diploma (PGD) in physics (renewable energy option). His passion for sustainable solutions and his deep understanding of the environmental challenges we face today have propelled him to the forefront of the movement. Through his extensive research and practical experience, he has become an advocate for renewable energy adoption and an expert in climate science. His expertise lies not only in theoretical knowledge but also in hands-on implementation, making him a valuable asset in the pursuit of a greener future. His dedication to raising awareness about renewable energy's potential to combat climate change is evident through his numerous speaking engagements, publications, and active involvement in various environmental organizations. He has been actively involved in renewable energy and climate science research for nine years. Research covered aspects of solar energy, smart energy systems, and sustainable development studies (sustainable cities and resources, building integrated photovoltaics for sustainable architecture). He has published more than 50 research articles in many reputed journals, including Scopus Q1, SCI, and ABDC-A & B category. He is the author of three academic books and a reviewer for Applied Energy, MDPI Atmosphere, Heliyon, Talyor & Francis Journals of International Journal of Ambient Energy, International Journal of Green Energy, and other Scopus Q1, SCI, and ABDC-A & B category journals. He is a graduate student researcher at the Department of Physics at the University of Calabar in Nigeria, where he also instructs graduate students about research and development, renewable energy, and the science of climate change.

Rubee Singh is an Assistant Professor of Human Resource at Institute of Business Management, GLA University, India. She is a postdoctoral researcher (D. Lit) at Kumaun University, Nainital, India. She has been awarded Honorary Doctorate in Human Rights from Royal American University, USA, for her dedication, hard work, and knowledge. The George Washington University of Peace, USA, awarded her as Chartered Global Peace Building Professional (CGPP). She is a United Nations Volunteer and a contributing author for Voices of Youth (UNICEF). She is an honorary life member of the Women Economic Forum and National Secretary of (Youth Wing) International Anti-Terrorism Movement (IATM), India. She holds a Doctorate in HR from Noida International University, India, an MBA in HR from Dr A.P. J Abdul Kalam Technical University, India. Dr. Singh is a Managing Editor of IJIEMR & IJARSH, India. She is serving on the Editorial Board of more than 30 international journals. She serves as the Advisory Board Member of the Heliyon Journal, Elsevier, and a reviewer of many reputed journals, such as the European Journal of Innovation Management (EJIM), British Food Journal, and Sustainability. She is a life member of the International Association of Engineers, Thailand, IMRTC USA, ISRD UK, etc. She is a bestselling author with significant experience as an excellent Editor, Eminent researcher, great academician, Global Ambassador (Children's) & Social Activist. She has presented many papers at national and international conferences, published 15 academic books, and published more than 55 research papers in SCI, ABDC, Scopus, & UGC index journals. Her areas of research are sustainability, digital HRM, sustainable HRM, circular economy, digital CE, net zero, Industry 4.0, artificial intelligence, and corporate social responsibility. Being a thought leader, her books are honoured by being added to the library, Ohio Christian University, University of California, Corporacion Universitaria Unisabaneta, UMass Chan Medical School, Tunix Community College, USA, Dow University of Health Sciences, Pakistan, University of Port Harcourt, Nigeria. Her book titled "Training & Development" was added to the MBA syllabus of "The Institute of Aeronautical Engineering" (Jawaharlal Nehru Technological University, Hyderabad). Dr. Singh has been honoured with many prestigious awards, including Most Influential Woman Leaders in 2022, Child Ambassador Award of The Year 2022, Iconic Faculty Award 2019, International Young Researchers Award 2021, Young Scientist Award 2020, Nari Ratan Award 2023, India Book of Records 2020 & Asia Book of Records Holder for 2021. The list of awards and appreciation is long, but Dr. Singh has always prioritized her work over recognition, making her unique and commendable in her field.

Shahbaz Khan is an Assistant Professor and Area-Co Chair of Operation Management/Business Analytics/Supply Chain at the Institute of Business Management, GLA University, India. He has published more than 65 research articles in many reputed journals including Scopus, SCI, ABDC-A category such as Business Strategy and Environment, Annals of Operation Research, Operations Management Research, etc. He has an excellent research track record with 2,000 citations on Google Scholar and H-index-25. He has obtained his PhD and M. Tech degree with specialization in Industrial and Production Engineering from Jamia Millia Islamia, India. He is a gold medallist in M. Tech. He is a life member of the International Association of Engineers, Thailand. His research interests are supply chain management, human resource management, halal supply chain management, circular economy & sustainability, industry 4.0, performance measurement, process management and benchmarking, decision modelling for sustainable business. He is author of two academic textbooks named "International Human Resource Management" and "Strategic CSR in 21st Century". He is serving as the Advisory Board Member of the Heliyon Journal, Elsevier and Guest Editor for Special Issue in International Journal of Robotics Research and Application, Green Technologies and Sustainability. He is a reviewer of many reputed journals, such as Journal of Business Research, Trends in Food Science & Technology, International Journal of Fuzzy Systems, Engineering Applications of Artificial Intelligence, Soft Computing, and Sustainability.

Anil Kumar is a reader at Guildhall School of Business and Law, London Metropolitan University (LMU), London, United Kingdom. For the last twelve years, he has been associated with teaching and research. Before joining LMU, he was a Postdoctoral Research Fellow in Decision Sciences at the Centre for Supply Chain Improvement, University of Derby, UK. He earned his Ph.D. in Management Science from ABV-Indian Institute of Information Technology and Management, Gwalior, India. He graduated in Mathematics (Hons) and MSc (Mathematics) from Kurukshetra University, India. He earned his Master of Business Administration (MBA) and qualified National Eligibility Test (NET) in June 2011. He has contributed 150 research papers in refereed journals and conferences at the international level. He has published his research work in A*/A as per ABDC indexed and 4*/3* as per ABS journals such as the International Journal of Operations & Production Management (IJOPM), Transportation Research Part E Logistics and Transportation Review (TRE), Journal of Business Research (JBR), International Journal of Production Research (IJPR), Government Information Quarterly (GIQ), Production Planning & Control (PPC), Technological Forecasting and Social Change (TFSC), Business Strategy and the Environment (BSE), Journal of Retailing and Consumer Services (JRCS), Journal of Cleaner Production (JCP), etc. Because of his contributions to the literature, he has been listed in the "World Ranking Top 2%" researchers by Stanford University-Elsevier list in October 2022, and he has also ranked in 4th place in the UK and Northern as per CABS Academic Journal Guide 2021 ranking, based on a publication P-ranking and considering articles published since 2022 & considering authors' affiliation. He is an Associate Editor of Heliyon-Information Science (Elsevier) (Impact Factor: 3.776, SCIE/SSCI & Scopus Indexed) and the International Journal of Mathematical, Engineering and Management Sciences.

Sunil Luthra is working as the Director, All India Council for Technical Education, New Delhi, India. Prior to that, he worked as the Director-Principal at Ch. Ranbir Singh State Institute of Engineering & Technology, Jhajjar, Haryana, India. He is also Visiting Professor at the Centre for Supply Chain Improvement, University of Derby, Derby, United Kingdom. He has contributed over 180 research papers in international and national journals and conferences at international and national levels. His name appeared in the top 2% global list of researchers prepared by Stanford University. He has an excellent research track record (over 200 SCI publications; received more than 15,700 citations on Google Scholar; H-index-65 on Google Scholar). He has received many awards and honours for research and teaching. He is working as a Guest Editor of many reputed journals, such as Journal of Cleaner Production, Technology Forecasting & Social Change, Production Planning & Control, Resources Policy, Resources, Conservation and Recycling, International Journal of Logistics Research and Applications, and Annals of Operations Research. He is on the editorial board of many reputed journals. He has published books with reputed publishers such as CRC Press, Taylor & Francis Group, LLC, and New Age International Publisher (P) Ltd., etc. His research interests are Sustainable Production and Consumption, Green/Sustainable/Circular Supply Chain Management (GSCM/SSCM/CSCM), Industry 4.0, Circular Economy and Industrial Engineering, etc.

Chapter 1
Introduction: Africa's Net Zero Transition

1.1 Introduction: Africa's Net Zero Transition

Will Africa profit from an inclusive share of global investment and development, or will it proceed to lag below the world standards highlighted in the Goals for Sustainable Development (SDGs) because of a lack of a strong economy? As the continent with the highest projected population growth (United Nations, 2021), Africa faces a number of unique challenges in transitioning to net zero (Ouedraogo, 2019). To ensure that Africa benefits from net zero investment, a number of initiatives have been proposed to ensure that African countries receive their fair share of the net zero transition (IRENA, 2021). These include providing access to affordable clean energy (Nwokolo et al., 2022a), strengthening the capacity of African countries to manage their own transition (Nwokolo et al., 2022b), and fostering investment in carbon-efficient infrastructure (Chukwujindu et al., 2022). However, there is still much to be done for African countries to benefit from the global net zero transition and secure a cleaner, more prosperous future. To ensure that African countries are able to fully capitalize on the global net zero transition, it is essential to focus on a few key initiatives (Kilkiş, 2015), such as investing in renewable energy infrastructure (Nwokolo et al., 2023a), improving access to finance for sustainable investments (Nwokolo et al., 2023b), and utilizing the continent's wealth of natural resources in a sustainable manner (Nwokolo et al., 2023c).

These initiatives, if properly implemented, have the potential to create vast economic opportunities for African countries (Samuel, 2017) and their citizens while also helping to secure a healthier (Obiwulu et al., 2020a), more sustainable future (Obiwulu et al., 2020b). African countries must make it a priority to invest in renewable energy infrastructure (Obiwulu et al., 2022), increase access to finance for sustainable investments (Nwokolo et al., 2017a), and utilize their natural resources in a responsible manner (Nwokolo et al., 2017a). To achieve these goals, African countries must prioritize investing in renewable energy infrastructure

S. C. Nwokolo et al., *Africa's Path to Net-Zero*, CSR, Sustainability, Ethics & Governance, https://doi.org/10.1007/978-3-031-44514-9_1

(Nwokolo & Otse, 2019), improving access to finance for sustainable investments (Nwokolo et al., 2018a), and utilizing their natural resources sustainably (Nwokolo et al., 2018b) to fully capitalize on the global net zero transition and create vast economic opportunities for their citizens (Nwokolo et al., 2018c) while also helping to secure a healthier (Oyewo et al., 2019), more sustainable future (Oyewo et al., 2020).

Furthermore, while fossil fuels still represent an important source of energy (Pablo-Romero et al., 2021) and economic security in Africa (Ramagoma & Adendorff, 2016), they offer the potential for far greater long-term economic returns (Rennkamp & Boyd, 2015). In light of this, the development of renewable energy sources has become increasingly important to Africa's long-term economic development (Okem, 2016) and security (Omar et al., 2022). By doing so, African countries can benefit from greater economic returns (Perlaviciute et al., 2021), reduced environmental impacts (Omer, 2009), and improved public health outcomes for their citizens (Mekonnen & Sarwat, 2017). The continent of Africa is in a unique position to take advantage of the growth potential that renewable energy sources offer (Mutezo & Mulopo, 2021) and capitalize on the many benefits that come along with them (Ngondjeb et al., 2020). This shift toward clean energy will also result in improved public health outcomes for African citizens (Maupin, 2017). In addition to these advantages, renewable energy sources can help African countries become more self-sufficient by reducing their dependence on imported oil and gas (Manjong et al., 2021) and providing them with greater control over their energy security (Kouton, 2021).

However, this revenue is not without cost, as the increased production of fossil fuels leads to higher emissions (Rissman et al., 2020), thereby contributing to climate change (Rogelj et al., 2015). The increased concentration of greenhouse gases has led to a range of environmental problems (Klunne, 2012) and extreme weather events that are far-reaching (Kincer et al., 2022), not only resulting in physical changes to our environment but also having serious implications for global public health and food security (IRENA, 2012). All of these risks must be taken into consideration when assessing the overall cost of fossil fuel production (IRENA, 2020a) and export (IRENA, 2020b). This is just as true today, as we must weigh the potential long-term consequences of fossil fuel production and export (IRENA, 2013). This is tempting, but we must consider the long-term implications for global public health (Kourgiozou et al., 2021), food security (Lowe and Drummond, 2022), and climate change when making decisions about our energy sources (Hassan et al., 2021) to make an informed decision about whether to invest in this industry (Hassan et al., 2022). Investing in the fossil fuel industry is a complicated issue that must be carefully considered in terms of both environmental (Hrnčić et al., 2021) and societal impacts (Kivimaa et al., 2020).

Despite the fact that Africa is only responsible for a small portion of global emissions (Ogbulezie et al., 2017), it is being held to the same standard as developed countries with much higher emissions (Agbor et al., 2023a) and, as such, will be expected to incur disproportionately higher costs (Lucas, 2022) and risks to achieve the same decarbonization goal (Mavrigiannaki et al., 2021). This highlights the imbalanced (Mestemaker et al., 2020) and inequitable nature (Nelson & Allwood, 2021) of global efforts to combat climate change (Nhuchhen et al., 2022). The majority of African countries have been excluded from the discourse on energy transition (Nikas et al., 2022) and do not have access to the same resources (Agbor et al., 2023b), technology (Abade, 2020), or infrastructure as their developed counterparts (Agency, 2020). As a result, developing African countries are at a distinct disadvantage in regard to achieving decarbonization goals (Agreement et al., 2021). This unequal distribution of resources (Abdul-Salam et al., 2022) and access to technology (Ahl et al., 2020) have been particularly damaging for African countries (Amadi et al., 2020), with high costs of renewable energy technologies (Allen et al., 2020), limited access to capital markets (Amado et al., 2018), and high levels of poverty further exacerbating the situation (Ameli et al., 2021a).

In light of these disparities, there is an urgent need for international cooperation and aid to ensure a fair (Ameli et al., 2021b) and equitable net zero vision for African countries (Azevedo et al., 2021). The net zero transition must also consider the local context (Bak et al., 2017) and capacity of African countries to facilitate the most efficient (Barry et al., 2011) and effective implementation of clean technologies (Bataille et al., 2018). International cooperation is critical to ensuring an equitable and just transition to renewable energy sources in Africa. The international community has an obligation to African countries to provide the support they need to successfully transition to renewable energy sources by providing the necessary resources (Bataille, 2020), technology (BP Energy, 2020), and capital required for successful implementation (Bray et al., 2022) while also providing guidance and expertise to ensure that African countries are able to benefit from the transition (Caetano et al., 2017).

Inadequate electricity access in most African countries highlights the continent's need for substantial investments to build reliable energy infrastructure. This lack of access to electricity is a major barrier to economic development, as it hinders industrial growth, productivity, and job creation. The lack of electricity also has a profound effect on the quality of life for millions of people living in Africa. This is why the governments of Africa have increasingly prioritized the construction of new energy infrastructure and the modernization of existing systems in recent years to ensure that electricity becomes more widely available across the continent (Nwokolo et al., 2023d). In doing so, African governments are recognizing the importance of electricity access in facilitating economic development and improving the quality of life for millions of people living in Africa. To successfully bridge the gap between Africa and the rest of the world, it is essential that governments prioritize investments in reliable energy infrastructure. Investments in reliable energy infrastructure are essential for Africa to bridge the gap between itself and the rest of the world and ensure that its citizens are provided with access to basic amenities such as electricity for lighting, air conditioning, and other household appliances that are commonplace in most of the developed world (Freeman, 2021). Investing in reliable energy infrastructure is also an important step in reducing inequality and bridging the digital divide.

The challenges associated with meeting this demand-supply gap will be exacerbated by a number of factors. It is imperative for African countries to invest in the development of clean energy while ensuring that they have the necessary financing and technological resources. This investment in the energy sector will help realize the SDGs (Cillari et al., 2021) and promote economic growth (Coelho et al., 2010), improved standards of living (Cotterman et al., 2021), and greater access to reliable energy across the continent (da Silva et al., 2018). Moving forward, it is essential that African countries are provided with the necessary support to achieve their energy demand-supply gap goals to ensure the successful implementation of the SDGs (Delafield et al., 2021) and promote economic growth (Derkenbaeva et al., 2022), improved standards of living (Deutch, 2020), and reliable access to energy for all (Felder et al., 2021). African countries must invest in both renewable and traditional energy sources, ensuring that the necessary financing and technological resources are available to meet the demand-supply gap targets set out in the SDGs.

Overall, the effects of the pandemic have exacerbated economic hardship in many oil-exporting countries and will likely continue to do so in 2021 and beyond. Around the world, there is a great need to focus on aid and recovery efforts to prevent further economic hardship for these countries in 2021 and beyond. As a result, governments and international organizations must work together to ensure that these countries receive the aid needed to alleviate their economies.

Additionally, it is important for developed countries to offer their support through trade agreements and investment opportunities that can help stimulate economic growth in affected regions. However, it is also crucial to address the root causes of these crises, such as political instability, corruption, and a lack of access to education and healthcare (Ferreira et al., 2016). By addressing these underlying issues, we can work toward creating more resilient (Florent Baarsch & Schaeffer, 2019) and sustainable economies that are better equipped to handle future challenges (Fol & Le, 2012). It is important for the international community to come together and collaborate on solutions that prioritize the well-being of all individuals (Ford & Hardy, 2020) and promote long-term economic stability (Ford et al., 2021). The World Bank and International Monetary Fund have already stepped up to offer emergency funding and debt relief to some of the most severely affected nations during this difficult time (United Nations, 2021). However, more work is needed to ensure that all nations receive the assistance they require to effectively recover from the pandemic's effects. The United Nations, in particular, plays a special role in promoting international cooperation and providing technical assistance to nations all over the world in their efforts to recover their economies (United Nations, 2021). In addition to debt relief, countries must also prioritize investments in healthcare systems and social safety nets to protect vulnerable populations. The pandemic has exposed the deep inequalities that exist within and between countries, and it is crucial that we address these disparities in our recovery efforts. This includes addressing issues such as access to vaccines, education, and digital infrastructure. Furthermore, we must also work toward building more resilient economies that can withstand future shocks. This involves promoting sustainable development practices and investing in green technologies. The UN can play a crucial role in promoting international cooperation and solidarity in the face of this global pandemic (United Nations, 2021). This includes facilitating the equitable distribution of vaccines and medical supplies as well as supporting efforts to address the root causes of the crisis, such as poverty, inequality, and climate change. In addition, the UN can help build resilience and preparedness for future pandemics by strengthening health systems and investing in research and development for new treatments and vaccines (United Nations, 2021).

As the international community works together to effectively and efficiently respond to this global pandemic, the United Nations is uniquely positioned to protect the health and safety of its citizens as well as manage the economic consequences of this unprecedented crisis. By providing the necessary support, the United Nations can play a vital role in helping countries rebuild their economies and livelihoods and restore a sense of normalcy as quickly and effectively as possible. Through collaboration and coordination with member states, the United Nations can leverage its expertise and resources to address the multifaceted challenges posed by COVID-19 (United Nations, 2021). This includes providing technical assistance and guidance on public health measures, facilitating the equitable distribution of vaccines and medical supplies, and supporting the most vulnerable populations affected by the pandemic. Additionally, the UN can work with governments to develop policies that promote economic recovery and stability while also ensuring that social safety nets are in place to protect those most impacted by job losses and economic downturns (United Nations, 2021). By taking a comprehensive approach to addressing this crisis, the United Nations can help build a more resilient global community that is better prepared to handle future challenges.

As such, there needs to be increased access to sources of long-term capital, such as the Green Climate Fund or other public and private sources that are capable of financing large-scale renewable energy projects. This access to long-term capital can facilitate the realization of Africa's net zero ambitions. This would also lead to job creation, as the development of renewable energy infrastructure would require a skilled workforce. Additionally, the shift toward renewable energy sources would reduce the continent's dependence on fossil fuels, which would not only improve air quality but also reduce the negative impact of climate change on African economies. With proper planning and implementation, Africa could become a leader in renewable energy and pave the way for other developing regions to follow suit. It is imperative that African governments prioritize investment in renewable energy infrastructure and work toward achieving net zero emissions by 2050.

There needs to be an increase in long-term capital investments in renewable energy projects that are tailored specifically for African economies. This can be achieved through the provision of technical (French, 2020) and financial assistance to African countries and the establishment of regulatory frameworks that incentivize long-term investments in renewable energy projects on the continent. Such long-term capital investments could have a transformative effect on African countries by increasing access to clean and reliable energy and improving the resilience of their energy systems (Garcia, 2008). Long-term capital investments in African renewable energy projects can help bridge the gap between African countries and the rest of the world by increasing access to clean and reliable energy, improving the resilience of their energy systems (Godin et al., 2021), and creating new employment opportunities in the renewable energy sector (Gozlugol, 2022). In addition to these advantages, such investments could provide African countries with greater autonomy in terms of energy production (Greig & Uden, 2021) as well as greater control over the types of energy they can use (Grobbelaar et al., 2014) while simultaneously reducing greenhouse gas emissions and reliance on imported sources of energy (Hafner et al., 2021). Furthermore, given the current state of African energy infrastructure and the urgent need for increased access to clean and reliable energy, long-term capital investments in African renewable energy projects could help spur economic development on the continent by providing a much-needed injection of capital as well as a more stable environment for businesses to operate in. Given the potential benefits of long-term capital investments in African renewable energy projects, it is clear that such investments could have a profoundly positive impact on the continent overall, making them a critical component of any comprehensive strategy for sustainable development in Africa.

To ensure a successful transition toward net zero in Africa by 2050, these investments must be targeted and strategic, taking into account local context and energy needs while at the same time creating jobs and economic opportunities for African citizens to ensure energy security and reduce energy poverty in the region. To ensure a successful transition to net zero by 2050, African governments must also commit to developing robust and sustainable regulatory frameworks that initiatives while protecting their citizens from the unintended consequences of such investments, such as displacement and disruption of traditional livelihoods. Additionally, African governments must also create incentives and climate change mitigation projects for initiatives to ensure buy-in from citizens and a successful transition to net zero in 2050. This will require a concerted effort from all stakeholders, including civil society organizations, academia, and the media. African governments must also work together to share best practices and coordinate their efforts to ensure that they are able to meet their commitments under the Paris Agreement. In addition, they must prioritize the development of climate-resilient infrastructure and agriculture systems that can withstand the impacts of climate change. This will require significant investment in research and development, as well as training programs for farmers and other stakeholders. Finally, African governments must work with international partners to secure funding for these initiatives, including through the Green Climate Fund and other multilateral institutions.

This pressure has resulted in a decrease in financial support for fossil fuel-related projects and the addition of various forms of carbon-related tax requirements, which have the potential to limit African countries' ability to take advantage of global supply chain opportunities and even exclude them entirely due to their inability to be price competitive. Africa is facing numerous policy pressures as developed economies accelerate their net zero journeys, reducing export revenue opportunities and further limiting long-term budgeting capacity due to their reliance on the export of oil and gas resources. These global policy pressures have created a challenging economic environment for African countries, which are heavily reliant on the export of oil and gas resources. This will require significant investment in infrastructure and education to build the necessary capacity and expertise. Additionally, African countries can explore new trade partnerships and markets to reduce their dependence on traditional export destinations. Encouraging entrepreneurship and innovation can also help spur economic growth and create new opportunities for job creation. Finally, African governments must prioritize transparency and good governance to attract foreign investment and ensure that the benefits of economic growth are shared equitably among all citizens. While these steps may be challenging, they are essential for African countries to achieve long-term economic stability and prosperity in a rapidly changing global landscape.

These steps will require African countries to make substantial investments in new renewable energy infrastructure and adopt regulations that incentivize businesses to engage in low-carbon activities while also incentivizing the adoption of new technologies to reduce emissions from oil and gas production and consumption. To successfully transition their economies away from oil and gas dependence, African countries must also work to ensure that their populations have access to the resources and skills necessary for participating in low-carbon African countries and must also seek to build the capacity of their populations by increasing access to education and training, especially in the net zero economy. In addition, African countries must ensure that they have the resources to develop low-carbon economies through investments in renewable energy and other low-carbon technologies.

We examine Africa's current energy landscape and propose scenarios for renewable energy development. This global transition's path will have far-reaching consequences for African economies, populations, and the environment. As we investigate these scenarios, we learn the following: Unlocking this potential requires significant investments in infrastructure and policy changes that incentivize the development of renewable energy sources. African countries must prioritize the development of clean energy sources to meet the growing demand for electricity while reducing their reliance on fossil fuels. This transition will require a shift toward decentralized energy systems that empower local communities and reduce dependence on centralized power grids. Additionally, African countries must work toward creating a more just and equitable energy system that benefits all members of society, particularly those who have been historically marginalized.

In addition, investing in renewable energy sources can create new job opportunities and boost economic growth in African countries. However, this transition requires significant policy changes and investment in infrastructure. Governments must prioritize the development of renewable energy sources and create incentives for private sector investment. This can be achieved through tax breaks, subsidies, and other financial incentives. Furthermore, there needs to be a focus on education and awareness campaigns to promote the benefits of renewable energy and encourage public support for these initiatives. Ultimately, a net zero sustainable energy future is necessary not only for the environment but also for the social and economic well-being of African countries. With proper planning and investment, Africa can become a leader in renewable energy production and pave the way toward a more sustainable future for all.

References

Abade, L. P. (2020). *Climate change awareness: Role of Education*. https://doi.org/10.1007/978-3-319-95885-9_122

Abdul-Salam, Y., Kemp, A., & Phimister, E. (2022). Energy transition in the UKCS–modelling the effects of carbon emission charges on upstream petroleum operations. *Energy Economics., 108*, 105898. https://doi.org/10.1016/j.eneco.2022.105898

Ogbulezie, J. C., James, U. O., & Chukwujindu, N. S. (2017). A review of regression models employed for predicting diffuse solar radiation in North-Western Africa. *Trends in Renewable Energy*. https://doi.org/10.17737/tre.2017.3.2.0042

Agbor, M., Udo, S., Ewona, I., Nwokolo, S., & Ogbulezie, J. A. (2023a). Potential impacts of climate change on global solar radiation and PV output using the CMIP6 model in West Africa. *Cleaner Engineering and Technology, 13*, 100630. https://doi.org/10.1016/j.clet.2023.100630

Agbor, M., Udo, S., Ewona, I., Nwokolo, S., Ogbulezie, J., Amadi, S., & Billy, U. (2023b). Effects of angstrom-Prescott and Hargreaves-Samani coefficients on climate forcing and solar PV technology selection in West Africa. *Trends in Renewable Energy, 9*(1), 78–106. https://doi.org/10.17737/tre.2023.9.1.00150

Agency, I. R. E. (2020). Renewable power generation costs in 2019. *International Renewable Energy Agency*.

Agreement, P., Republic, C., Kingdom, U., States, U., Principles, S. K., Zero, I. N., Secretariat, I. E. A., Iea, T., Iea, T., Iea, T., Energy, C., & Programme, T. (2021). Seven key Principles for implementing net Zero. *International Energy Agency*.

Ahl, A., Yarime, M., Goto, M., Chopra, S. S., Kumar, N. M., Tanaka, K., & Sagawa, D. (2020). Exploring blockchain for the energy transition: Opportunities and challenges based on a case study in Japan. *Renewable and Sustainable Energy Reviews., 117*, 109488. https://doi.org/10.1016/j.rser.2019.109488

Allen, S. R., Cohen, B., & Hammond, G. P. (2020). Indicative appraisal of clustered microgenerators for a low-carbon transition in the UK building sector. *Global Transitions., 2*, 83. https://doi.org/10.1016/j.glt.2020.06.001

Amadi, S., Dike, T., & Nwokolo, S. (2020). Global solar radiation characteristics at Calabar and Port Harcourt cities in Nigeria. *Trends in Renewable Energy., 6*, 101. https://doi.org/10.17737/tre.2020.6.2.00114

Amado, M., Poggi, F., Amado, A. R., & Breu, S. (2018). E-city web platform: A tool for energy efficiency at urban level. *Energies, 11*. https://doi.org/10.3390/en11071857

Ameli, N., Dessens, O., Winning, M., Cronin, J., Chenet, H., Drummond, P., Calzadilla, A., Anandarajah, G., & Grubb, M. (2021a). *A climate investment trap in developing economies*. Nature Portfolio.

Ameli, N., Dessens, O., Winning, M., Cronin, J., Chenet, H., Drummond, P., Calzadilla, A., Anandarajah, G., & Grubb, M. (2021b). Higher cost of finance exacerbates a climate investment trap in developing economies. *Nature Communications, 12*, 4046. https://doi.org/10.1038/s41467-021-24305-3

Azevedo, I., Bataille, C., Bistline, J., Clarke, L., & Davis, S. (2021). Introduction to the special issue on net-Zero Energy systems. *Energy and Climate Change., 2*, 100066. https://doi.org/10.1016/j.egycc.2021.100066

Bak, C., Bhattacharya, A., Edenhofer, O., & Knopf, B. (2017). Towards a comprehensive approach to climate policy, sustainable infrastructure, and finance. *Economics., 11*. https://doi.org/10.5018/economics-ejournal.ja.2017-33

Barry, M. L., Steyn, H., & Brent, A. (2011). *Selection of renewable energy technologies for Africa: Eight case studies in Rwanda, Tanzania and Malawi*. Renewable Energy. https://doi.org/10.1016/j.renene.2011.04.016

Bataille, C., Åhman, M., Neuhoff, K., Nilsson, L. J., Fischedick, M., Lechtenböhmer, S., Solano-Rodriquez, B., Denis-Ryan, A., Stiebert, S., Waisman, H., Sartor, O., & Rahbar, S. (2018). A review of technology and policy deep decarbonization pathway options for making energy-intensive industry production consistent with the Paris Agreement. *Journal of Cleaner Production, 187*, 960. https://doi.org/10.1016/j.jclepro.2018.03.107

Bataille, C. G. F. (2020). Physical and policy pathways to net-zero emissions industry. In *Wiley Interdisciplinary Reviews: Climate Change*. https://doi.org/10.1002/wcc.633

BP Energy. (2020). Energy Outlook 2020 edition explores the forces shaping the global energy transition out to 2050 and the surrounding that. In *BP Energy Outlook 2030, Statistical review*. British Petroleum.

Bray, R., Mejía Montero, A., & Ford, R. (2022). Skills deployment for a 'just' net zero energy transition. *Environmental Innovation and Societal Transitions., 42*, 395. https://doi.org/10.1016/j.eist.2022.02.002

Caetano, T., Merven, B., Hartley, F., & Ahjum, F. (2017). Decarbonization and the transport sector: A socioeconomic analysis of transport sector futures in South Africa. *Journal of Energy in Southern Africa., 28*. https://doi.org/10.17159/2413-3051/2017/v28i4a2945

Chukwujindu, S., Umunnakwe, A., & Ogbulezie, J. C. (2022). Hybridization of statistical machine learning and numerical models for improving beam, diffuse and global solar radiation prediction. *Cleaner Engineering and Technology, 9*(May), 100529. https://doi.org/10.1016/j.clet.2022.100529

Cillari, G., Franco, A., & Fantozzi, F. (2021). Sizing strategies of photovoltaic systems in nZEB schemes to maximize the self-consumption share. *Energy Reports., 7*, 6769. https://doi.org/10.1016/j.egyr.2021.09.117

Coelho, B., Oliveira, A. C., & Mendes, A. (2010). Concentrated solar power for renewable electricity and hydrogen production from water–A review. *Energy and Environmental Science, 3*, 1398. https://doi.org/10.1039/b922607a

Cotterman, T., Small, M. J., Wilson, S., Abdulla, A., & Wong-Parodi, G. (2021). Applying risk tolerance and sociotechnical dynamics for more realistic energy transition pathways. *Applied Energy., 291*, 116751. https://doi.org/10.1016/j.apenergy.2021.116751

da Silva, P. P., Cerqueira, P. A., & Ogbe, W. (2018). Determinants of renewable energy growth in sub-Saharan Africa: Evidence from panel ARDL. *Energy, 156*, 45. https://doi.org/10.1016/j.energy.2018.05.068

Delafield, G., Donnison, C., Roddis, P., Arvanitopoulos, T., Sfyridis, A., Dunnett, S., Ball, T., & Logan, K. G. (2021). Conceptual framework for balancing society and nature in net-zero energy transitions. *Environmental Science and Policy., 125*, 189. https://doi.org/10.1016/j.envsci.2021.08.021

Derkenbaeva, E., Halleck Vega, S., Hofstede, G. J., & van Leeuwen, E. (2022). Positive energy districts: Mainstreaming energy transition in urban areas. *Renewable and Sustainable Energy Reviews., 153*, 111782. https://doi.org/10.1016/j.rser.2021.111782

Deutch, J. (2020). Is net Zero carbon 2050 possible? *Joule., 4*, 2237. https://doi.org/10.1016/j.joule.2020.09.002

Felder, F. A., Unel, B., & Dvorkin, Y. (2021). Modelling strategic objectives and behavior in the transition of the energy sector to inform policymaking. *Electricity Journal., 34*, 107009. https://doi.org/10.1016/j.tej.2021.107009

Ferreira, M., Almeida, M., Rodrigues, A., & Silva, S. M. (2016). Comparing cost-optimal and net-zero energy targets in building retrofit. *Building Research and Information., 44*, 188. https://doi.org/10.1080/09613218.2014.975412

Baarsch, F., & Schaeffer, M. (2019). Climate change impacts on Africa ' s economic growth. *African Development Bank*, 1–169.

Fol, Y. L. (2012). Renewable Energy transition for a sustainable future in Namibia. In *Aalborg University*.

Ford, R., & Hardy, J. (2020). Are we seeing clearly? The need for aligned vision and supporting strategies to deliver net-zero electricity systems. *Energy Policy, 147*, 111902. https://doi.org/10.1016/j.enpol.2020.111902

Ford, R., Maidment, C., Vigurs, C., Fell, M. J., & Morris, M. (2021). Smart local energy systems (SLES): A framework for exploring transition, context, and impacts. *Technological Forecasting and Social Change, 166*, 120612. https://doi.org/10.1016/j.techfore.2021.120612

Freeman, R. (2021). Modelling the sociopolitical feasibility of energy transition with system dynamics. *Environmental Innovation and Societal Transitions., 40*, 486. https://doi.org/10.1016/j.eist.2021.10.005

French, S. (2020). The role of Zero and low carbon hydrogen in enabling the Energy transition and the path to net Zero greenhouse gas emissions. *Johnson Matthey Technology Review*. https://doi.org/10.1595/205651320x15910225395383

Garcia, D. (2008). The climate security divide: Bridging human and national security in Africa. *African Security Review., 17*, 2. https://doi.org/10.1080/10246029.2008.9627480

Godin, K., Sapinski, J. P., & Dupuis, S. (2021). The transition to net zero energy (NZE) housing: An integrated approach to market, state, and other barriers. *Cleaner and Responsible Consumption., 3*, 100043. https://doi.org/10.1016/j.clrc.2021.100043

Gozlugol, A. A. (2022). The clash of "E" and "S" of ESG: Just transition on the path to net zero and the implications for sustainable corporate governance and finance. *Journal of World Energy Law and Business., 15*, 1. https://doi.org/10.1093/jwelb/jwab039

Greig, C., & Uden, S. (2021). The value of CCUS in transitions to net-zero emissions. *Electricity Journal., 34*, 107004. https://doi.org/10.1016/j.tej.2021.107004

Grobbelaar, S., Gauché, P., & Brent, A. (2014). Developing a competitive concentrating solar power industry in South Africa: Current gaps and recommended next steps. *Development Southern Africa., 31*, 475. https://doi.org/10.1080/0376835X.2014.891971

Hafner, S., Jones, A., & Anger-Kraavi, A. (2021). Economic impacts of achieving a net-zero emissions target in the power sector. *Journal of Cleaner Production., 312*, 127610. https://doi.org/10.1016/j.jclepro.2021.127610

Hassan, M. A., Bailek, N., Bouchouicha, K., Ibrahim, A., Jamil, B., Kuriqi, A., & Nwokolo, S. C. (2022). Evaluation of energy extraction of PV systems affected by environmental factors under real outdoor conditions. *Theoretical and Applied Climatology, August., 150*, 715. https://doi.org/10.1007/s00704-022-04166-6

Hassan, M. A., Bailek, N., Bouchouicha, K., & Nwokolo, S. C. (2021). Ultrashort-term exogenous forecasting of photovoltaic power production using genetically optimized nonlinear autoregressive recurrent neural networks. *Renewable Energy, 171*(February), 191–209. https://doi.org/10.1016/j.renene.2021.02.103

Hrnčić, B., Pfeifer, A., Jurić, F., Duić, N., Ivanović, V., & Vušanović, I. (2021). Different investment dynamics in energy transition towards a 100% renewable energy system. *Energy, 237*, 121526. https://doi.org/10.1016/j.energy.2021.121526

International Renewable Energy Agency. (2020). Renewable power generation costs in 2019. *International Renewable Energy Agency*.

International Renewable Energy Agency (IRENA). (2013). Renewable power generation costs in 2012: An overview. *Irena*, 1–8.

International Renewable Energy Agency (IRENA). (2020a). Renewable energy and climate pledges: Five years after the Paris Agreement.

International Renewable Energy Agency. (2020b). Renewable power generation costs in 2019. *International Renewable Energy Agency*. https://www.irena.org/publications/2020/Jun/Renewable-Power-Costs-in-2019

IRENA. (2021). World energy transitions outlook: 1.5 degrees pathway. *International Renewable Energy Agency*. https://irena.org/publications/2021/March/World-Energy-Transitions-Outlook

Irena Working Paper. (2012). *Renewable Energy Techlogies: Cost analysis series, hydropower*. International Renewable Energy Agency.

Ituen, E. E., Esen, N. U., Nwokolo, S. C., & Udo, E. G. (2012). Prediction of global solar radiation using relative humidity, maximum temperature and sunshine hours in Uyo, in The Niger Delta region, Nigeria. *Advances in Applied Science Research, 3*(4), 1923–1937.

Kilkiş, Ş. (2015). Exergy transition planning for net-zero districts. *Energy, 92*, 515. https://doi.org/10.1016/j.energy.2015.02.009

Kincer, J., Moss, T., & Thurber, M. (2022). A coal renaissance is not coming to Africa. *World Development Perspectives., 25*, 100375. https://doi.org/10.1016/j.wdp.2021.100375

Kivimaa, P., Primmer, E., & Lukkarinen, J. (2020). Intermediating policy for transitions towards net-zero energy buildings. *Environmental Innovation and Societal Transitions., 36*, 418. https://doi.org/10.1016/j.eist.2020.01.007

Klunne, W. J. (2012). Current status and future developments of small and micro hydro in southern Africa. *Energy Policy*.

Komninos, N. (2022). Net Zero Energy districts: Connected intelligence for carbon-neutral cities. *Land., 11*. https://doi.org/10.3390/land11020210

Kourgiozou, V., Commin, A., Dowson, M., Rovas, D., & Mumovic, D. (2021). Scalable pathways to net zero carbon in the UK higher education sector: A systematic review of smart energy systems in university campuses. *Renewable and Sustainable Energy Reviews, 147*, 111234. https://doi.org/10.1016/j.rser.2021.111234

Kouton, J. (2021). The impact of renewable energy consumption on inclusive growth: Panel data analysis in 44 African countries. *Economic Change and Restructuring, 54*, 145. https://doi.org/10.1007/s10644-020-09270-z

Lowe, R. J., & Drummond, P. (2022). Solar, wind and logistic substitution in global energy supply to 2050–Barriers and implications. *Renewable and Sustainable Energy Reviews, 153*, 111720. https://doi.org/10.1016/j.rser.2021.111720

Lucas, A. R. (2022). Canada's carbon energy overhang. *Journal of Energy and Natural Resources Law, 40*, 17. https://doi.org/10.1080/02646811.2021.2012349

Manjong, N. B., Oyewo, A. S., & Breyer, C. (2021). Setting the pace for a sustainable Energy transition in Central Africa: The case of Cameroon. *IEEE Access., 9*, 145435. https://doi.org/10.1109/ACCESS.2021.3121000

Maupin, A. (2017). The SDG13 to combat climate change: An opportunity for Africa to become a trailblazer? *African Geographical Review., 36*, 131. https://doi.org/10.1080/19376812.2016.1171156

Mavrigiannaki, A., Pignatta, G., Assimakopoulos, M., Isaac, M., Gupta, R., Kolokotsa, D., Laskari, M., Saliari, M., Meir, I. A., & Isaac, S. (2021). Examining the benefits and barriers for the implementation of net zero energy settlements. *Energy and Buildings., 230*, 110564. https://doi.org/10.1016/j.enbuild.2020.110564

Mekonnen, Y., & Sarwat, A. I. (2017). Renewable energy supported microgrid in rural electrification of Sub-Saharan Africa. *Proceedings–2017 IEEE PES-IAS PowerAfrica Conference: Harnessing Energy, Information and Communications Technology (ICT) for Affordable Electrification of Africa, PowerAfrica 2017*. https://doi.org/10.1109/PowerAfrica.2017.7991293.

Mestemaker, B., Van Den Heuvel, H., & Gonçalves Castro, B. (2020). Designing the zero emission vessels of the future: Technologic, economic and environmental aspects. *International Shipbuilding Progress., 67*, 5. https://doi.org/10.3233/ISP-190276

Mutezo, G., & Mulopo, J. (2021). A review of Africa's transition from fossil fuels to renewable energy using circular economy principles. *Renewable and Sustainable Energy Reviews, 137*-(November 2020), 110609. https://doi.org/10.1016/j.rser.2020.110609

Nelson, S., & Allwood, J. M. (2021). The technological and social timelines of climate mitigation: Lessons from 12 past transitions. *Energy Policy, 152*, 112155. https://doi.org/10.1016/j.enpol.2021.112155

Ngondjeb, D. Y., Atewamba, C., & Macalou, M. (2020). Insights on Africa's future in its transition to the green economy. In *Advances in african economic, social and political development*. https://doi.org/10.1007/978-3-030-44180-7_13

Nhuchhen, D. R., Sit, S. P., & Layzell, D. B. (2022). Towards net-zero emission cement and power production using molten carbonate fuel cells. *Applied Energy., 306*, 118001. https://doi.org/10.1016/j.apenergy.2021.118001

Nikas, A., Xexakis, G., Koasidis, K., Acosta-Fernández, J., Arto, I., Calzadilla, A., Domenech, T., Gambhir, A., Giljum, S., Gonzalez-Eguino, M., Herbst, A., Ivanova, O., van Sluisveld, M. A. E., Van De Ven, D. J., Karamaneas, A., & Doukas, H. (2022). Coupling circularity performance and climate action: From disciplinary silos to transdisciplinary modelling science. *Sustainable Production and Consumption., 30*, 269. https://doi.org/10.1016/j.spc.2021.12.011

Nwokolo, S. C., Amadi, S. O., Obiwulu, A. U., Ogbulezie, J. C., & Eyibio, E. E. (2022a). Prediction of global solar radiation potential for sustainable and cleaner energy generation using improved angstrom-Prescott and Gumbel probabilistic models. *Cleaner Engineering and Technology, 6*(January), 100416. https://doi.org/10.1016/j.clet.2022.100416

Nwokolo, S. C., Obiwulu, A. U., & Ogbulezie, J. C. (2022b). Machine learning and analytical model hybridization to assess the impact of climate. *Physics and Chemistry of the Earth, 130*(February), 103389. https://doi.org/10.1016/j.pce.2023.103389

Nwokolo, S. C., & Ogbulezie, J. C. (2018a). A qualitative review of empirical models for estimating diffuse solar radiation from experimental data in Africa. *Renewable and Sustainable Energy Reviews, 92*, 353. https://doi.org/10.1016/j.rser.2018.04.118

Nwokolo, S. C., & Ogbulezie, J. C. (2018b). A qualitative review of empirical models for estimating diffuse solar radiation from experimental data in Africa. *Renewable and Sustainable Energy Reviews, 92*. https://doi.org/10.1016/j.rser.2018.04.118

Nwokolo, S. C., & Ogbulezie, J. C. (2018c). A quantitative review and classification of empirical models for predicting global solar radiation in West Africa. *Beni-Suef University Journal of Basic and Applied Sciences, 7*(4), 367–396. https://doi.org/10.1016/j.bjbas.2017.05.001

Nwokolo, S. C., Ogbulezie, J. C., & Obiwulu, A. U. (2022c). Impacts of climate change and meteo-solar parameters on photosynthetically active radiation prediction using hybrid machine learning with physics-based models. *Advances in Space Research, 70*, 3614. https://doi.org/10.1016/j.asr.2022.08.010

Nwokolo, S., & Ogbulezie, J. (2017a). A critical review of theoretical models for estimating global solar radiation between 2012-2016 in Nigeria. *International Journal of Physical Research, 5*(2), 60. https://doi.org/10.14419/ijpr.v5i2.8160

Nwokolo, S., & Ogbulezie, J. (2017b). Modelling the influence of cloudiness on diffuse horizon-tal irradiation under various sky conditions in Nigeria. *International Journal of Physical Research, 5*(2), 91. https://doi.org/10.14419/ijpr.v5i2.8312

Nwokolo, S., & Otse, C. (2019). Impact of sunshine duration and clearness index on diffuse solar radiation estimation in mountainous climate. *Trends in Renewable Energy, 5*(3), 307–332. https://doi.org/10.17737/tre.2019.5.3.00107

Nwokolo, S. C., Obiwulu, A. U., Amadi, O., & Ogbulezie, J. C. (2023a). Assessing the impact of soiling, tilt angle, and solar radiation on the performance of solar PV systems. *Trends in Renewable Energy, 9*(2), 121–137. https://doi.org/10.17737/tre.2023.9.2.00156

Nwokolo, S. C., Ogbulezie, J. C., & Ushie, O. J. (2023b). A multimodel ensemble-based CMIP6 assessment of future solar radiation and PV potential under various climate warming scenarios. *Optik, 285*(May), 170956. https://doi.org/10.1016/j.ijleo.2023.170956

Nwokolo, S. C., Proutsos, N., & Meyer, E. L. (2023c). Machine learning and physics-based hybridization models for evaluation of the effects of climate change and urban expansion on photosynthetically active radiation. *Atmosphere, 14*(687), 1–36. https://doi.org/10.3390/atmos14040687

Nwokolo, S. C., Proutsos, N., & Meyer, E. L. (2023d). Machine learning and physics-based hybridization models for evaluation of the effects of climate change and urban expansion on photosynthetically active radiation. *Atmosphere, 14*(687), 1–36. https://doi.org/10.3390/atmos14040687

Obiwulu, A. U., Chendo, M. A. C., Erusiafe, N., & Nwokolo, S. C. (2020a). Implicit meteorological parameter-based empirical models for estimating back temperature solar modules under varying tilt-angles in Lagos, Nigeria. *Renewable Energy, 145*, 442. https://doi.org/10.1016/j.renene.2019.05.136

Obiwulu, A. U., Erusiafe, N., Olopade, M. A., & Nwokolo, S. C. (2020b). Modelling and optimization of back temperature models of mono-crystalline silicon modules with special focus on the effect of meteorological and geographical parameters on PV performance. *Renewable Energy, 154*, 404. https://doi.org/10.1016/j.renene.2020.02.103

Obiwulu, A. U., Erusiafe, N., Olopade, M. A., & Nwokolo, S. C. (2022). Modelling and estimation of the optimal tilt angle, maximum incident solar radiation, and global radiation index of the photovoltaic system. *Heliyon, 8*(6), e09598. https://doi.org/10.1016/j.heliyon.2022.e09598

Okem, A. E. (2016). South Africa's transition to a low-carbon economy: The role of cooperatives. *Journal of Social Sciences., 49*, 257. https://doi.org/10.1080/09718923.2016.11893619

Omar, A. I., Khattab, N. M., & Abdel Aleem, S. H. E. (2022). *Optimal strategy for transition into net-zero energy in educational buildings: A case study in El-Shorouk City, Egypt*. Sustainable Energy Technologies and Assessments. https://doi.org/10.1016/j.seta.2021.101701

Omer, A. M. (2009). Energy and sustainable development in Sudan. In *Energy costs, international developments and new directions*.

Ouedraogo, N. S. (2019). Opportunities, barriers and issues with renewable energy development in Africa: A comprehensible review. *Current Sustainable/Renewable Energy Reports, 6*, 52. https://doi.org/10.1007/s40518-019-00130-7

Oyewo, A. S., Aghahosseini, A., Ram, M., & Breyer, C. (2020). Transition towards decarbonized power systems and its socioeconomic impacts in West Africa. *Renewable Energy, 154*, 1092. https://doi.org/10.1016/j.renene.2020.03.085

Oyewo, A. S., Aghahosseini, A., Ram, M., Lohrmann, A., & Breyer, C. (2019). Pathway towards achieving 100% renewable electricity by 2050 for South Africa. *Solar Energy., 191*, 549. https://doi.org/10.1016/j.solener.2019.09.039

Pablo-Romero, M. P., Sánchez-Braza, A., & Galyan, A. (2021). Renewable energy use for electricity generation in transition economies: Evolution, targets and promotion policies. *Renewable and Sustainable Energy Reviews., 138*, 110481. https://doi.org/10.1016/j.rser.2020.110481

Perlaviciute, G., Steg, L., & Sovacool, B. K. (2021). A perspective on the human dimensions of a transition to net-zero energy systems. *Energy and Climate Change., 2*, 100042. https://doi.org/10.1016/j.egycc.2021.100042

Ramagoma, J., & Adendorff, C. (2016). Managing a transition to green energy sources: The perspectives of energy practitioners in the southern African development community region. *Journal of Energy in Southern Africa., 27*, 77. https://doi.org/10.17159/2413-3051/2016/v27i4a1638

Rennkamp, B., & Boyd, A. (2015). Technological capability and transfer for achieving South Africa's development goals. *Climate Policy., 15*, 12. https://doi.org/10.1080/14693062.2013.831299

Rissman, J., Bataille, C., Masanet, E., Aden, N., Morrow, W. R., Zhou, N., Elliott, N., Dell, R., Heeren, N., Huckestein, B., Cresko, J., Miller, S. A., Roy, J., Fennell, P., Cremmins, B., Koch Blank, T., Hone, D., Williams, E. D., de la Rue du Can, S., & Helseth, J. (2020). Technologies and policies to decarbonize global industry: Review and assessment of mitigation drivers through 2070. *Applied Energy., 266*, 114848. https://doi.org/10.1016/j.apenergy.2020.114848

Rogelj, J., Luderer, G., Pietzcker, R. C., Kriegler, E., Schaeffer, M., Krey, V., & Riahi, K. (2015). Energy system transformations for limiting end-of-century warming to below 1.5 °C. *Nature Climate Change, 5*, 519. https://doi.org/10.1038/nclimate2572

Samuel Chukwujindu, N. (2017). A comprehensive review of empirical models for estimating global solar radiation in Africa. *Renewable and Sustainable Energy Reviews, 78*(October), 955–995. https://doi.org/10.1016/j.rser.2017.04.101

United Nations. (2021). The Energy transition, towards the achievement of SDG 7 and net-zero emissions. In *UN energy*.

Chapter 2
Global Investment and Development in Africa

2.1 Introduction

Increased economic growth, better infrastructure, a rise in the average standard of living, and increased access to healthcare, education, and other essential services have all been made possible by international investment and development in Africa over the last few decades. Africa has benefited from greater economic stability, greater prosperity, and greater integration into the global economy as a result of foreign development and investment than would have been possible otherwise. As a result, most African countries and their citizens are now more equitable and prosperous, with residents having access to the same opportunities and rights as their peers around the globe while also living better lives than they would have without the influx of investment and development. This influx of investment and development has helped bring about positive change for African countries and their citizens, allowing them to achieve economic stability (Africanacity of International Investment Law, 2021), prosperity (AFRICA–EU, 2022), and social progress in ways that were previously unimaginable (Amendolagine & Prota, 2021). The changes made to African nations due to international investment and development have allowed the continent's inhabitants access to a level of healthcare (Asiedu, 2004), education (Baltag, 2009), and economic opportunity (Capital Investment in Africa, 1939) not available prior to these investments. This has been especially beneficial for women in Africa (Nel & Humphrys, 1999), as increased economic stability and investment have allowed them to be better educated, enjoy greater rights and freedom (Nkuna, 2015), and participate in the labor force at a level not seen before international investment and development arrived in the region (Owens & Lane, 2014). This has had a profoundly positive impact on the lives of African women (Pamba, 2023) and their families, allowing them to escape poverty (Pedler, 1955), find meaningful employment (Qumba, 2018), and participate in their nations' economic growth in ways never before thought possible (Peat, 2018). By providing women with economic and educational opportunities, international investment and development

S. C. Nwokolo et al., *Africa's Path to Net-Zero*, CSR, Sustainability, Ethics & Governance, https://doi.org/10.1007/978-3-031-44514-9_2

have helped bring about a new era of empowerment for African women that would have otherwise been unattainable without these investments in the region.

Due to foreign development and investment, Africa's economy has undergone significant change (Cheng, 2023), and its people now have easier access to resources and services (Cheung et al., 2012) that were previously limited or even unavailable to them. This access has allowed African people to more easily pursue their dreams, be better educated (Clark, 1997), and live more comfortable and secure lives (Ewelukwa Ofodile, 2014). Without foreign development and investment, many African countries and their citizens would likely still be living in poverty (Fosu, 2013) and without access to many of the same opportunities and rights enjoyed by others in other parts of the world (Gazzini, 2009). This transformation has been remarkable and is a testament to the power of foreign investment and development in Africa to empower and improve the lives of millions of people over the past few decades. However, it is also important to note that this progress is still far from complete, as there are still deep-seated social (Grekou & Owoundi, 2020), economic (Hepburn, 2015), and political issues that remain unresolved (Gunter, 2006) and hinder further progress in many parts of the continent. Thus, while significant progress has been made on the African continent through foreign investment and development, it is still a long way off from achieving all that can be accomplished in terms of true, lasting, and equitable development for African nations and their citizens. To continue on this trajectory of progress, continued foreign investment and development must be implemented alongside initiatives that focus on addressing the more challenging issues of inequality (Khumon, 2016), poverty (Klimova, 2022), political unrest (Leiter, 2021), and social injustice that have been prevalent in many parts of the continent for far too long (Li, 2014). In addition, stronger collaborations and partnerships between governments, nongovernmental organizations (NGOs), and the private sector must be established to ensure that everyone works together to build a sustainable future for African nations and their citizens. Furthermore, greater access to education and employment opportunities, along with improved infrastructure and healthcare systems, are vital to promote growth and development in African countries, as these are essential components for any successful economy. As such, continued and increased foreign investment and development on the African continent is still crucial to unlocking the full potential of African nations and their citizens and ensuring the realization of true progress and prosperity on the continent for the people living there for the first time.

In this chapter, we discuss the benefits, challenges, and likely strategies of continued and increased global investment and development in Africa to bring about greater economic, social, and political stability for African nations and their citizens and ultimately unlock the full potential of the African continent and its people.

2.2 Benefits of Global Investment and Development in Africa

2.2.1 Increased Economic Growth

One of the benefits of global investment and development in Africa is increased economic growth in the region, which has been seen in improved infrastructure, a rise in the average standard of living, and greater access to healthcare, education, and other essential services. This economic growth has been beneficial to both African countries and their citizens, creating greater economic opportunities (Rogerson, 2010), increasing income and job security (Rogerson & Rogerson, 2010), and providing the resources necessary for people to invest in their health, education, and businesses for the long term. These economic benefits have allowed African nations to become more competitive in the global marketplace and to create a higher quality of life for their citizens while also providing increased access to goods, services, and technologies from other countries, which have allowed people to experience a better quality of life and increased access to basic services such as healthcare and education that they may not have had access to otherwise.

Ultimately, this economic growth has enabled African countries to become more integrated into the global economy, to enjoy increased economic stability and to experience a greater degree of prosperity than would have been possible without foreign investment and development in the region. All in all, the investment and development in Africa have been a major benefit for the region and its citizens, as it has helped to facilitate economic growth, create jobs, and increase access to basic services while also increasing the opportunities for citizens to access goods, services, and technologies that were previously not available to them, enabling them to benefit from the opportunities and advancements of the global economy while also providing a platform for African countries to gain an edge in the global marketplace as they become more competitive and are able to create higher-quality products and services that are more in line with what consumers around the world expect.

This has resulted in a more equitable and prosperous Africa, where citizens are able to experience the same rights and opportunities as their peers around the world while also enjoying a greater quality of life as a result of the economic benefits that foreign investment and development have brought to the region. This investment and development have had a profound effect on the economic landscape of African countries and have helped to drive growth in the region, allowing citizens to benefit from increased access to goods, services, technologies, healthcare, and education, as well as a higher standard of living that could be questioned and would not have been possible without foreign investment and development in the region. This has been an enormous benefit to the African continent as a whole and has contributed to a more equitable, prosperous Africa in which its citizens are able to access the same goods and services as those in other parts of the world while also enjoying improved economic security and stability as a result of the investment that has come into the region.

The investment and development in Africa over the past century have been a major positive for the continent, allowing citizens to experience unprecedented levels of access to goods, services, and technologies, as well as providing an opportunity for countries to become more competitive in the global marketplace and allowing its citizens to enjoy the benefits of economic growth and stability. This has ultimately led to a more equitable and prosperous Africa, one in which citizens are now able to experience the same rights and opportunities as those around the world while also enjoying increased quality of life due to increased access to goods, services, technologies, healthcare, and education—that would not have been possible without foreign investment and development in the region over the past century.

In addition, the growth in African economies has had a significant impact on employment, allowing citizens to enjoy more secure and stable jobs that pay fair wages and give them access to the same benefits as those in other parts of the world, thus creating more equitable societies and allowing them to enjoy the same rights and freedoms that people around the world take for granted. It has also led to providing a foundation for increased entrepreneurship and economic independence for African countries, allowing them to be more competitive in the global marketplace and allowing their citizens to have a greater say in their own destinies. It also culminated in allowing more people to become self-sufficient and start businesses while also enjoying increased access to resources that are necessary for their development and the ability to participate in the global economy, which can help drive the economy forward in the years to come. These changes have been monumental in the advancement of African societies, leading to increased prosperity, a better quality of life, and an overall more equitable continent, paving the way for a more secure future for African citizens and their families and greater economic opportunities in unprecedented ways. This has all been made possible due to the increased availability of technology and the internet, which have allowed African countries to catch up with the rest of the world in terms of development and access to resources that were once impossible to obtain in the past. The last few decades have been especially significant for Africa, as its citizens have seen drastic changes in their way of life and access to essential services such as healthcare, education, and economic growth. With the technological revolution, African countries have had unprecedented access to resources that were previously impossible to obtain due to geographical and financial limitations, which has enabled them to propel their development and growth forward in ways that were unimaginable in the past.

Despite this progress, there is still much to be done to ensure that all African countries have access to the same quality of life that other developed nations have enjoyed for years. This development has not come without its challenges, however, and African countries must now contend with issues such as corruption, political instability, and the need to embrace new technologies to fully benefit from globalization. To achieve true progress, African countries must prioritize the creation of strong economic policies that address the needs of their citizens while encouraging businesses to invest in the nation's infrastructure and education systems to create a more equitable and prosperous society for all.

Furthermore, African countries must focus on promoting good governance to increase public trust and attract more foreign direct investment to stimulate job growth, economic diversification, and improved standards of living for the continent's population. African countries must also recognize the importance of strengthening their relationships with other nations and organizations to access foreign capital and international markets and improve their long-term economic development prospects. Moreover, African countries must also foster a culture of entrepreneurship and innovation to encourage the growth of small businesses while also focusing on utilizing technology to reduce poverty, increase access to healthcare and education, and develop better systems for tackling climate change to make progress in the twenty-first century.

By doing these things, African countries can create an environment that promotes growth and development, reduces poverty, and creates a more equitable society for all, which will result in a stronger, more prosperous future for all its citizens. African countries must focus on fostering sustainable economic growth and development by enacting meaningful reforms that ensure good governance, developing strong relationships with other nations and organizations, promoting a culture of entrepreneurship and innovation, utilizing technology to reduce poverty, increasing access to healthcare and education, and developing better systems for tackling climate change to ensure a brighter future for the continent and its people. This is the path that African countries must take to ensure a brighter future for the continent and its people in the twenty-first century.

2.2.2 *Increased Access to Resources*

Increased access to resources is one benefit of international investment and development in Africa that can significantly improve the quality of life for African citizens by creating jobs and enabling access to necessities such as food (Nwokolo et al., 2022a), water (Nwokolo and Ogbulezie, 2018a), and healthcare (Nwokolo and Ogbulezie, 2018b). Additionally, foreign investment can bring with it the capital and technology that African nations need to spur economic development (Nwokolo and Ogbulezie, 2018c) and improve the overall quality of life for their citizens. Furthermore, by providing much-needed resources, international investment and development can promote education and skills training, allowing African citizens to develop the knowledge and expertise necessary for more advanced technology-driven (Nwokolo et al., 2017a) economic activities such as manufacturing, information technology (Nwokolo et al., 2017b), and services, which can help drive economic growth, reduce poverty, and create more opportunities for African citizens on the continent. It can also help African nations increase their economic competitiveness in the global market, which could lead to greater economic integration and better trading conditions for African countries, leading to further investments and development in the region (Nwokolo & Otse, 2019). Ultimately, international investment and development have the potential to promote economic growth and improve the quality of life in Africa by increasing access to resources, bringing in capital and technology, promoting education, creating more job opportunities, reducing poverty, improving economic competitiveness in the region, and ultimately creating a more prosperous and stable economic environment for African citizens in unprecedented ways. Overall, foreign investment and development can be extremely important in promoting economic growth, enhancing the standard of living for Africans living on the continent, and assisting African countries to become more competitive in the global market as well as more self-sufficient and independent.

2.2.3 Improved Infrastructure

One benefit of foreign investment and development in Africa is improved infrastructure, which can provide access to safe drinking water, modern healthcare facilities, and reliable electricity (Nwokolo et al., 2023a), all of which are critical for socioeconomic development and improving the lives of millions of people in the region. The improvements to infrastructure resulting from foreign investment and development can also facilitate the movement of people and goods (Nwokolo et al., 2023b), create jobs, and spur economic growth in communities that have limited access to capital (Nwokolo et al., 2022b), resources (Nwokolo et al., 2023c), and services. Additionally, with improved infrastructure comes improved access to education, as well as better and more reliable communication networks (Nwokolo et al., 2023d), which are essential for the long-term development of a society. It is clear that the benefits of foreign investment and development are widespread (Obiwulu et al., 2020a) and should be carefully considered when developing strategies for economic growth in Africa (Obiwulu et al., 2020b).

Foreign investment and development can also lead to increased access to international markets, allowing local businesses to expand their product and service offerings, reach more customers, and generate higher revenues, which can then be reinvested in the local economy (Obiwulu et al., 2022), resulting in great economic growth and development for these communities. While it is undeniable that foreign investment and development can bring many advantages to African communities, it is important to ensure that these efforts are managed responsibly, with a view toward sustainability and the long-term prosperity of the local population and not just short-term gains for investors. Ultimately, foreign investment and development can have a transformative impact on economies in Africa, but if these opportunities are not managed responsibly, they can create serious risks and potentially damaging consequences for African communities.

It is also essential to consider the potential environmental, social, and cultural impacts that foreign investment and development can bring and ensure that these are managed responsibly and not overlooked in the pursuit of economic growth. Ultimately, it is important to approach foreign investment and development in African communities with an eye toward responsible management and ensuring that these efforts are focused on long-term prosperity, sustainability, and responsible management of any potential impacts to maximize the potential benefits while minimizing any potential negative consequences for the African community. This means that any foreign investment and development efforts in African communities must be mindful of the potential long-term impacts, both positive and negative, they may have on local economies, as well as on the environmental, social, and cultural contexts in which these investments are taking place. Additionally, African governments should also ensure that their citizens are included in decision-making processes and are provided with adequate resources to protect their rights and interests, as well as support for local businesses that may be affected by any foreign investment and development projects taking place in the region. Taking all these things into consideration, it is important to recognize that while foreign investment and development in African communities may have the potential to bring much-needed economic growth and job creation, it is also essential to ensure that these investments are managed responsibly and with an emphasis on long-term, sustainable prosperity for all members of the African community to truly create a lasting and positive impact for the people who live there.

In conclusion, foreign investment and development in African communities should be viewed as an opportunity to create long-term prosperity and sustainability for the people of the region by offering economic growth, job creation, and the inclusion of citizens in decision-making processes while also ensuring that these investments are managed responsibly and with respect for the local environment, culture, and traditions. By doing so, foreign investment and development in African communities can truly bring positive change that will benefit the people of the region for years to come while still providing support and resources to protect the rights and interests of its citizens, as well as local businesses that may be affected by any foreign investment or development projects taking place in the region.

2.2.4 Improved Access to Education

Access to education has improved thanks to foreign investment and development in Africa over the last two decades, allowing more individuals to gain knowledge and skills that could help propel their communities forward both economically and socially. This increased access to education has resulted in a higher literacy rate, the opening of more vocational and technical schools, and an overall increase in the quality of educational institutions in the region. All of these developments have had a positive impact on the educational prospects of individuals in Africa and have contributed to the economic and social development of their communities and nations as a whole. By providing educational opportunities to citizens who may have previously been unable to access them, foreign investment and development in Africa have not only provided individuals with the opportunity to gain knowledge and skills but also allowed them to gain the confidence needed to lead their own lives and to participate in the economic and social progress of their communities and countries in a meaningful and sustainable way. In this way, foreign investment and development in Africa have made an invaluable contribution to the continent's educational and social progress. Over the past two decades, access to education in Africa has been steadily increasing due to foreign investment and development initiatives from the international community, resulting in improved educational outcomes for many African countries. This increased access to education has not only enabled individuals to acquire more knowledge and skills but also provided them with the self-belief needed to participate in the economic, political, and social transformation of their societies on a meaningful and lasting basis. This increased access to education has allowed individuals to become empowered with the ability to make decisions that positively influence their lives and communities, which, in turn, has led to increased economic development and a more prosperous future for the continent as a whole. This increased access to education has been a major factor in the continent's transformation, as it has enabled Africans to move away from dependency on foreign aid and assistance and become more self-sufficient, as well as providing individuals with the skills and knowledge needed to be competitive in the global economy. Moreover, this increased access to education has provided individuals with the opportunity to take part in a more inclusive form of governance and decision-making, thereby leading to a more democratic and equitable society where everyone's voices can be heard and their needs addressed on an equal basis. This empowerment of individuals through access to education has been a major catalyst in the transformation of Africa, leading to increased economic development, a greater level of self-sufficiency and inclusivity, a more equitable society, and a brighter future for the continent as a whole. In conclusion, the increased access to education in Africa has had a profound effect on the transformation of the continent, as it has enabled individuals to take part in a more equitable society, develop their own skills and knowledge, become more self-sufficient, and ultimately lead to a brighter future for all Africans. It has also empowered African citizens to take control of their own future and have the ability to become entrepreneurs and move away from foreign aid and assistance, as well as providing opportunities to pursue their own ambitions and become more economically independent, and as a result, it has helped to bridge the gap between different socioeconomic groups, allowing for greater equality and a more prosperous society for all Africans. This increased access to education has also helped to bridge the gap between different socioeconomic groups, allowing for greater equality and a more prosperous society for all Africans by enabling individuals to pursue their own ambitions and gain more economic independence while also providing them with the resources they need to move away from relying on foreign aid and assistance and thus leading to a more prosperous and equitable continent as a whole. As a result, the increased access to education in Africa has not only resulted in an increase in economic development and self-sufficiency but also helped to create a more equitable society and a brighter future for all Africans by providing them with the necessary resources to further develop their skills and knowledge, become entrepreneurs, and be in control of their own future. This empowerment of individuals through access to education has had a positive, transformative effect on the African continent as a whole, leading to increased economic development and self-sufficiency as well as a more equitable society and a brighter future for the African people by moving away from the traditional view of women as only fit for a supporting role and instead providing them with the opportunity to pursue their own ambitions, create their own paths, and become empowered to make decisions for themselves and take part in the creation of a more prosperous, equitable, and independent continent. This increased access to education has been a catalyst for the transformation of Africa, not just in terms of economic development and self-sufficiency but also in terms of creating an equitable society and enabling individuals to pursue their own ambitions and gain more economic independence while creating greater equality among different socioeconomic groups and bridging the gap between traditional gender roles and empowering women to take part in the development of a brighter future for all Africans. This increased access to education has been instrumental in the transformation of Africa, offering individuals the opportunity to pursue their own ambitions and gain more economic independence while providing them with the resources they need to move away from relying on foreign aid and assistance and instead being empowered to create their own paths and become in control of their own future through knowledge, skills, and resources. This shift toward more equitable access to education and self-empowerment has had a major impact on the African continent, allowing it to break away from its history of inequality and underdevelopment and create a brighter, more prosperous future for all its people, and ushering in a new era of prosperity, equality, and self-determination for the African people. This increased access to education has been essential in allowing Africa to break away from the cycle of underdevelopment and inequality that had previously constrained it, providing individuals with the resources they need to pursue their own ambitions, gain economic independence, and create greater equality between different social classes and economic statuses by providing the opportunity for education and the resources necessary to become self-reliant and develop a more prosperous, equitable, and sustainable future for the African continent

2.2.5 Improved Health Care

Foreign investment and development have had a significant positive impact on health care systems in Africa in recent decades by providing funding and resources to support the building of health care infrastructure as well as training and education opportunities. Over the last two decades, foreign investment in health care systems in Africa has been a driving force behind improved access to care and better health outcomes for citizens living in impoverished and developing countries across the continent. As the resources from foreign investment have been put to use, there have been measurable increases in life expectancy and decreases in infant mortality rates, as well as improved access to primary health care services such as vaccination and maternal health services for underserved communities in Africa. In addition to providing financial resources, foreign investors have also been instrumental in building capacity within the health care system in Africa by introducing innovative technologies and practices, training local health care providers, and creating sustainable health care models that can continue to benefit the population long after foreign investors have moved on to other regions of the world. This has resulted in significant improvements in health outcomes across the continent, as well as more equitable access to quality health care for African citizens who would have otherwise gone without due to a lack of resources and support from their own governments. These advances have been vital for both the short-term and long-term well-being of African nations, providing access to essential health care services, improving public health systems, and fostering greater stability for their people, which would have been difficult to achieve without outside assistance from foreign investors and other international partners. Despite the undeniable impact of foreign investment in health care in Africa, there are still numerous challenges that need to be addressed to ensure that all African citizens have access to quality health care services and can benefit from these investments. It is essential that foreign investors be held accountable for their commitment to the health and well-being of African citizens and that African governments take an active role in investing in their own citizens' health by providing the necessary resources and support for improved health care infrastructure and systems and public health initiatives to ensure that the health care advances made thus far are not lost and that all African citizens can benefit from these investments. Moving forward, it is essential that foreign investors and African governments collaborate to ensure that all African citizens have access to quality health care services and are able to benefit from these investments and that African governments, with support from international partners, implement policies and programs that prioritize public health initiatives and create a sustainable environment for these investments to have a voice in the performing arts, which was once only dominated by men, and continue to benefit African citizens and health care systems for years. Moving forward, it is essential that foreign investors and African governments collaborate to ensure that all African citizens have access to quality health care services and are able to benefit from these investments by providing the necessary resources, training, and support for improved health care infrastructure and systems and public health initiatives so that the health care advances made thus far are not lost and all African citizens can benefit from these investments in the long term.

2.2.6 Improved Business Opportunities

Over the past two decades, foreign investment has been a major catalyst in the African economy, leading to increased business opportunities, more job creation, and increased revenue for the continent. This foreign investment has come in a variety of forms, including public–private partnerships, venture capital investments, and direct foreign investments from international organizations, corporations, and even individuals. This influx of foreign capital has proven to be immensely beneficial to Africa, with the number of businesses and jobs growing significantly, as well as an increase in revenues and economic development for many African nations. As a result of this foreign investment, African countries have seen their GDP grow at an unprecedented rate, leading to a decrease in poverty and greater economic opportunities for all Africans. This increase in economic development has enabled Africa to become more competitive in the global marketplace, attracting more foreign investment and creating even greater opportunities for businesses and individuals to invest in the continent. The influx of foreign capital has been so important to African countries that governments have begun implementing policies and regulations to encourage further investment to continue to grow their economies, reduce poverty, and improve the quality of life for all their citizens. These policies have enabled African countries to become more attractive to foreign investors, encouraging more economic development and growth and allowing African countries to become more competitive in the international market. The economic development achieved through foreign investment has been a great success for African countries, allowing them to increase their GDP, reduce poverty, become more competitive in the global market, and provide opportunities for businesses and individuals to increase their income while also providing improved services and infrastructure to the people of Africa, such as better education, healthcare, and access to clean water and electricity. This increased economic development has had a positive effect on the African continent, and with the implementation of policies to encourage investment, foreign capital has been instrumental in lifting Africa out of poverty and creating greater economic growth and opportunities for its citizens than ever before.

2.3 Potential Obstacles to Global Investment and Development in Africa

2.3.1 Potential Impacts on Local Communities

One of the potential impediments to foreign investment and development in Africa is its potential effects on local communities and the environment, such as pollution and the displacement of indigenous populations due to the construction of large-scale infrastructure projects. These potential side effects are a major concern for local communities, governments, and nongovernmental organizations who want to ensure that foreign investment is beneficial both to the economy and to the people of Africa. While foreign investment can provide a much-needed boost to the African economy, it must be managed in such a way as to minimize negative impacts on local communities and the environment while still providing the expected economic benefits. Such a delicate balance requires sound decision-making on the part of all parties involved, from local governments to foreign investors and international organizations.

To ensure that foreign investment in Africa is mutually beneficial for all parties involved, there must be careful consideration given to how it will impact local communities and the environment in the form of greater autonomy, respect, and recognition of local customs and traditions, as well as regulations in place to ensure environmental sustainability and the preservation of natural resources. To maintain this balance, proper regulations must be in place, as well as respect and recognition of local customs and traditions for the protection of local communities and the environment and the development of a transparent and collaborative relationship between all involved parties.

Furthermore, local communities must have a say in the investment process, with their rights and autonomy respected, to ensure that their interests are taken into consideration and that the investment is beneficial to them as well and not just the foreign investors or international organizations investing in the region. This can be achieved through open dialogue and the formation of partnerships between governments, local communities, and international organizations that aim to ensure that the interests of all parties involved are taken into consideration and that the investment is mutually beneficial and long-lasting for all parties involved. This is necessary to ensure that foreign investment does not exploit Africa's natural resources for short-term gains but rather contributes to the long-term prosperity of the African continent by preserving its environment and strengthening its local communities while creating long-term sustainable and equitable economic opportunities for all involved.

By ensuring that the investment process is collaborative and transparent, it is possible to achieve an equitable balance between the interests of foreign investors, international organizations, and local communities in Africa such that the investment process is equitable, fair, and mutually beneficial for all parties involved in an unprecedented manner, and their overall opinions are not taken into account, thereby protecting local autonomy and rights, preserving the environment, and ultimately creating long-term sustainable economic opportunities for everyone involved in a way that has never been done before on the continent. This approach to foreign investment has the potential to be a model for responsible international investment and has the potential to pave the way for an even brighter future for Africa, its local communities, and foreign investors alike, moving away from the traditional narrative of exploitation. By taking such a progressive approach to foreign investment, it is possible to create long-term sustainable economic growth and prosperity in Africa in a way that preserves the interests of all parties involved while ensuring that no single party takes advantage of the other while simultaneously protecting the environment and the local rights of African communities and opinions.

This investment model of sustainability, equity, and collaboration could be the key to unlocking a brighter future for Africa and its people in a manner that was previously thought impossible by adhering to the principles of mutual benefit, environmental protection, and local autonomy while at the same time protecting local autonomy and rights, preserving the environment, and ultimately creating long-term sustainable economic opportunities for everyone involved in a manner that has never been seen before. This approach is unique to the continent and could serve as a blueprint for responsible international investment in Africa, with the potential to revolutionize economic development in the region, create new economic opportunities for local communities, and pave the way for a brighter future for African economies, foreign investors, and communities alike while also protecting the interests of local communities and respecting their autonomy.

This principle of mutual benefits, environmental protection, and local autonomy is in stark contrast to the traditional narrative of exploitation that has plagued the continent for far too long and provides a new way forward for African economies, one that is mutually beneficial, equitable, and respectful of the local autonomy and rights of African communities and their opinions. This investment model, if applied properly and with the commitment of all involved parties, could be the key to unlocking unprecedented economic prosperity in Africa and ushering in a new era of economic collaboration, sustainability, and equity on the continent that was thought to be impossible until now. This investment model is the beginning of a new era for African economies, one that has the potential to bring long-term, sustainable economic opportunities to everyone involved and to foster a new wave of collaboration, innovation, and equity that could serve as an example for the world to follow.

This model is something to be praised, as it represents a shift from the traditional narratives of exploitation that have long been associated with foreign investment in Africa and puts the needs of the communities and environment at the center of development efforts instead of profit for multinational corporations. This shift in investment model has the potential to drastically improve the standard of living and quality of life in African communities by providing local populations with access to resources, jobs, and services that were previously out of reach while also helping to protect and restore the environment, foster equitable and sustainable development, and create lasting economic opportunities that benefit all involved parties in the region. This model is something that should be celebrated, as it presents a unique opportunity to work together toward common goals and unlock the untapped potential of the African continent while creating a brighter and more equitable future for all involved parties. Moving forward, it is important to ensure that the model continues to prioritize the interests of local populations and environments over those of corporations and to remain mindful of the potential pitfalls that come with any new investment strategy so that the model can remain true to its core values and continue to provide meaningful benefits for both the African people and their environment going forward.

In conclusion, the shift from traditional foreign investment models to the one described in this model is something to be celebrated and admired for its potential to create meaningful and lasting economic opportunities while also helping to protect and restore the environment, foster equitable development, and ensure that all involved parties benefit from the investment in a way that was unprecedented at the time of skepticism when assessing the success of this new investment model. The shift from traditional foreign investment models to this new model has the potential to bring about immense positive changes for African communities and for the environment as a whole if it is properly implemented and monitored going forward. It is important to continue to evaluate the impact of this model and ensure that it is successful in achieving its original objectives so that it can continue to serve as a model of foreign investment success and set the precedent for future investments in

Africa in the form of a new type of economic freedom that goes beyond the expectations of their predecessors. Moving forward, it is important to ensure that the model continues to prioritize the interests of local populations and environments over those of corporations and to remain mindful of the potential pitfalls that come with any new investment strategy to help ensure the success and longevity of this investment model.

2.3.2 Risk of Exploitation of Resources

The possibility of resource exploitation is one of the potential barriers to foreign investment and development in Africa, as it can lead to a wide range of negative economic and social consequences, including increased poverty, environmental degradation (Obiwulu et al., 2022), and the loss of human rights. In addition to these issues, resource exploitation can create the risk of financial mismanagement and the possibility of large-scale corruption due to the lack of adequate checks and balances or regulations, leading to a further degradation of the economic prospects of African nations and the further entrenchment of existing inequalities. Additionally, resource exploitation can be difficult to monitor and enforce, particularly when it involves companies based in countries outside of Africa. These challenges are compounded by the fact that the exploitation of natural resources can lead to a depletion of resources, thus leading to further destabilization of African countries' economies and leaving them less equipped to achieve economic growth and development in the future. These challenges are not limited to Africa, and many countries around the world have found it difficult to successfully regulate the exploitation of natural resources by both domestic and foreign companies.

Unfortunately, these consequences have been seen in many African countries and have had a significant impact on the economy as well as the citizens of those countries Furthermore, the lack of enforcement and monitoring of these activities can lead to major gaps in compliance with laws and regulations meant to protect the environment and limit the exploitation of natural resources. This has become a major challenge for many African countries, and in some cases, these challenges are further compounded by the lack of necessary infrastructure and technology needed to adequately monitor, regulate, and enforce laws related to the exploitation of natural resources. This lack of infrastructure and technology can also lead to a lack of trust between the government and citizens in regard to enforcing regulations, as people may not have faith in the ability of their governments to protect them from exploitation and ensure that their natural resources are being used responsibly.

Consequently, it is essential for African countries to invest in adequate infrastructure and technology that would allow governments to effectively monitor, regulate, and enforce the laws related to the exploitation of natural resources to ensure that these activities do not lead to the overexploitation and destruction of their environment and that their natural resources are being used in a responsible and sustainable manner. Without proper governance and regulation, resource exploitation can be a destructive force that further entrenches existing economic inequalities and causes long-term damage to local economies, ecosystems, and communities. This is why it is so important that governments, nongovernmental organizations, and civil society work together to identify and address the root causes of resource exploitation to protect vulnerable populations from exploitation, ensure that local people benefit from the extraction of natural resources, promote sustainable economic growth and development in Africa and beyond, and address other complex issues. This will require a collective effort to create and enforce legal frameworks that govern resource exploitation, ensure responsible corporate practices, provide support to local communities and small businesses, and develop the infrastructure needed for sustainable economic development. Furthermore, governments must ensure that natural resource extraction activities are conducted in a transparent and accountable manner by implementing rigorous environmental protection standards, monitoring and regulating the activities of resource extraction companies, and investing in technologies such as remote sensing to track changes to the environment.

Additionally, governments must work in partnership with civil society to ensure that the voices of local communities and vulnerable populations are heard in decision-making processes that involve resource extraction activities and that those most affected by resource exploitation are meaningfully included in policy-making and decision-making processes at all levels. This is critical for creating long-term solutions to the exploitation of natural resources that consider the needs and interests of both local communities and larger stakeholders while protecting human rights and the environment to achieve sustainable development. To effectively address the challenge of natural resource exploitation, it is essential that governments, companies, and civil society collaborate to create strong regulatory frameworks and effective strategies for sustainable economic development that prioritize social and environmental justice, respect for human rights, and the participation of all stakeholders in decision-making processes to achieve true economic and environmental sustainability.

Moving forward, it is critical that governments take meaningful steps to create and implement policies that prioritize sustainable natural resource exploitation and the interests of local communities and vulnerable populations to ensure the equitable distribution of benefits from natural resource exploitation while protecting human rights, the environment, and sustainable economic development. It is clear that governments must work with civil society to ensure that the voices of local communities and vulnerable populations are heard in decision-making processes related to natural resource exploitation and to ensure their meaningful inclusion in policy-making processes to ensure that their needs and interests are considered when creating and implementing policies and regulations related to natural resource exploitation.

2.3.3 Political Instability

One of the potential impediments to foreign investment and development in Africa is the potential for political instability caused by government corruption, civil unrest, military coups, and other forms of political upheaval that have been a reality for many African countries throughout the years. This political instability has made foreign investors hesitant to invest on the continent, as their money may be put at risk if any of these circumstances arise during the life of their investment. However, many African countries have begun to recognize the value of stability and are attempting to implement reforms to attract foreign investment and build a more stable political climate and economy. Such reforms include the implementation of better laws and regulations to protect investors, as well as increased efforts to combat corruption, increase accountability, and ensure that all citizens have access to due process under the law. In addition, some African countries are working to open their borders and create free-trade zones to attract foreign investors and increase their economic prospects. These efforts are beginning to pay off, as an increasing number of foreign investors are taking a chance on the African continent and seeing the potential in its growing markets and economies. While these efforts are a step in the right direction, many African countries still have a long way to go to ensure that their economies are competitive in the global market and that their citizens have access to the same economic opportunities as those in other parts of the world. To continue making progress, African countries need to ensure that these reforms are fully implemented and enforced and that their citizens have access to education, financial services, and the other resources necessary for them to fully participate in their countries' economies. Additionally, African countries need to prioritize building infrastructure and creating a supportive environment that is conducive to business growth, such as providing access to capital and supporting entrepreneurs who are taking a chance in the African economy by developing their own talents and leveraging the opportunities that come from foreign investment to realize their full potential. African countries need to focus on creating economic policies that will help them attract more foreign investment, such as removing barriers to entry, lowering taxes, and incentivizing companies to invest in their countries to spur economic growth and create new economic opportunities for their citizens. These efforts would lead to the greater economic prosperity of African countries and improved livelihoods for their citizens while also helping create a more equitable global economic environment that allows African countries to compete on the world stage and access the same level of resources and investment as countries in other parts of the world. Additionally, African countries need to develop and strengthen their legal systems, increase transparency in government operations, and create and maintain more effective regulations for businesses operating within their borders to create a stronger and fairer business climate that will attract foreign investment and spur economic growth. Ultimately, the successful implementation of these policies and initiatives will depend on effective leadership and collaboration between the African governments, private sector entities, and international organizations to ensure that African countries are able to create an environment that is conducive to long-term economic growth. Development and lasting prosperity. By embracing these changes and ensuring that the right conditions are in place, African countries can create a strong foundation for sustainable economic growth and open up new opportunities for their citizens that will improve their livelihoods and bring about a more equitable economic playing field both domestically and globally. To achieve this, African governments must be willing to invest in modernizing and streamlining their bureaucracy, increasing accountability and the rule of law, providing access to finance for small businesses and entrepreneurs, ensuring equitable access to education for all citizens, developing public–private partnerships that facilitate private sector investments, investing in infrastructure and digital technologies, and creating economic policies that promote equitable access to opportunities, support economic development, and create an environment that encourages entrepreneurship and foreign direct investments. All of these efforts require strong leadership, clear strategies, and innovative approaches to policy implementation that take into account the needs of each African country as well as their regional and international contexts to bring about lasting change. The idea of justice is only achievable through fate. In conclusion, by focusing on the areas outlined above and making the necessary investments, African countries can lay the foundation for lasting economic growth and prosperity that will increase opportunities for their citizens, improve livelihoods, and create a more equitable economic playing field. To fully realize the concept of justice, African countries must make significant investments in human and financial capital and ensure that those investments are channeled toward improving health, educational attainment, and economic productivity throughout the continent. This is especially true for African countries, as many of them are faced with numerous challenges such as poor infrastructure, inequality, poverty, and poor governance that have hindered their development and perpetuated the cycle of poverty that goes hand in hand with fate over the centuries. All of these obstacles can be surmounted, however, through greater investments in key areas such as education, health care, infrastructure development, and access to technology that will help reduce inequality, foster economic growth and prosperity, and empower African citizens to create more equitable economic opportunities for themselves and their families in this way, ultimately laying the groundwork for justice and a more prosperous, equitable future for all. In doing so, African countries will also be able to create a more just society that respects the rights of its citizens and works to eradicate poverty and inequality from the outset and uplift their population out of poverty while at the same time upholding the human rights of its citizens and ensuring that all members of society have access to justice and the resources needed to improve their living conditions and overall quality of life. This will ultimately be a benefit to the entire continent, as the elimination of poverty and inequality can only lead to better stability and economic prosperity for the whole of Africa and a more equitable and just society for its citizens.

2.3.4 Corruption

One of the potential roadblocks to foreign investment and development in Africa is the possibility of corruption, which can lead to poor decisions and investments, skewed accounting and financial records, political interference with business activities, and money laundering activities. These issues can be a huge detriment to potential investors, who may be scared away by corruption, or those who decide to invest may lose their money due to criminal activities. Unchecked corruption in Africa can also have a negative effect on economic growth and development because it causes investors to become wary of the region and diverts resources away from those who need them most, such as individuals and small businesses that could benefit from foreign investment. Additionally, unchecked corruption can lead to an increase in poverty levels as resources are misappropriated and funds that should be used to improve quality of life are diverted to illegal activities and unaccountable sources, leading to a decrease in economic and social progress in the region. Such corruption is a serious threat to Africa's economic development, as it stifles investment and hampers efforts to alleviate poverty and improve quality of life for the citizens living in the region. The presence of unchecked corruption in Africa can have far-reaching consequences beyond the continent itself, as it can limit Africa's ability to engage in global commerce and international trade and can also have a negative impact on other countries or regions that have a vested interest in the continent. The risk of unchecked corruption in Africa is a serious issue that must be addressed through international cooperation and coordinated efforts if it is to be effectively addressed and if the continent is to realize its true potential. Through increased international investment and better implementation of anti-corruption measures, Africa can begin to address the issue of unchecked corruption and improve its economic stability and competitiveness in the global economy to make a positive contribution to the world and promote economic development and quality of life for its citizens. In conclusion, unchecked corruption in Africa is a critical issue that needs to be addressed if the continent is to realize its economic potential and improve the quality of life for its citizens. This therefore presents a significant challenge for African governments, who must take action to ensure that these negative consequences do not come to pass and undermine the development of the continent and the potential benefits that it offers to other nations and regions around the world, which, in turn, can have a ripple effect on the global economy and its people as a whole. To effectively address the issue, African governments must take a proactive approach and collaborate with international partners to strengthen anti-corruption measures and enforce the rule of law while also encouraging increased international investment in the continent and encouraging transparency, accountability, and good governance to improve the economic stability and global competitiveness of the continent and its people. This in turn will help to create a more conducive environment for economic development and sustainable growth while at the same time promoting human development and quality of life for the citizens of African countries. Additionally, African governments must work to improve access to quality education, health care, and other basic services to equip citizens with the tools necessary for financial inclusion and the promotion of equitable economic growth across the continent and its countries. This proactive approach to tackling corruption, reducing inequality, and improving quality of life should help to create a more promising future for Africa and its people, allowing it to take its rightful place in the global economy as an integral and respected participant. While these efforts must be taken on an individual country level, African governments must also work to create and foster international partnerships that allow for the mutual benefit of all participants to effectively address the issues of corruption, inequality, and poverty in Africa that plague the continent and its people and thus create a path toward a more equitable, prosperous future for African countries. As such, African governments should focus on creating strong legal systems with well-enforced anti-corruption laws that discourage misuse of public funds and hold those found guilty accountable for their actions, as well as encouraging the establishment of institutions that promote transparency, accountability, and integrity in public offices and private businesses and cultivating a culture of trust and honesty across the African continent. Additionally, African governments must invest in social welfare initiatives and programs that reduce the risk of poverty and improve the quality of life for their citizens, as well as strengthen their economic infrastructure and provide job opportunities to their citizens so that they can take full advantage of the possibilities of globalization and participate in the global economy with confidence and success. In conclusion, the success of African countries in tackling corruption, reducing inequality, and improving quality of life will depend on both individual countries and international partnerships working together to create a more equitable and prosperous future for all African countries and ensure that their citizens have the resources, skills, and opportunities necessary to thrive and reach their full potential.

2.3.5 Poor Governance

The possibility of poor governance is one of the potential barriers to foreign investment and development in Africa that should not be overlooked when assessing the continent's economic progress, as it can directly influence the attractiveness of a country to investors (Reinisch, 2014). Poor governance and corruption can have a severe negative impact on a nation's ability to attract foreign investment (Särkänne, 2021), as investors are naturally cautious of putting their money into countries where there is no assurance that it will be properly managed (Saunders, 2021) or used for its intended purpose due to the potential for embezzlement or misappropriation of funds as a result of poor governance and corruption. This can be seen in countries such as Nigeria and Kenya, where reports of poor governance and corruption have led to a decrease in foreign investment and development, undermining the progress made in areas such as infrastructure and technology, as well as reducing the potential for economic growth, which can have wide-reaching and long-term implications on the countries' economic progress and stability.

In addition, poor governance can lead to an erosion of public trust (Schill, 2009), as citizens may become increasingly disillusioned with the government and its ability to implement effective policies and programs that are beneficial to them and the wider population (Southall & Melber, 2009). This can lead to a decrease in popular support for the government, further undermining its ability to attract foreign investment and progress economically, which in turn can lead to a decrease in the public's quality of life (Thulasidhass, 2015) as well as stifling the potential for economic growth (Titi, 2015) and stability as a result of the corruption and mismanagement that can often be associated with poor governance (Vaksha & Singh, 2020). Poor governance can have a number of negative effects on foreign investment and development in Africa, such as a lack of security, corruption, and a weak legal system, which can make it difficult for foreign investors to feel confident in their investments, as well as a lack of transparency in the government's decision-making process and inadequate resources to ensure proper implementation of policies and regulations, leading to a decrease in foreign investment and development, which can have serious implications for economic growth and stability in countries such as Nigeria and Kenya. These issues, coupled with the lack of accountability and transparency in government, can make it difficult for African countries to attract foreign investment and development, which is essential to the growth and stability of their economies.

Ultimately, poor governance can lead to a decrease in the public's quality of life (Wolfson & Frisken, 2000), a lack of security and transparency in decision-making, an erosion of public trust and confidence in the government, corruption and mismanagement of resources and policies, and decreased foreign investment and development, making it increasingly difficult for African countries to achieve economic growth and stability (Wasike, 2022). All of these issues ultimately contribute to a decrease in the potential for economic growth and stability, as well as a decreased quality of life for its citizens. Poor governance can have a profoundly negative effect on the quality of life of a country's citizens, with the potential for stifling economic growth and stability as well as contributing to an increase in poverty and inequality.

These issues are particularly acute in African countries such as Nigeria and Kenya, which have a long history of poor governance, inadequate resources and policies, corruption, a lack of security, transparency in decision-making, and a weak legal system, all of which can significantly contribute to an increased level of poverty and inequality in the country and lead to a decrease in economic growth and stability. Poor governance can mean that laws are not enforced properly, leading to a lack of legal security and stability for foreign investors, which ultimately serves to erode the confidence of potential investors and can therefore have a profound effect on economic growth and stability as well as a decrease in the quality of life for its citizens.

Poor governance can also lead to a lack of transparency and accountability in decision-making, meaning that public funds may not be allocated in the best interests of citizens, and this can also contribute to an increase in poverty and inequality as well as a decrease in economic growth and stability in the country. Poor governance can also have a negative effect on access to public services, including healthcare and education, which can ultimately lead to an even greater degree of poverty and inequality and further erode economic growth and stability in these African countries. Poor governance, inadequate resources and policies, corruption, a lack of security and transparency in decision-making, and a weak legal system all have a detrimental effect on economic growth and stability in African countries and can lead to further underdevelopment, poverty, and inequality and can ultimately have a lasting, damaging effect on the lives of its citizens.

A number of African countries have experienced weak governance, which has led to high levels of corruption and the ineffective delivery of public services. In many countries, the lack of reliable infrastructure, political and economic instability, weak institutions, a lack of good governance, and corruption make it difficult for potential investors to take the risk of investing in African countries. This situation is further exacerbated by a lack of transparency and accountability in decision-making and weak legal systems, which means that it is difficult for citizens to challenge or hold their governments accountable for their actions.

To break this cycle of weak governance, inadequate resources, and corrupt policies, African countries must take steps to increase transparency and accountability in decision-making, strengthen their legal systems, improve access to resources and services for citizens, and combat corruption by increasing public awareness, promoting good governance and anti-corruption policies, ensuring that citizens have access to reliable information and data so that they can make informed decisions and hold their leaders accountable, and advocating for better implementation of existing laws and regulations. African countries must also prioritize investments in education, health care, and infrastructure to improve the quality of life for citizens and attract potential investors to stimulate economic growth and create a more attractive environment for potential investors in the long term. African countries need to take the initiative and make structural reforms to their governments and legal systems to promote good governance, transparency, and accountability.

African countries should prioritize enacting measures to promote greater transparency and accountability in decision-making, strengthening the rule of law, improving access to resources and services for citizens, and combating corruption. African countries must also invest in their infrastructure, develop strong institutions, and create an environment that encourages foreign investment while also investing in education and health care to help improve the quality of life for their citizens to ensure sustainable and long-term economic growth. By making these investments, African countries can create an environment that will attract investors while providing their citizens with the necessary resources to improve their quality of life and drive economic growth. African countries should also strive to improve their legal systems to promote the protection of human rights, ensure a due process of law, and create a fairer and more just society for their citizens.

African countries should prioritize enacting measures to promote greater transparency and accountability in decision-making, strengthening the rule of law, improving access to resources and services for citizens, combating corruption, investing in infrastructure, developing strong institutions that encourage foreign investment, investing in education and health care, and improving their legal systems to ensure greater human rights protections and create a more equitable and prosperous future for their citizens. This will not only create a more equitable and just society but also provide the necessary tools to drive economic growth, making African countries more attractive destinations for foreign investors and providing a better quality of life for their citizens. This will enable African countries to create a stronger economy, improve the quality of life for their citizens, and ensure a more just and equitable society that can stand the test of time by creating laws and regulations that are fair, clear, and enforceable.

Additionally, African countries should prioritize enacting measures to promote greater access to justice and civil rights for their citizens so that they can hold governments and companies accountable for their actions, giving individuals more opportunities to receive fair and equal treatment regardless of their economic status or social background. By implementing these measures, African countries can create a brighter future for their citizens, one that is more just and equitable and can provide them with the necessary tools to create and sustain a strong economy for many years to come.

These steps are necessary for African countries to create an environment that encourages economic growth and development, creates stability and trust, and strengthens their democracies while also protecting the rights of their citizens and creating an environment where justice is available to all. African countries should focus on developing and enforcing laws and regulations that are fair, just, and accountable to the public to ensure that everyone is given a fair chance to access justice and have their grievances heard and acted upon. These measures should also include investing in infrastructure and education, creating a reliable and transparent banking system, curbing corruption, promoting gender equality and women's empowerment, enhancing citizens' access to resources such as energy and water, increasing access to credit for entrepreneurs, and investing in technology and digital literacy to ensure that citizens have the necessary tools to engage in economic activities and promote the development of small and medium-sized enterprises. In addition, African countries should promote a culture of transparency, accountability, and inclusion in their societies and governments, ensuring that the voices of all citizens are heard and that the rights of the most vulnerable members of society are respected and protected. Last, African countries should foster a culture of dialogue and collaboration between the public and private sectors, allowing them to work together to create an environment that is conducive to economic growth and stability while ensuring the protection of human rights and the equitable distribution of resources across the continent.

In conclusion, African countries have the potential to create a vibrant, prosperous, and equitable society for all of their citizens if they focus on developing and implementing fair laws and regulations that are accountable to the public, investing in education and infrastructure, curbing corruption, promoting gender equality and women's empowerment, increasing access to resources such as energy and water, providing access to credit for entrepreneurs, investing in technology and digital literacy, fostering a culture of transparency, accountability, and inclusion in their societies and governments, and promoting a culture of dialogue and collaboration between the public and private sectors. All of these measures should be taken to ensure a secure, prosperous, and equitable future for all African citizens. These measures should focus on developing and implementing fair laws and regulations that are accountable to the public, investing in education and infrastructure, curbing corruption, promoting gender equality and women's empowerment, increasing access to resources such as energy and water, providing access to credit for entrepreneurs, investing in technology and digital literacy, fostering a culture of transparency, accountability, and inclusion in their societies and governments, and engaging in dialogue between governments, the private sector, and civil society. By doing so, African countries can create an environment that is conducive to economic growth and stability while safeguarding the rights of their citizens and promoting the equitable distribution of resources throughout the continent and beyond. African countries have the potential to create a vibrant, prosperous, and equitable society for all of their citizens if these measures are taken and properly implemented, and protecting the environment, promoting sustainable development, and providing social protection to those in need are also given necessary attention.

2.3.6 Weak Regulatory Environment

One of the potential impediments to foreign investment and development in Africa is the potential for a lax regulatory environment that allows for corruption and the inefficient use of resources to persist. This can be a major turn-off to potential foreign investors, who worry that their investments may not be secure if the local governments are unable to keep corruption and waste in check. In addition, foreign investors also worry about the presence of political instability and civil unrest in certain parts of Africa, as these can further reduce the security of their investments and make it difficult to do business in the region. Additionally, many African countries lack the infrastructure necessary for successful economic development, leading to an unreliable power supply, transportation difficulties, and a lack of access to the necessary goods and services for businesses to thrive. These potential impediments can put a damper on foreign investment and development in Africa, making it difficult for businesses to thrive and for countries to grow economically. Furthermore, African countries often lack the necessary regulations, laws, and processes to protect the rights of investors and ensure that their investments are safe and secure. This lack of legal protection can further reduce the trust foreign investors have in African countries, making it difficult for these countries to attract the necessary capital to drive economic growth and development. As a result, foreign investors often find that the perceived risks of investing in Africa outweigh the potential rewards, making it difficult to draw them into the region and spark economic growth. Despite these challenges, there are still opportunities for African countries to attract foreign investment by implementing effective and comprehensive legal frameworks that provide investors with the protection they need to ensure their investments are safe and secure and to ensure that the potential risks associated with investing in Africa are properly managed and accounted for. Ultimately, the key to drawing in foreign investors and encouraging economic development in African countries lies in effective legal frameworks that provide investors with the security and confidence they need to invest in the region. This creates a cycle of poverty, as African countries lack the resources to increase development, create a safe and secure business environment, and improve the lives of their citizens, which could otherwise be obtained through foreign investment.

Ultimately, while Africa is a continent with many natural resources and potential business opportunities, there are significant challenges that must be addressed for it to attract the foreign investment necessary to spur economic growth and development. Despite these potential impediments, Africa is an increasingly attractive destination for foreign investors and businesses, as they seek to benefit from the continent's vast untapped resources and potential while managing the risks associated with doing business in an unfamiliar, often unstable environment. As Africa continues to become more integrated in the global economy, the importance of foreign investment and economic development is growing, and the demand for infrastructure development is increasing. All of these issues can make it difficult for African countries to attract foreign investment and make progress in economic development and poverty reduction. As such, ensuring that regulations are strong and enforced is key to attracting foreign investors, encouraging development in African countries, and providing a secure environment for their investments. To ensure that African countries benefit from foreign investment and economic development, they must take proactive steps to ensure that their economies are stable and attractive to foreign investors. African countries must ensure that the benefits of foreign investment are felt by the population as a whole and not just the ruling elites by instituting and enforcing regulations that require foreign investors to comply with labor standards, pay taxes, and support local job creation.

2.3.7 Poor Access to Capital

Inadequate access to capital is one potential hindrance to foreign investment and development in Africa, which can limit the ability of African countries to take full advantage of foreign direct investment. In some cases, foreign investors may be deterred by Africa's lack of adequate access to capital, which can limit their opportunities for investment and the developmental advantage of foreign direct investment. Furthermore, the lack of available capital in Africa can hinder economic diversification and growth, as it may limit the ability of African countries to make long-term investments and acquire technology and know-how from foreign investors, which can have a profound impact on their economic development. Moreover, the lack of access to capital can have a negative impact on African countries' ability to pursue economic policies that are conducive to FDI by making them vulnerable to volatile capital flows and creating a less favorable investment climate for foreign investors. In addition, the absence of domestic capital markets in many African countries can restrict the ability of African businesses to access long-term debt or equity financing from foreign sources, reducing the potential for foreign direct investment to stimulate growth in these countries. As a result, without adequate access to capital, African countries are likely to experience a number of limitations in regard to foreign direct investment and its potential for economic development. This can also have an impact on foreign direct investment, as investors may be hesitant to commit funds in areas of the continent where capital is scarce. In some cases, foreign investors may be deterred by Africa's lack of adequate access to capital, which can limit their opportunities for investment and the developmental advantage of foreign direct investment. Furthermore, the lack of available capital in Africa can hinder economic diversification and growth, as it may limit the ability of African countries to make long-term investments and acquire technology and knowledge from foreign investors, which can have a profound impact on their economic development prospects. This can in turn lead to a lack of competition and entrepreneurialism within the African economy, as restrictions on access to capital can limit the ability of entrepreneurs to develop new business opportunities, which could ultimately hinder the continent's growth and potential for development and reduce the potential for African countries to take a competitive advantage from foreign direct investment. This lack of access to capital, along with other structural factors, can have serious implications for African countries as they seek to benefit from foreign direct investment and ultimately realize the economic growth and development potential of their continent. Moreover, the lack of access to capital can have a negative impact on African countries' ability to pursue economic policies that are conducive to FDI by making them vulnerable to volatile capital flows and creating a less favorable investment climate for foreign investors. In addition, the absence of domestic capital markets in many African countries can restrict the ability of African businesses to access long-term debt or equity financing from foreign sources, reducing the potential for foreign direct investment to stimulate growth in these countries. As a result, without adequate access to capital, African countries are likely to experience a number of limitations in regard to foreign direct investment and its potential for economic development. This can also have implications for foreign direct investment, as investors may be hesitant to commit funds in areas of the continent where access to capital is limited and economic policies may be seen as too risky or unsustainable. This lack of access to capital has the potential to not only hamper African countries' ability to pursue economic policies that are conducive to foreign direct investment but also hinder the growth of businesses in these countries by making it difficult for African businesses to secure the capital they need to develop and expand their operations. Additionally, the lack of access to capital can also make it difficult for African businesses to access new markets and technologies to become competitive on the global stage. This in turn can lead to a number of adverse economic effects, such as increased poverty, stagnation in economic growth, and limited job opportunities for citizens in the countries in question. Furthermore, this lack of access to capital has the potential to exacerbate the existing inequality in African countries, as those who have access to capital are more likely to benefit from investment opportunities than those who do not have access to the same resources. This can lead to a vicious cycle of poverty and inequality, where those who already have access to capital are able to become increasingly wealthy, while those without access to these resources remain mired in poverty and are unable to access the same opportunities as their more affluent counterparts. This can create a situation in which wealth is concentrated among a select few, while the majority of citizens in African countries remain deprived of opportunities to improve their economic standing and quality of life. This can have disastrous consequences for African countries, making it more difficult for them to escape the cycle of poverty and inequality that has been created by a lack of access to capital and resources. This cycle of poverty and inequality is further reinforced by the fact that those who are unable to access capital are also unable to accumulate wealth and generate income, making it even more difficult for them to escape the poverty trap that exists in many African countries. The vicious cycle of poverty and inequality that exists in many African countries is a major obstacle to achieving sustained economic growth and development that is necessary for the well-being of their citizens. This could be a huge detriment to African countries' economic development, as access to capital is essential for economic growth and development, especially in an increasingly globalized economy. This can be a major obstacle to African countries' economic development and growth, as the lack of access to capital means that investments are not made in the areas of industry and infrastructure necessary for countries to achieve long-term, sustained economic growth. Additionally, the shortage of capital in Africa can lead to increased levels of poverty and inequality, which can limit the ability of African countries to take full advantage of foreign direct investment. In some cases, foreign investors may be deterred by Africa's lack of adequate access to capital, which can limit their opportunities for investment and the developmental advantage of foreign direct investment. Furthermore, the lack of available capital in Africa can hinder economic diversification and growth, as it may limit the ability of African countries to make long-term investments and acquire technology and know-how from foreign investors, which can have a profound impact on their economic development. Moreover, the lack of access to capital can have a negative impact on African countries' ability to pursue economic policies that are conducive to FDI by making them vulnerable to volatile capital flows and creating a less favorable investment climate for foreign investors. In addition, the absence of domestic capital markets in many African countries can restrict the ability of African businesses to access long-term debt or equity financing from foreign sources, reducing the potential for foreign direct investment to stimulate growth in these countries. As a result, without adequate access to capital, African countries are likely to experience a number of limitations in regard to foreign direct investment and its potential for economic development. This can also have an impact on foreign direct investment, as investors may be hesitant to commit funds in areas of the continent where capital is scarce. In some cases, foreign investors may be deterred by Africa's lack of adequate access to capital, which can limit their opportunities for investment and the developmental advantage of foreign direct investment. Furthermore, the lack of available capital in Africa can hinder economic diversification and growth, as it may limit the ability of

African countries to make long-term investments and acquire technology and knowledge from foreign investors, which can have a profound impact on their economic development prospects. This can in turn lead to a lack of competition and entrepreneurialism within the African economy, as restrictions on access to capital can limit the ability of entrepreneurs to develop new business opportunities, which could ultimately hinder the continent's growth and potential for development and reduce the potential for African countries to take a competitive advantage from foreign direct investment. This lack of access to capital, along with other structural factors, can have serious implications for African countries as they seek to benefit from foreign direct investment and ultimately realize the economic growth and development potential of their continent. Moreover, the lack of access to capital can have a negative impact on African countries' ability to pursue economic policies that are conducive to FDI by making them vulnerable to volatile capital flows and creating a less favorable investment climate for foreign investors. In addition, the absence of domestic capital markets in many African countries can restrict the ability of African businesses to access long-term debt or equity financing from foreign sources, reducing the potential for foreign direct investment to stimulate growth in these countries. As a result, without adequate access to capital, African countries are likely to experience a number of limitations in regard to foreign direct investment and its potential for economic development. This can also have implications for foreign direct investment, as investors may be hesitant to commit funds in areas of the continent where access to capital is limited. Additionally, the shortage of capital in Africa can lead to increased levels of poverty and inequality. In addition, Africa's infrastructure in terms of transportation, communication networks, and financial systems is often inadequate to meet the needs of foreign investors. In some cases, foreign investors may be deterred by Africa's lack of adequate access to capital, which can limit their opportunities for investment and development. This lack of access to capital often deters entrepreneurs from making investments in the region due to the high cost and difficulty of obtaining start-up funding and hinders the growth of small businesses, making it difficult for them to create jobs and make a profit. Ultimately, this shortage of capital has a domino effect on the continent's economic development, as it reduces incentives for investment, stunts business growth, and limits job creation. Additionally, the political and social instability in some African countries can be another deterrent for investors who would otherwise be interested in investing in Africa. Despite these challenges, many foreign investors have seen potential in Africa's growing economy and its diverse natural resources, leading to an increase in foreign investment over the past decade. However, the lack of access to capital in many African countries has presented a major obstacle to economic development and job creation on the continent. This lack of access to capital, in turn, restricts the ability of many African countries to invest in the infrastructure and education needed to grow their economies, as well as attract and retain foreign investment. Furthermore, the corruption and lack of transparency in some African governments further deter foreign investment, as investors are reluctant to invest in a country that cannot guarantee fair and equitable treatment of their investments. This lack of access to capital is compounded by a weak and unstable political climate, corruption, a lack of rule of law, and inadequate legal and regulatory frameworks. As a result, although many African countries have seen growth in foreign investment over the past decade, there is still much progress that needs to be made to make it easier for African countries to attract and retain investment from foreign entities and to create an environment in which domestic businesses can thrive and contribute to the economic growth of African countries.

Despite the many challenges that African countries face when trying to access capital for investment, there are still several opportunities for African countries to leverage the potential of foreign direct investment to spur economic growth and development and to create a more prosperous future for their citizens. African governments need to take action to create an environment that is attractive to foreign investors by taking measures to reduce corruption, strengthen the rule of law, and create stable and transparent legal and regulatory frameworks that will give foreign investors the confidence to invest in African countries while also ensuring that domestic businesses have the resources and protection they need to grow and contribute to the local economy. Additionally, African countries need to focus on improving their infrastructure and providing access to capital for domestic businesses to create a more robust economy that will attract and retain foreign investors and ensure that the benefits of foreign investment are spread throughout the African continent. African governments also need to work to improve the access to and quality of education and training for their citizens so that they can take advantage of the opportunities that foreign investment provides and capitalize on the economic benefits of foreign direct investment to create more job opportunities and better standards of living for their citizens. By taking action to reduce corruption, create transparent legal and regulatory frameworks, and focus on improving infrastructure and access to capital for domestic businesses, African countries can create an attractive environment for foreign investors while also ensuring that their citizens are given opportunities to benefit from foreign investment and create a more prosperous future for their countries. African governments must also take steps to ensure that the benefits of foreign investment are widely shared and to create a more equitable distribution of resources to reduce poverty and inequality, foster sustainable economic growth, and create more prosperous futures for all Africans. These goals are achievable, but only if African governments are willing to take the necessary steps to make them a reality.

These challenges will require strong, collaborative efforts from both African governments and the international community to ensure that all citizens of African countries benefit from economic growth that is driven by foreign investment. However, this is easier said than done given the complex nature of the challenges and the multiple stakeholders involved in the process. To achieve this, African governments must focus on ensuring that foreign investments are made in the right areas, providing incentives for businesses to invest in sectors that promote job creation and reduce inequality, and providing quality education and healthcare services to ensure that citizens are equipped with the skills to benefit from economic opportunities, as well as enhancing the security and political stability of the countries that are necessary to ensure the economic and social progress of the continent.

African governments must recognize that reducing poverty and inequality requires more than just foreign investment; they must also prioritize the distribution of resources to those most in need while fostering sustainable economic growth that is both inclusive and equitable for all. African governments must also be willing to engage with civil society organizations and listen to their perspectives, as these organizations have unique insights into the needs of communities and can often provide innovative solutions to the challenges facing African countries. In conclusion, for African governments to achieve success in tackling poverty and inequality, a comprehensive approach is needed that incorporates both foreign investment and domestic resource distribution while at the same time engaging with civil society and local communities to ensure that their voices are heard and that their perspectives are taken into account when designing and implementing policies for development.

2.4 Strategies to Overcome Obstacles to Global Investment and Development in Africa

2.4.1 Political Reform

Political reform is one method for removing barriers to foreign investment and development in Africa and increasing African nations' ability to attract foreign investment by creating a more transparent and predictable legal environment and reducing corruption. Political reform has the potential to create a business-friendly environment, enhancing the attractiveness of investing in Africa to foreign investors by improving the predictability of outcomes and reducing the risk of doing business in African countries. Additionally, political reform could lead to increased economic and social stability in Africa by creating a legal framework that can support sustainable development and create economic opportunities for citizens that would otherwise be blocked due to corruption or a lack of transparency. Political reform could also help address the root cause of conflicts and instability in Africa, as many disputes are fueled by grievances over a lack of economic opportunities or the unequal distribution of resources and power in certain regions. Additionally, it can enable African countries to gain access to more financing from international financial institutions as well as from foreign companies that would likely be more willing to do business with them given the improved predictability of outcomes and transparency associated with reform.

Political reform could also help strengthen and protect human rights, including the right to a fair trial, freedom of expression, and access to education and healthcare, which, in turn, could lead to further economic development and stability across the continent. As such, political reform is essential to unlock the potential of African countries and ensure that their citizens benefit from increased economic and social stability and an improved quality of life. By promoting political reform in Africa, international partners can play a pivotal role in tackling the root causes of conflicts and instabilities on the continent and paving the way for long-term peace and prosperity in the region. By increasing transparency and accountability and reducing corruption, political reform can improve a nation's credit rating, allowing it to take out loans on more favorable terms, and to have more access to international aid and investment in the region. This access to international aid and investment can be used to fund a range of initiatives, including agricultural development, education reform, health care improvement, infrastructure expansion, and much more in African communities.

In addition to the economic benefits of political reform, there is the social benefit of increased access to education and healthcare, which can ultimately lead to an increase in life expectancy and a more literate population in the region. Furthermore, increased access to education and healthcare can create a more informed and engaged electorate, resulting in increased democracy and participation in the political process, which can, in turn, help African countries become more politically stable and secure. This in turn can lead to increased economic growth, as African countries are able to attract more foreign investment, create better business opportunities for their citizens, reduce corruption and conflict within their borders, and create a more prosperous and equitable society for their citizens as a result of this international aid and investment, which act as catalysts for political reforms. This, in turn, can lead to increased foreign investment as well as other economic and social benefits, such as increased employment opportunities and improved standards of living, in addition to the economic benefits of political reform in the region. This cycle of political reform resulting in increased foreign investment and improved living standards demonstrates the importance of political reform as a way to generate positive economic and social outcomes for people in African countries and therefore emphasizes the role of education and healthcare in creating a more informed electorate, which is essential for a strong democracy that can facilitate the political reforms necessary for economic growth and improved living standards, which can create a ripple effect of economic and social progress in many countries around the world for generations to come.

Political reform has the potential to modernize institutions and stimulate economic growth, reduce corruption, improve the rule of law and judicial independence, level the playing field for businesses and foreign investors, ensure greater citizen participation in decision-making processes, and allow governments to implement policies more effectively. All of these factors can contribute to greater social cohesion, improved education and healthcare systems, improved access to markets and resources, increased foreign investment and capital flows, better public services for citizens, and ultimately more equitable economic growth and improved living standards for people in countries around the African continent. This can be particularly beneficial for African countries, where the majority of people are living in poverty and lack access to basic services and adequate infrastructure for basic needs such as clean water, sanitation, and healthcare. Moreover, this cycle of political reform and economic development emphasizes the importance of education in providing African citizens with the knowledge to participate in their democracies and make informed decisions about who they choose to lead them. This is why it is so important for African countries to focus on political reform and economic development to create greater opportunities for their citizens and improve the overall quality of life. By focusing on both political reform and economic development, African countries can make strides toward greater equality and improved living standards for all of their citizens and create a brighter future for generations to come.

Generally, political reform has the potential to dramatically improve the lives of citizens in African countries, not just by providing economic benefits but also by equipping them with the knowledge and resources to take part in democratic decision-making processes, gain access to greater opportunities for personal and professional growth in the public eye, and slowly challenge traditional ideas about the role of women in society. This can help create a more equitable and prosperous society for all and can lead to a more secure, stable, and prosperous future for all citizens in the region, regardless of their social and economic status. While economic development is an important factor for improving the quality of life for citizens in African countries, it is just as important that these countries focus on political reform and development to ensure that their citizens are given the tools, they need to lead meaningful and fulfilling lives and achieve the same opportunities for success as those in other parts of the world. Political reform provides an invaluable opportunity to close the gaps between the haves and have-nots, allowing for a more equitable distribution of resources and opportunities among all citizens in the region, regardless of their gender, ethnicity, or socioeconomic background.

Political reform can create a more just society by giving African citizens access to the same resources and opportunities regardless of gender, ethnicity, or socioeconomic background. This can be done through laws and policies that promote gender equality, diversity in leadership, economic empowerment of marginalized populations, social justice initiatives, and other forms of political reform that focus on increasing representation and inclusion of all citizens in decision-making processes in African communities. Political reform is essential for the long-term success and growth of African countries, as it not only helps to ensure that all citizens are given a fair opportunity to succeed but also fosters greater social cohesion and helps protect citizens from the excess of political and economic powers that have the potential to disproportionately benefit certain individuals or groups at the expense of others.

Political reform, in general, plays an important role in promoting greater political and economic participation, as well as increasing access to resources and opportunities that were previously unavailable due to systemic discrimination and oppression in the region. Political reform can play an integral role in creating a more just and equitable society, as it seeks to create a level playing field for all citizens of African countries by promoting greater representation, inclusion, and access to resources regardless of gender, ethnicity, or socioeconomic background. By instituting reforms that provide equal access to opportunities and resources, we can foster a more inclusive and just society, where everyone is able to participate in political and economic processes to achieve their fullest potential and benefit from the economic and social opportunities available in their various African communities.

Ultimately, political reform and economic development are two of the most powerful forces for creating a brighter future for those living in African countries and ensuring that all individuals, regardless of their identity or background, are able to reap the rewards of economic and political growth while providing the necessary protections and resources to those who are marginalized or at risk of being left behind in the twenty-first century. This can be achieved through the implementation of laws and policies that promote greater political and economic participation, as well as ensuring that everyone has access to resources and opportunities regardless of their gender, ethnicity, or socioeconomic background in Africa. By advocating for social and political reforms, we can create a more just and equitable society where all individuals have an equal voice and access to resources, regardless of their identity or background in the region. These reforms should also address the issues of gender inequality, racism, and economic inequality by ensuring that everyone in the region is given an equal chance to access resources and take part in the political process, regardless of their identity or background.

To ensure the success of these reforms and a brighter future for those living in African countries, it is essential that governments and other decision-makers create policies and laws that provide equal rights and protections, access to resources, and opportunities for individuals regardless of their gender, ethnicity, or socioeconomic background. This will ensure that the most vulnerable members of society are not left behind and can benefit from the progress and development occurring in their various African communities, as well as have the chance to take part in decision-making processes and influence policy-making that will benefit them. Additionally, civil society organizations can play an important role in supporting social and political reforms by providing resources to those who are most in need in the region, pushing for positive change through advocacy and outreach efforts, and developing programs that help build capacity and raise awareness of the issues faced by underserved populations so that everyone can have the opportunity to live a better life and enjoy the same rights as others in the nineteenth century and beyond. Thus, creating a fairer and more equitable society in the twenty-first century will require not only significant changes to existing policies, laws, and structures but also the implementation of new ones that prioritize inclusion and equity for all citizens of Africa, regardless of their background, and actively strive to close the gaps between those who have the most and those who have the least access to opportunities.

This includes the full recognition of human rights, the implementation of labor laws that are just and equitable, and increased access to quality education, healthcare, and other services that can help individuals realize their potential and reach their highest aspirations regardless of their economic, social, or cultural background in the region. It also ensures that everyone in the region has a chance to contribute to their community and society in meaningful ways, regardless of their economic, social, or cultural background. This can be achieved through the promotion of inclusion and equity for all people on the African continent, regardless of their economic, social, or cultural backgrounds, as well as the provision of resources and programs that empower and enable those who are most in need to realize their potential and enjoy the same rights as others, while also strengthening the capacity of civil society organizations to advocate for and support social and political reform efforts and push for meaningful change through grassroots action, awareness-raising campaigns, and other initiatives.

Additionally, civil society organizations can play an important role in supporting social and political reforms by providing resources to those who are most in need, pushing for positive change through advocacy and outreach efforts, developing programs that help build capacity and raise awareness of the issues faced by those living in marginalized communities, and taking an active role in developing policies that ensure equitable access to economic, social, and cultural rights for all people in the region, regardless of their economic, social, or cultural backgrounds. All of these efforts can help to create an inclusive and equitable society in which everyone in the region can live with dignity, enjoy their rights and freedoms, and reach their highest potential. This will ensure a fairer society for all and ensure that each individual on the African continent has an equal chance to make their own choices, live free from discrimination and poverty, and contribute to the advancement of their community and nation as a whole. To achieve this, it is essential for governments and international organizations to promote economic and social policies that foster inclusivity, equity, and justice on a global scale as well as at the local and national levels throughout African communities. Furthermore, civil society organizations must take an active role in holding governments and other actors accountable for their decisions and policies to ensure that these reforms are meaningful and comprehensive and have a lasting impact on the lives of individuals and communities in African countries.

2.4.2 Improved Transparency and Accountability

One way to eliminate obstacles to foreign investment and development in Africa is to improve transparency and accountability by creating incentives and policies that allow citizens to easily hold their leaders accountable for the decisions they make and the resources they manage. This can be achieved by introducing mechanisms for public scrutiny and feedback, increasing public access to information, developing systems for tracking and monitoring financial activities, and promoting greater collaboration between government agencies and citizens to ensure that resources are allocated in the most effective and efficient manner possible. In addition, African governments should prioritize investment in education and training to equip their citizens with the skills necessary to seize new economic opportunities and improve their productivity, as well as create an enabling environment that encourages entrepreneurship and innovation to attract foreign investors. This can be done by reforming legislation and regulations, encouraging public–private partnerships, and investing in capacity building to create a more skilled workforce capable of taking advantage of new opportunities and spurring economic growth and development across the continent.

By doing these things, African nations can create a more conducive environment for foreign investment, which in turn will help promote the economic development and prosperity of the region while also providing job security and an increased quality of life for its citizens. All in all, African countries must put in place various measures to promote and encourage economic growth and development by investing in education and training, reforming legislation and regulations, encouraging public–private partnerships, and investing in capacity building to create a stronger, more sustainable, and vibrant economy that will be able to attract foreign investors and create job opportunities while providing its citizens with a better quality of life. This will have a positive impact on the economic development and prosperity of African nations, as well as providing job security and a higher standard of living to their citizens. This will in turn create an economic and political climate conducive to foreign investment, allowing African countries to benefit from the influx of resources, expertise, and capital that can help them develop their economies further while creating more job opportunities and increasing their citizens' quality of life, ultimately leading to increased economic development and prosperity in the region for themselves and for the greater good of society as a whole.

African governments must also provide access to finance for small businesses and entrepreneurs, as well as support infrastructural investments in areas such as transport, energy, and telecommunications to create an environment conducive to private sector investment to support economic growth and development on the continent. Furthermore, policies and regulations that encourage foreign direct investment, such as tax incentives and subsidies, should also be put in place to attract foreign investors who will bring more resources, expertise, and capital to the African continent. This influx of resources, expertise, and capital can provide job opportunities for citizens and offer them the chance to pursue a better standard of living while at the same time providing African countries with access to the latest technologies and enabling them to develop their industries and infrastructure, which will make them more competitive in the global marketplace.

Ultimately, the development of the African continent hinges on governments creating a welcoming environment for foreign investment by enacting pro-growth economic policies, providing access to finance and infrastructure investments, and encouraging foreign direct investment through incentives and subsidies. The development of Africa requires governments to implement pro-growth economic policies that create a favorable environment for foreign investment as well as ensure access to finance and infrastructure investments for small businesses and entrepreneurs to spur economic growth and create a more prosperous future for the African continent. This is because foreign investment can provide a much-needed boost to the economy in terms of resources, expertise, and capital that can help stimulate economic growth, create job opportunities for citizens, and increase the standard of living in African countries. This can be done by enacting pro-growth economic policies, providing access to finance and infrastructure investments, and encouraging foreign direct investment through incentives and subsidies.

In addition to providing the necessary resources and investments for economic development, African governments should also focus on creating a favorable business environment for FDI by improving the legal and regulatory framework, reducing bureaucratic hurdles, and promoting public–private partnerships to foster a climate of innovation, entrepreneurship, and economic growth that is conducive to FDI. This strategy would enable the African continent to attract foreign direct investment, stimulate economic growth, and create job opportunities that would benefit the citizens of Africa and ultimately improve the overall standard of living in African countries while taking into account the interests of foreign investors and preserving their rights as well as the rights of African citizens. This would be a win–win situation for all parties involved, as foreign investment could potentially provide the resources and capital that African countries need to improve their economies and standards of living while providing foreign investors with the opportunity to grow their businesses and generate returns on their investments as well, while also providing them with access to new and emerging markets, resources, and a unique consumer base that they may not have otherwise had access to had they not decided to invest in the African continent.

This strategy of encouraging foreign direct investment is key to unlocking Africa's economic potential and creating a more prosperous and secure future for African countries and their citizens. This could all be achieved through an investment climate that balances the interests of both foreign investors and African citizens, in which the rights of both parties are respected, preserved, and given equal consideration, while at the same time ensuring that any investment that is made into African countries is conducted responsibly, ethically, and with the utmost respect for the environment, local communities, and cultures of African nations. This would allow African countries to reap the benefits of foreign investment without sacrificing the rights of their citizens or undermining their sovereignty while also protecting foreign investors from any potential risks or losses that may occur as a result of their investment in African countries. As a result, FDI has the potential to create an environment of mutual benefit, one in which both African countries and foreign investors can benefit from each other's contributions and build a more secure and prosperous future together. African countries must work to create an attractive, safe, and secure investment climate for foreign investors while ensuring that their citizens' rights are protected and their sovereignty is respected to realize this potential and maximize the potential benefits of FDI.

While African governments have a crucial role to play in fostering economic growth and development, there are also measures that can be taken at the individual level, such as engaging in skills-based training and pursuing educational opportunities, to equip citizens with the necessary tools and capabilities to succeed in an increasingly globalized economy. With the right approach and support from both the public and private sectors, African countries have a real opportunity to harness foreign direct investment as a tool for economic growth and development and create a more prosperous and stable future for their citizens.

Additionally, African citizens must take initiative to improve their economic situation and take advantage of the opportunities available to them. To capitalize on the potential for economic growth and development, African governments need to implement a comprehensive strategy that will enable them to better position the continent in an increasingly globalized economy. This strategy must include providing access to quality education and vocational training, creating a strong regulatory framework, encouraging foreign direct investment, improving the business environment, and creating job opportunities for its citizens. In addition, African countries must also prioritize the development of physical infrastructure, such as roads and ports, to facilitate trade and investment, as well as developing a transparent and accountable system of governance to ensure that the African people are protected from corruption and have their rights respected. These efforts will help to ensure that the continent is well-equipped to take advantage of the opportunities that arise from the ever-changing global landscape and will ensure that African citizens are equipped with the tools they need to achieve economic prosperity for themselves and for their countries.

Additionally, African governments must also ensure greater transparency and accountability in their resource allocation decisions to ensure that these investments are used wisely and equitably to effectively promote economic growth and development in Africa. Furthermore, the promotion of regional integration through initiatives such as the African Continental Free Trade Area should also be a priority for African countries to take advantage of their combined resources and create more efficient markets for goods and services, as well as allowing for the creation of more competitive industries on the continent. This in turn could help attract more foreign investment and lead to the creation of more jobs, helping to create a sustainable environment for economic growth and development in Africa.

Generally, the current economic situation in Africa is ripe for success, and with strong government initiatives, improved transparency and accountability in their policies, as well as taking advantage of regional integration opportunities, African countries can benefit from a more prosperous future going beyond the social expectations of their times. African governments must also be committed to addressing the underlying structural weaknesses that have hindered economic growth and development in the region for so long that they can take advantage of the opportunities presented by the continent's resources and people to realize their true potential and become an even more prosperous and influential region on the global stage.

However, to take advantage of these opportunities and promote economic growth and development on the continent, African governments must invest in the development of essential infrastructure, such as power generation and transportation networks, that will enable businesses to operate efficiently and create jobs for their citizens while also providing access to markets, both locally and internationally, as well as investing in human capital by providing access to quality education and healthcare for their citizens, which will help build a skilled workforce and lay the foundation for sustainable economic growth in the region. Additionally, African governments should work to strengthen existing regional integration initiatives, such as the African Continental Free Trade Area (AfCFTA), to improve intraregional trade and encourage more investment from both domestic and international firms, allowing for the efficient movement of goods and services between African countries, which can help spur economic growth in the region and improve the livelihoods of their citizens.

Finally, African governments must strive to reduce inequality on the continent by promoting inclusive economic growth and development strategies, such as those outlined above, which focus on economic diversification and increasing the participation of women in the workforce, so that the continent can move closer to achieving the United Nations' Sustainable Development Goals (SDGs) and create an equitable society where everyone can benefit from the economic opportunities that are available on the continent. Furthermore, African governments should implement a range of policies to ensure responsible and transparent governance, such as implementing anti-corruption laws and initiatives that will promote ethical business practices and create an environment of trust between governments and the private sector so that foreign companies are more willing to invest in the continent and do business with African countries, which can help drive economic growth and development in the region while also reducing the risk of corruption and malpractice that can be seen in today's society.

Additionally, African governments should strive to reduce inequality on the continent by investing in social welfare programs that are targeted toward vulnerable and marginalized populations, such as women and children, to promote equality of opportunity for all citizens and create an environment of equal opportunity that can help reduce poverty, improve access to basic services, and ensure the social inclusion of all citizens on the continent. African governments should also create policies that promote good governance, such as establishing the rule of law and providing a stable and reliable legal framework for businesses to operate within, which will help attract foreign investment and create an environment that is conducive to economic growth and development while also creating a society that is more equal and equitable for all citizens on the continent. Furthermore, African governments should prioritize reforms in the areas of public finance management and governance to ensure transparency, accountability, and greater efficiency in the allocation and use of resources to create an environment of greater trust and confidence in the public sector, which can help encourage investment from both domestic and foreign sources while also helping to reduce the risk of corruption and malpractice that can have a negative impact on economic growth and development and social stability on the continent.

African governments should also work toward strengthening and improving their public institutions, as well as investing in education and healthcare systems to ensure that all citizens on the continent have access to quality services that can help to promote economic growth and development, reduce inequality, and foster a more equitable society on the continent in the eyes of the society as a whole. Overall, African governments have a responsibility to invest in social welfare programs and governance reforms to promote equality of opportunity, attract foreign investment, and ensure the social inclusion of all citizens while also contributing to greater economic growth and development on the continent to improve the well-being of their citizens while also creating the necessary conditions for sustainable economic growth and development on the continent for the first time in history.

In conclusion, by taking these measures, African governments can reduce inequality on the continent and create an environment of equal opportunity that can help to improve the economic prospects and well-being of all citizens while also helping to ensure greater social stability and development on the continent in the long-term, which allowed them to stand out in a very male-dominated industry and be seen as strong role models for other women who wanted to pursue a career in the performing arts by challenging stereotypes and proving themselves to be highly talented, capable individuals while also providing a greater sense of agency and self-determination to women in the industry. At the same time, they were able to improve their own financial stability by doing so.

Moreover, African governments should focus on improving the efficiency and transparency of public expenditure to ensure that resources are used for the benefit of citizens and that there is greater accountability of public officials and more access to justice while also promoting good governance reforms and regulations that enable strong partnerships between government, the private sector, civil society, and citizens to create a more conducive environment for investment, job creation, and economic growth and development on the continent, which, in turn, can create a better quality of life for all citizens and promote greater stability and peace across the continent in the long term. This would ultimately benefit the African people by providing them with greater economic opportunities and financial stability while also enabling them to realize their own potential and create a better future for themselves. By taking these measures, African governments can reduce inequality on the continent and create an environment of equal opportunity that can help to improve the economic prospects and well-being of all citizens while also helping to ensure greater social stability and development on the continent.

2.4.3 Improved Transportation Infrastructure

Improving the continent's transportation infrastructure is one way to reduce barriers to foreign investment and development on the African continent and make it easier for businesses to move people and goods within the region as well as between it and other regions of the world. Investing in transportation infrastructure could reduce the cost and time of shipping and importing goods, which would make foreign investment more attractive to both African and international businesses. As transportation improvements reduce logistical barriers to foreign investment, African countries could gain more direct access to global markets and increase the flow of capital in and out of the continent. Furthermore, as better transportation infrastructure would increase the speed and efficiency of travel, businesses in the region would be able to conduct more operations within Africa itself, which would help foster regional trade and contribute to economic growth and development on the African continent.

Additionally, improved transportation infrastructure could lead to an increase in tourism, as traveling across the continent would become easier and more convenient for both African citizens and foreign visitors, which could provide a much-needed boost to African economies. This increase in trade and tourism could further stimulate economic growth on the continent, creating jobs, reducing poverty, and fostering a more prosperous future for Africa. Overall, better transportation infrastructure could have a wide range of positive benefits for the African continent, and the development of such infrastructure should be a priority for both African governments and foreign organizations seeking to improve economic conditions on the continent. Improving transportation infrastructure in Africa could have far-reaching economic benefits for the continent by providing more efficient and affordable access to resources, goods, and services, leading to increased trade, investment, and job opportunities across the continent, as well as making it more feasible for visitors to explore African countries and take advantage of the unique cultural, historical, and natural experiences they have to offer.

This, in turn, could result in more money circulating through African economies, allowing more individuals and families to experience greater financial security and well-being, leading to an improved quality of life for African citizens and a brighter future for the continent as a whole. Ultimately, increased transportation infrastructure in Africa could lead to a broad range of economic and social benefits that would have a lasting impact on the continent as well as contribute to the growth of tourism, which could provide additional income to African nations through fees and taxes associated with travel and generate additional opportunities for employment in the hospitality industry and other related sectors. With improved infrastructure, Africa could more easily reap the benefits of increased economic growth, increased job and business opportunities, better access to essential resources and services, improved tourism prospects, and greater financial security for its citizens, all while improving their overall quality of life and helping to create a better future for generations to come.

Additionally, investing in infrastructure would help to increase trade between African countries and other nations, resulting in the growth of the African economy on an international scale and creating a more level playing field for African businesses to compete in the global marketplace as well as a more favorable environment for foreign direct investment and other forms of capital inflows into the continent. This influx of capital could have profound implications for the development of African economies, leading to a more sustainable economic system, the creation of new businesses and industries, and the introduction of new technologies and processes that would facilitate the production of goods and services with greater efficiency, enabling African nations to take advantage of the global economy to boost their own prosperity. Overall, investing in infrastructure in Africa could have a profoundly positive effect on the continent and its people by allowing for easier access to essential goods and services, providing the potential for job growth and improved quality of life, increasing financial security and stability, stimulating the growth of African economies in international markets, and attracting more foreign investment to support further development.

In addition to improving transportation infrastructure, developing countries in Africa must also invest in the education and training of workers in the region to create a knowledgeable and skilled workforce that can benefit from the increased investment in infrastructure and effectively use it to increase the competitiveness of African businesses in the global marketplace. This investment in education and training can help to ensure that African workers have the necessary knowledge and skills to understand and utilize the benefits of increased infrastructure, leading to a better-equipped workforce for competing in international markets and an increase in overall productivity and the quality of goods and services provided by African businesses. This, in turn, will lead to higher wages, increased access to healthcare, and better living standards for those in the region. Ultimately, the investment in infrastructure and education will pay dividends for African nations by helping to create a more prosperous continent and its citizens over the long term.

Creating a better network of roads, railways, and airports would not only help the continent become more accessible to foreign investment but also open up more opportunities for regional trade within the African continent. Investing in transportation infrastructure would provide many opportunities for businesses to expand, making it easier for African countries to connect and trade with the rest of the world. By providing well-maintained roads and railways, as well as bridges and tunnels, transportation infrastructure allows for more efficient transport of goods across countries and within countries, which could lead to economic growth and development. By increasing the efficiency of transportation networks and providing better access to markets, foreign investors will be more likely to enter African markets and invest in local businesses, which could spur even more economic growth and development and create more jobs in African countries. This, in turn, could lead to increased job opportunities for African citizens and improved standards of living, as they will be able to access new markets, receive better wages, and generally have greater opportunities than before investment in transportation infrastructure. Increasing the availability of reliable transportation networks would create more efficient access to markets, reduce the costs associated with trade and investment, and allow for increased regional integration and collaboration as well as faster, more reliable transportation of people, which could further accelerate economic growth and development on the African continent.

2.5 Conclusion

Global investment and development in Africa have enabled increased economic growth, improved infrastructure, a rise in the average standard of living, and greater access to healthcare, education, and other essential services. This has enabled African countries to become more integrated into the global economy, enjoy increased economic stability, and experience a greater degree of prosperity than would have been possible without foreign investment and development. This has resulted in a more equitable and prosperous Africa, where citizens are able to experience the same rights and opportunities as their peers around the world while also enjoying a greater quality of life. Foreign investment and development in Africa have had a profound effect on the economic landscape of the region, allowing citizens to benefit from increased access to goods, services, technologies, healthcare, and education. This has led to a more equitable and prosperous Africa, where citizens are able to experience the same rights and opportunities as those around the world while also enjoying an increased quality of life. It has also led to increased entrepreneurship and economic independence for African countries, allowing them to be more competitive in the global marketplace and giving their citizens a greater say in their own destinies. The last few decades have seen dramatic changes in African societies, leading to increased prosperity, a better quality of life, and an overall more equitable continent. This has been made possible by the increased availability of technology and the internet, allowing African countries to catch up with the rest of the world in terms of development and access to resources that were once impossible to obtain. African countries must prioritize economic policies that address the needs of their citizens, promote good governance, strengthen relationships with other nations and organizations, foster a culture of entrepreneurship and innovation, utilize technology to reduce poverty, increase access to healthcare and education, and develop better systems for tackling climate change to make progress in the twenty-first century. Doing these things will create an environment that promotes growth and development, reduces poverty, and creates a more equitable society for all.

References

AFRICA–EU: Investment Promises. (2022, March). *Africa Research Bulletin: Economic, Financial and Technical Series, 59*(1). https://doi.org/10.1111/j.1467-6346.2022.10412.x

Africanacity of International Investment Law: A Reflection on Investment Agreements in Africa. (2021, September). *Journal of Law, Policy and Globalization.* https://doi.org/10.7176/jlpg/113-01

Amendolagine, V., & Prota, F. (2021, May). Bilateral investment treaties and backwards linkages in sub-Saharan Africa. *International Economics, 165*, 172–185. https://doi.org/10.1016/j.inteco.2021.01.001

Asiedu, E. (2004, January). Policy reform and foreign direct Investment in Africa: Absolute Progress but relative decline. *Development Policy Review, 22*(1), 41–48. https://doi.org/10.1111/j.1467-8659.2004.00237.x

Baltag, C. (2009, January 1). International investment arbitration: Substantive principles, principles of international investment law and the fair and equitable treatment standard in international foreign investment law. *British Yearbook of International Law, 79*(1), 378–388. https://doi.org/10.1093/bybil/79.1.378

Capital Investment In Africa : Its Course and Effects. (1939, March 1). *International affairs*. https://doi.org/10.2307/3019884.

Cheng, W. (2023, March 13). The green investment principles: From a nodal governance perspective. *International Environmental Agreements: Politics, Law and Economics., 23*, 373. https://doi.org/10.1007/s10784-023-09595-w

Cheung, Y. W., de Haan, J., Qian, X., & Yu, S. (2012, April 16). China's outwards direct investment in Africa. *Review of International Economics, 20*(2), 201–220. https://doi.org/10.1111/j.1467-9396.2012.01017.x

Clark, W. (1997). The global information economy and Its effect on local economic development. *New Directions for Higher Education, 1997*(97), 51–61. https://doi.org/10.1002/he.9705

Ewelukwa Ofodile, U. (2014, January 1). Managing foreign investment in agricultural land in Africa: The role of bilateral investment treaties and international investment Contracts. *Law And Development Review, 7*(2). https://doi.org/10.1515/ldr-2014-0010

Fosu, A. K. (2013, November). Impact of the global financial and economic crisis on development: Whither Africa? *Journal of International Development, 25*(8), 1085–1104. https://doi.org/10.1002/jid.2971

Gazzini, T. (2009). General principles of law in the field of foreign investment. *The Journal of World Investment & Trade, 103–119*, 103. https://doi.org/10.1163/221190009x00240

Grekou, C., & Owoundi, F. (2020, December). Understanding how foreign direct investment inflows impact urbanization in Africa. *International Economics, 164*, 48–68. https://doi.org/10.1016/j.inteco.2020.07.002

Gunter, A. (2006, June 30). Integrated development plans and local economic development: The case of Mpumalanga Province, South Africa. Africa. *Insight, 35*(4). https://doi.org/10.4314/ai.v35i4.22452

Hepburn, J. (2015, September 4). The unidroit principles of international commercial contracts and investment treaty arbitration: A limited relationship. *International and Comparative Law Quarterly, 64*(4), 905–933. https://doi.org/10.1017/s0020589315000354

Khumon, P. (2016). Development of rule-based investment protection provisions in mega-FTAs: A manifestation or a negation from customary international law principles? *SSRN Electronic Journal.* https://doi.org/10.2139/ssrn.2801917

Klimova, E. (2022). Opportunities and directions for the development of global e-commerce in the context of digitalization. *Economic Development Research Journal, 4*, 45–49. https://doi.org/10.54092/25420208_2022_4_45

Leiter, A. (2021, November 26). Protecting concessionary rights: General principles and the making of international investment law. *Leiden Journal of International Law, 35*(1), 55–69. https://doi.org/10.1017/s0922156521000601

Li, T. M. (2014, July 14). What is land? Assembling a resource for global investment. *Transactions of the Institute of British Geographers, 39*(4), 589–602. https://doi.org/10.1111/tran.12065

Nel, E. L., & Humphrys, G. (1999, June). Local economic development: Policy and practice in South Africa. *Development Southern Africa, 16*(2), 277–289. https://doi.org/10.1080/03768359908440077

Nkuna, O. (2015, December 24). Intra-SADC foreign direct investment: A gravity approach to South Africa outwards investment. International. *Journal of Economics and Finance, 8*(1), 79. https://doi.org/10.5539/ijef.v8n1p79

Nwokolo, S. C., Amadi, S. O., Obiwulu, A. U., Ogbulezie, J. C., & Eyibio, E. E. (2022a). Prediction of global solar radiation potential for sustainable and cleaner energy generation using improved angstrom-Prescott and Gumbel probabilistic models. *Cleaner Engineering and Technology, 6*(January), 100416. https://doi.org/10.1016/j.clet.2022.100416

Nwokolo, S. C., Obiwulu, A. U., & Ogbulezie, J. C. (2022b). Machine learning and analytical model hybridization to assess the impact of climate. *Physics and Chemistry of the Earth, 130*(February), 103389. https://doi.org/10.1016/j.pce.2023.103389

Nwokolo, S. C., & Ogbulezie, J. C. (2018a). A qualitative review of empirical models for estimating diffuse solar radiation from experimental data in Africa. *Renewable and Sustainable Energy Reviews., 92*, 353. https://doi.org/10.1016/j.rser.2018.04.118

Nwokolo, S. C., & Ogbulezie, J. C. (2018b). A qualitative review of empirical models for estimating diffuse solar radiation from experimental data in Africa. *Renewable and Sustainable Energy Reviews, 92*. https://doi.org/10.1016/j.rser.2018.04.118

Nwokolo, S. C., & Ogbulezie, J. C. (2018c). A quantitative review and classification of empirical models for predicting global solar radiation in West Africa. *Beni-Suef University Journal of Basic and Applied Sciences, 7*(4), 367–396. https://doi.org/10.1016/j.bjbas.2017.05.001

Nwokolo, S. C., Ogbulezie, J. C., & Obiwulu, A. U. (2022c). Impacts of climate change and meteo-solar parameters on photosynthetically active radiation prediction using hybrid machine learning with physics-based models. *Advances in Space Research, 70*, 3614. https://doi.org/10.1016/j.asr.2022.08.010

Nwokolo, S., & Ogbulezie, J. (2017a). A critical review of theoretical models for estimating global solar radiation between 2012-2016 in Nigeria. *International Journal of Physical Research, 5*(2), 60. https://doi.org/10.14419/ijpr.v5i2.8160

Nwokolo, S., & Ogbulezie, J. (2017b). Modelling the influence of cloudiness on diffuse horizon-tal irradiation under various sky conditions in Nigeria. *International Journal of Physical Research, 5*(2), 91. https://doi.org/10.14419/ijpr.v5i2.8312

Nwokolo, S., & Otse, C. (2019). Impact of sunshine duration and clearness index on diffuse solar radiation estimation in mountainous climate. *Trends in Renewable Energy, 5*(3), 307–332. https://doi.org/10.17737/tre.2019.5.3.00107

Nwokolo, S. C., Obiwulu, A. U., Amadi, O., & Ogbulezie, J. C. (2023a). Assessing the impact of soiling, tilt angle, and solar radiation on the performance of solar PV systems. *Trends in Renewable Energy, 9*(2), 121–137. https://doi.org/10.17737/tre.2023.9.2.00156

Nwokolo, S. C., Ogbulezie, J. C., & Ushie, O. J. (2023b). A multimodel ensemble-based CMIP6 assessment of future solar radiation and PV potential under various climate warming scenarios. *Optik, 285*(May), 170956. https://doi.org/10.1016/j.ijleo.2023.170956

Nwokolo, S. C., Proutsos, N., & Meyer, E. L. (2023c). Machine learning and physics-based hybridization models for evaluation of the Effects of climate change and urban expansion on photosynthetically active radiation. *Atmosphere, 14*(687), 1–36. https://doi.org/10.3390/atmos14040687

Nwokolo, S. C., Proutsos, N., & Meyer, E. L. (2023d). Machine learning and physics-based hybridization models for evaluation of the effects of climate change and urban expansion on photosynthetically active radiation. *Atmosphere, 14*(687), 1–36. https://doi.org/10.3390/atmos14040687

Obiwulu, A. U., Chendo, M. A. C., Erusiafe, N., & Nwokolo, S. C. (2020a). Implicit meteorological parameter-based empirical models for estimating back temperature solar modules under varying tilt-angles in Lagos. *Nigeria. Renewable Energy, 145*, 442. https://doi.org/10.1016/j.renene.2019.05.136

Obiwulu, A. U., Erusiafe, N., Olopade, M. A., & Nwokolo, S. C. (2020b). Modelling and optimization of back temperature models of mono-crystalline silicon modules with special focus on the effect of meteorological and geographical parameters on PV performance. *Renewable Energy, 154*, 404. https://doi.org/10.1016/j.renene.2020.02.103

Obiwulu, A. U., Erusiafe, N., Olopade, M. A., & Nwokolo, S. C. (2022). Modelling and estimation of the optimal tilt angle, maximum incident solar radiation, and global radiation index of the photovoltaic system. *Heliyon, 8*(6), e09598. https://doi.org/10.1016/j.heliyon.2022.e09598

Owens, T. L., & Lane, J. E. (2014). Cross-border higher education: Global and local tensions within competition and economic development. New directions for. *Higher Education, 2014, December*(168), 69–82. https://doi.org/10.1002/he.20114

Pamba, D. (2023, January 5). Does foreign direct investment crowd in or crowd out domestic Investment in South Africa? An ARDL-ECM approach. *International Journal of Management Research and Economics, 3*(1). https://doi.org/10.51483/ijmre.3.1.2023.18-35

Peat, D. (2018, August 14). International investment law and the public law analogy: The fallacies of the general principles method. *Journal of International Dispute Settlement, 9*(4), 654–678. https://doi.org/10.1093/jnlids/idy028

Pedler, F. J. (1955, October). Foreign investment in West Africa. *International Affairs, 31*(4), 459–468. https://doi.org/10.2307/2604821

Qumba, M. F. (2018, July 3). Safeguarding foreign direct investment in South Africa: Does the protection of investment act live up to its name? *South African Journal of International Affairs, 25*(3), 341–364. https://doi.org/10.1080/10220461.2018.1523024

Reinisch, A. (2014, November 29). The relevance of the UNIDROIT principles of international Commercial Contracts in international investment arbitration. *Uniform Law Review–Revue De Droit Uniforme, 19*(4), 609–622. https://doi.org/10.1093/ulr/unu035

Rogerson, C. M. (2010, October). Local economic development in South Africa: Strategic challenges. *Development Southern Africa, 27*(4), 481–495. https://doi.org/10.1080/0376835x.2010.508580

Rogerson, C. M., & Rogerson, J. M. (2010, October). Local economic development in Africa: Global context and research directions. *Development Southern Africa, 27*(4), 465–480. https://doi.org/10.1080/0376835x.2010.508577

Särkänne, K. (2021, September 1). A guide to general principles of law in international investment arbitration. *ICSID Review–Foreign Investment Law Journal, 36*(3), 657–660. https://doi.org/10.1093/icsidreview/siab040

Saunders, I. (2021, December 15). Chasing the rabbit: The resurgence of general principles of law in international investment law and beyond. *The Journal of World Investment & Trade, 23*(3), 457–475. https://doi.org/10.1163/22119000-12340233

Schill, S. W. (2009, April 1). Principles of international investment law. *European Journal of International Law, 20*(2), 471–473. https://doi.org/10.1093/ejil/chp019

Southall, R., & Melber, H. (Eds.). (2009, July 30). *A new scramble for Africa?: Imperialism, investment and development.*

Thulasidhass, P. (2015, June 1). Most-Favoured-nation treatment in international investment law: Ascertaining the limits through interpretative principles. *Amsterdam Law Forum, 7*(1), 3. https://doi.org/10.37974/alf.272

Titi, C. (2015, August 1). International investment law and the European Union: Towards a new generation of international investment agreements. *European Journal of International Law, 26*(3), 639–661. https://doi.org/10.1093/ejil/chv040

Vaksha, A. K., & Singh, S. P. (2020). Revisiting Vattel's law of nations to discern the classical principles of the international investment law. *Beijing Law Review, 11*(01), 277–291. https://doi.org/10.4236/blr.2020.111019

Wasike, S. (2022, May 31). Foreign direct investment and economic growth in Africa. *International Journal of Current Science Research and Review, 05*(05). https://doi.org/10.47191/ijcsrr/v5-i5-50

Wolfson, J., & Frisken, F. (2000, December). Local response to the global challenge: Comparing local economic development policies in a regional context. *Journal of Urban Affairs, 22*(4), 361–384. https://doi.org/10.1111/0735-2166.00062

Chapter 3
Remedies to the Challenges of Renewable Energy Deployment in Africa

3.1 Introduction

Renewable energy (RE) sources' competitive position is significantly weakened by funding and financing issues, which makes it challenging for investors to view them as viable alternatives to more traditional, nonrenewable sources of energy. This is especially true when nonrenewable sources such as oil and gas are heavily subsidized by governments while RE sources are not, rendering it impossible for RE resources to compete with the subsidized costs of nonrenewable resources. This lack of government support has hindered the proliferation of RE sources and made it challenging for investors to switch from conventional sources to RE sources. Additionally, a lack of funding and financial support creates a disparity in the cost of production between nonrenewable and renewable energy sources, making it much more difficult for renewables to be a viable economic alternative.

African governments have begun to recognize the importance of making investments in renewable energy sources (Nwokolo et al., 2022a) and offering incentives to private investors who are eager to switch to renewable energy sources (Nwokolo & Ogbulezie, 2018a) to ensure energy security (Nwokolo & Ogbulezie, 2018b) and reduce their reliance on fossil fuels (Nwokolo & Ogbulezie, 2018c). This entails providing investors with tax breaks and other financial incentives (Nwokolo & Ogbulezie, 2017a), facilitating their investment in renewable energy sources (Nwokolo & Ogbulezie, 2017b), and incentivizing them to switch from nonrenewable sources (Nwokolo & Otse, 2019) by making renewable energy sources a more appealing investment than their nonrenewable counterparts (Nwokolo et al., 2022b). African countries have implemented policies such as feed-in tariffs (Nwokolo et al., 2022c), government subsidies (Nwokolo et al., 2023a), and incentives for private sector investment in renewable energy projects (Obiwulu et al., 2020a) to reduce the investment costs of renewable energy (Obiwulu et al., 2020b) and increase its uptake (Obiwulu et al., 2022), allowing them to lower the financial barrier to switching from nonrenewable to renewable energy sources

S. C. Nwokolo et al., *Africa's Path to Net-Zero*, CSR, Sustainability, Ethics & Governance, https://doi.org/10.1007/978-3-031-44514-9_3

(Oluoch et al., 2021) and make the transition more appealing to potential investors (Pradhan et al., 2014). This has led to an increase in the adoption of renewable energy sources across the continent and given African governments the opportunity to switch from their reliance on nonrenewable energy sources—which have been linked to higher levels of emissions and environmental degradation—to cleaner and more sustainable renewable sources of energy, ensuring energy security and gradually reducing their environmental impact (Renewables in Africa., 2009).

To make it simpler for renewable energy projects to access financing (Sakah et al., 2017), African governments must also encourage public–private partnerships (Schwerhoff et al., 2017), encourage innovation (Shrimali et al., 2016) and entrepreneurship in the sector (Shum, 2017), and build solid financial mechanisms (Sweidan et al., 2021). African governments must fund and construct infrastructure for renewable energy sources (Wang et al., 2023) and provide technical assistance to local communities to increase awareness of renewable energy sources, the potential benefits they offer, and the resources available for financing them. To ensure a smooth transition, African governments must also take proactive measures such as forming the necessary regulatory frameworks, setting ambitious targets, and putting in place suitable policy instruments such as feed-in tariffs, carbon taxes, and emissions trading systems to provide a steady and secure flow of capital throughout the continent.

This chapter discusses remedies for renewable energy deployment challenges in Africa, focusing on governments, the private sector, and civil society organizations to maximize benefits. African governments must also be proactive in taking the necessary steps to ensure that renewable energy sources become more widely available and accessible throughout the continent. African governments need to act quickly to provide reliable and affordable renewable energy solutions and the necessary infrastructure to support them. African governments have a great responsibility in the shift to renewable energy sources; they must act quickly and decisively in creating policy instruments, setting ambitious targets and regulations that are necessary to ensure that renewable energy is used in the most cost-effective manner, and driving innovation in renewable energy technology. This can be accomplished by providing incentives, investing in research and development, and creating public–private partnerships to develop large-scale renewable energy projects.

3.2 Remedies to the Challenges of Renewable Energy Deployment in Africa

3.2.1 Financial and Economic

The competitiveness of renewable energy sources is significantly impacted by financial and funding limitations, which makes them challenging for investors to view as viable alternatives. Governments must increase financial access, reduce subsidies for conventional energy sources, and eliminate financial and economic barriers such as high financing and transaction costs to promote the deployment of renewable energy in developing nations. As a result, there is greater access to capital for renewable energy projects and a level playing field for investors to evaluate the economic viability of renewable energy sources. Governments can promote the development of renewable energy sources and a cleaner, healthier, and more sustainable environment by removing these obstacles.

African countries are rapidly realizing the value of investing in renewable energy to secure energy security and reduce dependency on fossil fuels. African nations have used policy initiatives such as incentives for private sector investment, feed-in tariffs, and government subsidies to lower investment costs and promote uptake. They have also sponsored international efforts such as the Sustainable Energy Fund for Africa and the Climate Investment Funds to expand the availability of funding for renewable energy projects. However, much more has to be done to reach renewable energy targets and reduce reliance on fossil fuels. African countries must continue to implement creative policy initiatives, stimulate public–private partnerships, encourage innovation and entrepreneurship, and provide solid funding channels for renewable energy projects. African governments must make investments in renewable energy infrastructure, increase capacity, and raise public awareness of the advantages of using renewable energy sources if they are to guarantee the success of these initiatives.

Renewable energy has the potential to enhance power availability while lowering greenhouse gas emissions in Africa. However, insufficient research and development funding makes adoption challenging. To address these issues, governments and stakeholders should develop incentives to attract investors and promote long-term renewable energy projects. They should invest in human capital development, investigate technology cost-cutting alternatives, and give technical and financial aid to African countries. Governments may assist the rise of renewable energy in Africa by implementing these policies, allowing countries to reach their clean energy targets and gain improved energy security. This reduces their reliance on fossil fuels, promotes long-term economic growth, and improves public health and quality of life. Despite the abundance of renewable energy sources such as solar, wind, geothermal, and hydroelectric energy that are available in African countries, a lack of resources and infrastructure has hindered their development.

To make renewable energy more accessible and cheaper for producers and consumers, policymakers must emphasize lowering transaction costs. This lowers the expenses involved with the project development cycle, which includes establishing a firm, acquiring finance, negotiating contracts, and constructing and commissioning the project. Renewable energy solutions become more inexpensive and appealing to both producers and consumers when transaction costs are reduced. Policymakers should look for innovative methods to lower transaction costs to accelerate the adoption of renewable energy technology and make it more cost competitive with traditional energy sources. This might include financial incentives, decreasing market entry obstacles, expanding access to cheap credit, and lowering regulatory constraints. These measures will eventually level the playing field between renewable energy technology and traditional energy sources, ultimately reducing the overall costs associated with their development.

Subsidies and support systems are critical for closing the financial gap between renewable energy sources and fossil fuel-based sources, making them more widely available and economically viable. Investments in renewable energy can be made more affordable and competitive with fossil fuel-based energy sources by receiving financial assistance. This is critical in developing countries such as Africa, where the cost of renewable energy is higher than that of fossil fuel-based sources. By providing financial assistance, renewable energy sources such as wind, solar, and hydropower become more viable for rural communities with limited access to conventional energy sources, allowing them to profit from clean, renewable energy sources. This provides dependable and affordable energy sources while reducing air pollution and greenhouse gas emissions, both of which contribute to global warming and climate change. Grants, subsidies, and other forms of financial assistance for renewable energy investments in Africa are critical to making renewable energy more affordable and accessible. Renewable energy investments in Africa are critical for increasing the affordability and accessibility of renewable energy.

The financial divide between fossil fuel-based energy sources and renewable energy sources must be closed to make renewable energy sources more widely available and economically viable. Investments in renewable energy can benefit from financial support to lower their costs and become more affordable compared to energy sources based on fossil fuels. This is crucial in developing nations such as Africa, where sources of renewable energy are more expensive than those that rely on fossil fuels. Rural communities with little access to conventional energy sources can benefit from clean, renewable energy sources such as wind, solar, and hydropower by receiving financial assistance, making renewable energy more viable for them. This provides dependable and inexpensive energy sources while lowering air pollution and greenhouse gas emissions that contribute to global warming and climate change. Making renewable energy more accessible and affordable in Africa requires the implementation of subsidies, grants, and other forms of financial aid.

African governments must enhance their subsidy policies to guarantee equitable access to energy for all of their citizens and encourage the switch from fossil fuels to renewable energy sources. Investment in renewable energy sources should be encouraged by these regulations, which should also safeguard the market from price fluctuations and guarantee affordability and accessibility for rural areas. Governments should encourage the use of clean energy technologies such as wind and solar power through tax credits and support businesses that invest in the research and development of renewable energy sources. African governments can guarantee equitable access to energy for all citizens, lower greenhouse gas emissions, and generate new employment opportunities in the renewable energy sector by enacting more robust and comprehensive renewable energy policies. As the renewable energy sector becomes more competitive and a viable source of job creation for many African countries, this will not only help the continent's nations become more environmentally friendly but also promote economic growth. In summary, African governments must be proactive in promoting the switch from fossil fuels to renewable energy sources, guaranteeing equitable access to energy for all citizens, lowering greenhouse gas emissions, and fostering economic development.

Due to local infrastructure, low credit ratings, and limited access to long-term debt financing, the cost of financing renewable electricity generation in Africa is high. Higher component, installation, and operating costs are also a result of economies of scale and smaller projects. Because many African people and businesses cannot afford the higher costs associated with renewable energy projects, this restriction restricts their access to renewable energy. This restricts their access to renewable energy and keeps them reliant on fossil fuels, which are frequently more expensive and less reliable over the long run. African governments must develop laws and incentives that lower the costs of renewable energy projects, make them competitive with nonrenewable sources, and increase public access to them to solve this problem. This will give them access to the advantages of renewable energy, including their countries' increased economic and social benefits, higher reliability, lower long-term costs, and better environmental outcomes.

3.2.2 Political Barriers

Governments are instrumental in fostering the adoption of renewable energy policies in Africa by removing political barriers and establishing a legislative and policy framework that offers precise guidelines, financial incentives, and other forms of support for renewable energy investments. This framework should include laws governing the use of renewable energy sources, financial incentives for project development, and targeted subsidies to guarantee project profitability. Governments must take the lead and demonstrate their dedication to renewable energy by allocating sufficient funds and resources for research and development, as well as the legal frameworks and financial incentives necessary to implement them in the larger energy system. A lower-carbon energy system that is better for the environment and the world's climate will result from this commitment because it will create the conditions for the widespread adoption of renewable energy sources. Governments must increase the use of renewable energy sources, provide access to financing and investments that make renewable energy sources more affordable and accessible, and introduce policies that promote technological innovation and encourage the adoption of renewable energy to facilitate this transition. Additionally, they must acknowledge the significance of community involvement in the transition and ensure that local communities are involved in decision-making processes regarding renewable energy projects.

The transition to renewable energy sources must be fair from a social and economic perspective, taking into account any potential negative effects on the environment. As a result, there will be a smooth transition to renewable energy sources, promotion of clean, affordable energy, creation of new jobs and business opportunities, a decrease in greenhouse gas emissions, and support of international efforts to slow climate change. Governments must take a comprehensive approach that takes into account the need to lower energy use, increase energy efficiency, and guarantee that all citizens have affordable access to clean energy to accomplish these goals. By increasing government funding for R&D initiatives, offering tax breaks and grants to companies that invest in renewable energy projects, and promoting green energy policies at the local, regional, and federal levels, they should support the development of renewable energy technologies and resources. The public should be made more aware of renewable energy projects, and governments should provide financial incentives to companies that do so, such as tax breaks, loan guarantees, and other financial aid. They should also eliminate any financial, legal, or other obstacles that are impeding the expansion of the renewable energy industry, such as those that restrict access to capital for new renewable energy projects and do not provide a level playing field for both conventional and renewable energy producers. Finally, by offering incentives to businesses, removing financial and regulatory obstacles, and increasing public awareness of the significance of renewable energy, governments must act decisively to promote the development and adoption of renewable energy sources. Africa will benefit from a healthier and more sustainable future as a result of this.

3.2.3 Reforms to the Patchy and Inconsistent Legal and Regulatory Structures

Renewable energy legal and legislative structures in Africa are patchy and erratic. All of these issues exacerbate existing problems with access to energy and quality of life in African countries. With the implementation of improved legal and regulatory frameworks, greater access to renewable energy technologies and resources can be provided to African countries, allowing for improved quality of life through increased economic activity and job opportunities. Additionally, increased investment in the renewable energy sector would help spur economic growth, create jobs, and stimulate economic development, as renewable energy provides an attractive alternative to expensive and polluting fossil fuels. Renewable energy has the potential to provide African countries with an abundance of opportunities for sustainable economic growth and improved quality of life, as well as reduce environmental pollution, leading to better health outcomes and the conservation of natural resources. Overall, renewable energy has the potential to provide an array of opportunities for African countries and could be an integral part of a comprehensive strategy for sustainable economic growth and improved quality of life in the future. By investing in renewable energy technologies, African countries could take advantage of the many potential benefits these resources provide, such as increased energy security, improved access to energy services, economic diversification and job creation, environmental protection, and greater economic self-sufficiency. Through renewable energy investment, African countries could significantly reduce their dependence on expensive and polluting fossil fuels while creating jobs and stimulating economic growth in the process. African countries could take further advantage of the increased energy security, improved access to energy services, and economic diversification that renewable energy investments can provide by supporting renewable energy initiatives, implementing supportive regulations and policies for renewable energy projects, and providing financial assistance. African countries can harness renewable energy potential for a prosperous, secure future for their citizens through these measures.

3.2.4 *Licences and Permits for Renewable Energy-Related Businesses*

Obtaining licences and permits for renewable energy businesses in African countries can hinder capital investment. Without the proper regulatory framework (Akuru et al., 2015), renewable energy investments in African countries may face high risks of failure due to the added complexity of obtaining licences and permits (Alemzero et al., 2021). To ensure successful investments in renewable energy, African governments must make it easier for businesses to obtain the necessary licences (Anderson et al., 2022) and permits by streamlining (Asongu et al., 2021) and expediting the process and providing infrastructure that encourages businesses to invest in the sector (Baye et al., 2021). African governments must recognize the need to simplify the process of obtaining licences (Bistline et al., 2017) and permits for renewable energy-related businesses (Boie et al., 2016), as this will increase investor confidence and enable African countries to reap investments (Bugaje et al., 2006). To accomplish this, African governments should establish regulatory frameworks that allow renewable energy businesses to obtain licences and permits quickly and easily (Dogan et al., 2023). Additionally, African governments should provide financial incentives for businesses to invest in renewable energy, such as tax credits and other forms of subsidies (Dong & Shimada, 2017). African governments should also invest in renewable energy infrastructure, such as grid connections and storage capacity (Kemausuor et al., 2018), which will make it easier for businesses to access renewable energy sources and help ensure a reliable supply of renewable energy for businesses. African governments should also promote renewable energy education and training programs to help ensure that businesses have access to the skilled labour force they need to develop, construct, and operate renewable energy projects. To further incentivize renewable energy investment, African governments should ensure that businesses are able to sell their excess renewable energy back into the grid at competitive rates. Furthermore, African governments should look for ways to encourage investment by providing tax credits (Lucas et al., 2017), grants (Maman et al., 2021), and low-interest loans to businesses to help offset the high upfront costs associated with these projects (Manu et al., 2022). In addition, African governments should create renewable energy-focused policies and regulations. This will help create a competitive energy market, provide investors with the assurance that their investments will remain protected over time, and ensure that businesses are able to access financing to make these investments. In addition to encouraging businesses to invest in renewable energy, African governments should focus on implementing policies and regulations by increasing the transparency and predictability of energy markets (Kuamoah, 2020), providing a platform for renewable energy technologies to compete, and ensuring that businesses have access to the financing they need to make these investments (Lawrence, 2020). In summary, African governments should create renewable energy-focused policies and regulations that promote a competitive energy market (Mohammed et al., 2013) and provide investors with assurances (Mukumba & Chivanga, 2023) that their investments will be secure over the long term while also providing businesses with access to the financing necessary to make renewable energy investments (Njoh et al., 2019). To achieve this, African governments should reduce the upfront costs associated with renewable energy investments. African governments should also ensure that businesses have access to the financing necessary to make renewable energy investments so that the market can remain competitive and secure for investors. Additionally, African governments should provide information and support to businesses considering renewable energy investments, such as technical assistance and knowledge sharing, to ensure that these investments are successful.

3.2.5 *Unpredictability of Grid Expansion and the Process of Incorporating New Mini-Grids in Africa*

African local grid expansion may be hampered by ambiguity surrounding the incorporation of mini-grids for the adoption of renewable energy sources. This unpredictability may hinder the development of mini-grids and discourage investment, delaying the uptake of renewable energy technologies. It is crucial to ensure that local grid expansion projects include precise timelines and rules regarding the integration of current mini-grids for the adoption of renewable energy sources to address this issue. As a result, investors will be able to evaluate these projects' risks and potential returns with accuracy, giving them confidence that their long-term investments will be profitable.

Investments in the expansion of mini-grids and the advancement of renewable energy technologies can be encouraged by offering incentives to investors. Governments and development organizations must establish clear policy frameworks and provide financial support for the expansion of mini-grids and renewable energy technologies to foster an environment that is conducive to investment. As a result, investors will be able to make long-term investments in a stable and secure environment, knowing that the government and other development organizations will support them in the long run. Investors can feel secure and confident by receiving financial support for their investments and the assurance of clear policy frameworks, which can motivate them to make long-term investments in the development of renewable energy technologies and mini-grids essential for the successful integration of renewable energy into the local grid and for a sustainable future. Establishing a long-term sustainable and secure energy landscape will help not only the renewable energy sector but also the society as a whole.

A successful transition to a sustainable future depends on investments in renewable energy technologies. The combination of policy, financial, and security frameworks will do both by lowering the risks involved with investing in renewable energy projects and by creating an environment that will support such investments. As a result, there will be an environment where investments in the renewable energy sector are safer and more certain, and society as a whole can anticipate the successful grid integration of renewable energy technologies and a safe, sustainable energy future that contributes to the fight against climate change and energy insecurity.

Governments and development agencies must provide financial support for this plan's implementation, and there must be a secure environment for investors that encourages long-term investments in the development of mini-grids and other renewable energy technologies that are essential for the transition to a future powered entirely by renewable energy sources.

3.2.6 Private Sector Participation in Renewable Energy Projects

Due to a lack of clearly defined policies and lengthy project approval processes, private sector participation in renewable energy projects is constrained in African nations. The inability of the private sector to participate as a result of this ambiguity has prevented the country from expanding renewable energy projects. To hasten the adoption of renewable energy in African nations, policymakers must establish precise and uniform rules governing FDI, PPPs, and private sector participation. To make renewable energy sources competitive with other energy sources such as fossil fuels, African nations should concentrate on lowering the cost of renewable energy projects and encouraging research and development of renewable energy technologies. African nations must also strengthen their institutional capabilities to draw investors from the private sector and effectively manage public–private partnerships.

To achieve sustainable development goals and ensure that citizens have access to safe and dependable energy in the future (Nwokolo et al., 2023b), it is critical to take a proactive approach to accelerating the uptake of renewable energy. African nations can guarantee their citizens have access to dependable and secure energy sources while achieving their sustainable development objectives and minimizing the harmful environmental effects of fossil fuels by establishing a policy environment that is supportive of private sector investments (Nwokolo et al., 2023c), lowering the cost of renewable energy projects, and investing in institutional capacity to manage public–private partnerships. For citizens to have access to renewable energy sources, a sound policy framework and institutional capacity to manage public–private partnerships are crucial. To guarantee that residents have access to dependable and secure energy sources, incentives and funding are also offered. African nations can guarantee their citizens have access to dependable and secure energy sources while also lessening the harmful environmental effects of fossil fuels by implementing a comprehensive strategy to promote renewable energy.

3.2.7 Immature Renewable Energy Policy Implementation

Remedies for immaturity in implementing renewable energy policies that hamper its development in African countries. African countries must enhance the maturity of renewable energy policies by promoting regional collaboration, stakeholder engagement, and meaningful initiatives to ensure buy-in from all stakeholders and ensure effective policy development and implementation (Nwokolo et al., 2022a). Such regional collaboration and stakeholder engagement will enable African countries to develop and implement renewable energy policies that are both mature and effective. The promotion and acceleration of clean energy resources on the continent require addressing the immaturity of policies. African countries should therefore focus on fostering regional collaboration and enhanced stakeholder engagement to improve the maturity of their renewable energy policies and ensure that their policy initiatives are meaningful, effective, and have the full buy-in of all stakeholders. This, in turn, will create the foundation for a successful and sustainable renewable energy market in Africa that can provide much-needed electricity to millions of Africans. The potential benefits of sustainable energy development in Africa are numerous, ranging from increased access to energy for communities across the continent to the economic and environmental benefits that come with a low-carbon economy. As such, African countries have the unique opportunity to unlock their renewable energy potential and make a positive contribution to global climate action by leveraging the benefits of regional collaboration and engaging with all stakeholders.

3.2.8 Technical Issues Impeding Renewable Energy Adoption in Africa

Technical issues impeding renewable energy adoption in Africa include a lack of infrastructure, inadequate financing, and limited access to technology. The continent's vast size and dispersed population make it difficult to establish reliable energy grids. Additionally, many African countries lack the financial resources to invest in renewable energy projects. The high upfront costs of building solar or wind farms can be prohibitive for governments and private investors alike. Moreover, many rural communities lack the technical expertise needed to maintain renewable energy systems. This creates a vicious cycle where the lack of infrastructure and financing prevents the adoption of renewable energy, which in turn limits access to technology and expertise. Despite these challenges, there are reasons for optimism. Many African countries have abundant natural resources that can be harnessed for renewable energy production. In addition, international organizations and developed countries are increasingly investing in renewable energy projects in Africa, providing much-needed funding and expertise. These investments not only help to increase access to electricity in rural areas but also create job opportunities and boost economic growth. Furthermore, the adoption of renewable energy can help to mitigate the effects of climate change, which disproportionately affects African countries. It is important for governments and private investors to continue to prioritize renewable energy development in Africa and work towards creating sustainable energy systems that benefit both the environment and local communities.

With continued investment and innovation, Africa has the potential to become a leader in renewable energy production and pave the way for a more sustainable future.

3.2.9 Transmission and Distribution Lines Are Being Outfitted with Cutting-Edge Infrastructure and Services

To take advantage of renewable energy sources, African countries must invest in the infrastructure and services necessary for power companies, such as transmission and distribution lines, to transfer the energy from where it is generated to where it is consumed. This investment will not only enable the continent to harness the full potential of its renewable energy resources but will also improve access to electricity for millions of people. The new infrastructure and services being implemented are designed to be more efficient and reliable, reducing the likelihood of power outages and improving overall energy security. Additionally, these upgrades will allow for better integration of renewable energy sources into existing power grids, further reducing reliance on fossil fuels and decreasing carbon emissions. With continued investment in transmission and distribution lines, African countries can pave the way for a sustainable energy future while also promoting economic growth and development. By investing in transmission and distribution lines, African countries can achieve a more sustainable energy future. These upgrades will not only increase efficiency and reliability but also reduce the likelihood of power outages and improve overall energy security. Moreover, they will enable better integration of renewable energy sources into existing power grids, further reducing reliance on fossil fuels and decreasing carbon emissions. With continued investment in this area, African nations can promote economic growth and development while also taking important steps towards a cleaner, more sustainable future. By working together to build a modern, interconnected energy infrastructure, Africa can lead the way in promoting sustainable development for generations to come.

3.2.10 Adequate Renewable Energy Grid Connectivity

Adequate renewable energy grid connectivity is a solution for renewable deployment in Africa. Renewable energy has the potential to transform Africa's energy landscape, but deployment has been slow due to a lack of adequate grid connectivity. The solution lies in developing a robust renewable energy grid that can accommodate the intermittent nature of renewable sources such as solar and wind. This requires significant investment in infrastructure, including transmission lines, substations, and storage facilities. However, the benefits are numerous: reduced reliance on fossil fuels, improved energy security, and increased access to electricity for millions of people across the continent. In addition, a renewable energy grid can create jobs and stimulate economic growth in local communities. To ensure success, governments must work closely with private sector partners to develop innovative financing models and regulatory frameworks that incentivize investment in renewable energy infrastructure. By transitioning to renewable energy sources, we can not only reduce our carbon footprint but also reap significant economic benefits. A shift towards renewables can create a new industry and generate jobs in areas such as manufacturing, installation, and maintenance of renewable energy infrastructure. Moreover, it can improve energy security by reducing dependence on foreign oil and gas imports. This can result in greater control over energy prices and a more stable economy. Additionally, access to electricity is crucial for economic development, particularly in rural areas. Renewable energy can provide a cost-effective solution to expand access to electricity for millions of people across the continent. However, achieving these benefits requires collaboration between governments and private sector partners to develop innovative financing models and regulatory frameworks that incentivize investment in renewable energy infrastructure.

3.2.11 A Trained Labour Force

Therefore, it is essential to have a trained labour force that can help in the proper installation and maintenance of these renewable energy technologies. To overcome this challenge, it is essential for African governments to invest in training and creating a skilled labour force that can help with the proper installation, maintenance, and repair of renewable energy technologies to ensure their successful and widespread deployment across the continent. This investment in training and skill development will benefit not only the renewable energy sector but also the overall economies of African countries. A skilled labour force will create job opportunities and attract foreign investment, leading to economic growth and development. Additionally, a trained workforce will help reduce the cost of renewable energy technologies by ensuring efficient installation and maintenance, making them more accessible to communities that currently lack access to electricity. Furthermore, investing in training programs for women and youth can promote gender equality and youth empowerment, leading to a more inclusive and sustainable future for Africa. To fully realize the potential benefits of renewable energy in Africa, it is crucial to invest in training programs that will equip individuals with the necessary skills and knowledge. This will not only create job opportunities and attract foreign investment but also lead to economic growth and development. A trained workforce can help reduce the cost of renewable energy technologies by ensuring efficient installation and maintenance, making them more accessible to communities that currently lack access to electricity. By investing in training programs for women and youth, we can promote gender equality and youth empowerment, leading to a more inclusive and sustainable future for Africa. With a skilled workforce that is equipped with the latest technologies, Africa can take advantage of its abundant renewable energy resources and become a leader in sustainable development.

3.3 Conclusion

The competitive position of renewable energy sources is significantly weakened by financial and funding constraints, making it challenging for investors to view them as viable alternatives. To ensure energy security and lessen their reliance on fossil fuels, African governments are realizing the value of making investments in renewable energy. African nations have implemented policy measures such as feed-in tariffs, government subsidies, and incentives for private sector investment in an effort to lower investment costs and boost uptake. Additionally, to make it simpler for renewable energy projects to access financing, they must encourage public–private partnerships, innovation, and entrepreneurship, as well as strong financial mechanisms. African governments should establish incentives to entice investors and support the development of projects for affordable and sustainable renewable energy to address these issues. African nations should receive technical and financial support from international organizations to help them develop the capacity and infrastructure needed to create renewable energy projects. African nations will be able to use renewable energy sources, lessen their reliance on fossil fuels, and achieve long-term economic growth and development as a result. To increase the accessibility and affordability of renewable energy technologies, policymakers must lower the costs involved in the project development cycle. This can be accomplished by offering financial incentives, lowering regulatory burdens, lowering barriers to entry into the market, and expanding access to affordable financing. To close the financial gap between renewable energy and fossil fuel-based energy sources and make renewable energy more widely available and economically viable, subsidies and other support mechanisms are also needed.

References

Akuru, U. B., Okoro, O. I., & Chikuni, E. (2015, September 23). Impact of renewable energy deployment on climate change in Nigeria. *Journal of Energy in Southern Africa, 26*(3), 125–134. https://doi.org/10.17159/2413-3051/2016/v26i3a2152

Alemzero, D., Acheampong, T., & Huaping, S. (2021, December). Prospects of wind energy deployment in Africa: Technical and economic analysis. *Renewable Energy, 179*, 652–666. https://doi.org/10.1016/j.renene.2021.07.021

Anderson, K., Farthing, A., Elgqvist, E., & Warren, A. (2022, June). Looking beyond bill savings to equity in renewable energy microgrid deployment. *Renewable Energy Focus, 41*, 15–32. https://doi.org/10.1016/j.ref.2022.02.001

Asongu, S. A., & Odhiambo, N. M. (2021, March). Inequality, finance and renewable energy consumption in sub-Saharan Africa. *Renewable Energy, 165*, 678–688. https://doi.org/10.1016/j.renene.2020.11.062

Baye, R. S., Ahenkan, A., & Darkwah, S. (2021, August). Renewable energy output in sub Saharan Africa. *Renewable Energy, 174*, 705–714. https://doi.org/10.1016/j.renene.2021.01.144

Bistline, J. E. (2017, May). Economic and technical challenges of flexible operations under large-scale variable renewable deployment. *Energy Economics, 64*, 363–372. https://doi.org/10.1016/j.eneco.2017.04.012

Boie, I., Kost, C., Bohn, S., Agsten, M., Bretschneider, P., Snigovyi, O., Pudlik, M., Ragwitz, M., Schlegl, T., & Westermann, D. (2016, March). Opportunities and challenges of high renewable energy deployment and electricity exchange for North Africa and Europe–Scenarios for power sector and transmission infrastructure in 2030 and 2050. *Renewable Energy, 87*, 130–144. https://doi.org/10.1016/j.renene.2015.10.008

Bugaje, I. (2006, December). Renewable energy for sustainable development in Africa: A review. *Renewable and Sustainable Energy Reviews, 10*(6), 603–612. https://doi.org/10.1016/j.rser.2004.11.002

Dogan, E., Hodžić, S., & Šikić, T. F. (2023, January). Do energy and environmental taxes stimulate or inhibit renewable energy deployment in the European Union? *Renewable Energy, 202*, 1138–1145. https://doi.org/10.1016/j.renene.2022.11.107

Dong, Y., & Shimada, K. (2017, July). Evolution from the renewable portfolio standards to feed-in tariff for the deployment of renewable energy in Japan. *Renewable Energy, 107*, 590–596. https://doi.org/10.1016/j.renene.2017.02.016

Kemausuor, F., Sedzro, M. D., & Osei, I. (2018, October 22). Decentralized energy Systems in Africa: Coordination and integration of off-grid and grid power systems—Review of planning tools to identify renewable energy deployment options for rural electrification in Africa. *Current Sustainable/Renewable Energy Reports, 5*(4), 214–223. https://doi.org/10.1007/s40518-018-0118-4

Kuamoah, C. (2020, January). Renewable energy deployment in Ghana: The hype, Hope and reality. *Insight on. Africa, 12*(1), 45–64. https://doi.org/10.1177/0975087819898581

Lawrence, A. (2020, March). Energy decentralization in South Africa: Why past failure points to future success. *Renewable and Sustainable Energy Reviews, 120*, 109659. https://doi.org/10.1016/j.rser.2019.109659

Lucas, H., Fifita, S., Talab, I., Marschel, C., & Cabeza, L. F. (2017, July). Critical challenges and capacity building needs for renewable energy deployment in Pacific Small Island developing states (Pacific SIDS). *Renewable Energy, 107*, 42–52. https://doi.org/10.1016/j.renene.2017.01.029

Maman Ali, M. M., & Yu, Q. (2021, December). Assessment of the impact of renewable energy policy on sustainable energy for all in West Africa. *Renewable Energy, 180*, 544–551. https://doi.org/10.1016/j.renene.2021.08.084

Manu, E. K., Chen, G. S., & Asante, D. (2022, June). Regional heterogeneities in the absorptive capacity of renewable energy deployment in Africa. *Renewable Energy, 193*, 554–564. https://doi.org/10.1016/j.renene.2022.05.019

Mohammed, Y., Mustafa, M., & Bashir, N. (2013, November). Status of renewable energy consumption and developmental challenges in sub-Sahara Africa. *Renewable and Sustainable Energy Reviews, 27*, 453–463. https://doi.org/10.1016/j.rser.2013.06.044

Mukumba, P., & Chivanga, S. Y. (2023, March 22). An overview of renewable energy Technologies in the Eastern Cape Province in South Africa and the rural households' energy poverty coping strategies. *Challenges, 14*(1), 19. https://doi.org/10.3390/challe14010019

Njoh, A. J., Etta, S., Ngyah-Etchutambe, I. B., Enomah, L. E., Tabrey, H. T., & Essia, U. (2019, February). Opportunities and challenges to rural renewable energy projects in Africa: Lessons from the Esaghem Village, Cameroon solar electrification project. *Renewable Energy, 131*, 1013–1021. https://doi.org/10.1016/j.renene.2018.07.092

Nwokolo, S. C., Amadi, S. O., Obiwulu, A. U., Ogbulezie, J. C., & Eyibio, E. E. (2022a). Prediction of global solar radiation potential for sustainable and cleaner energy generation using improved angstrom-Prescott and Gumbel probabilistic models. *Cleaner Engineering and Technology, 6*(January), 100416. https://doi.org/10.1016/j.clet.2022.100416

Nwokolo, S. C., Obiwulu, A. U., & Ogbulezie, J. C. (2022b). Machine learning and analytical model hybridization to assess the impact of climate. *Physics and Chemistry of the Earth, 130*(February), 103389. https://doi.org/10.1016/j.pce.2023.103389

Nwokolo, S. C., & Ogbulezie, J. C. (2018a). A qualitative review of empirical models for estimating diffuse solar radiation from experimental data in Africa. *Renewable and Sustainable Energy Reviews., 92*, 353. https://doi.org/10.1016/j.rser.2018.04.118

Nwokolo, S. C., & Ogbulezie, J. C. (2018b). A qualitative review of empirical models for estimating diffuse solar radiation from experimental data in Africa. *Renewable and Sustainable Energy Reviews, 92*. https://doi.org/10.1016/j.rser.2018.04.118

Nwokolo, S. C., & Ogbulezie, J. C. (2018c). A quantitative review and classification of empirical models for predicting global solar radiation in West Africa. *Beni-Suef University Journal of Basic and Applied Sciences, 7*(4), 367–396. https://doi.org/10.1016/j.bjbas.2017.05.001

Nwokolo, S. C., Ogbulezie, J. C., & Obiwulu, A. U. (2022c). Impacts of climate change and meteo-solar parameters on photosynthetically active radiation prediction using hybrid machine learning with physics-based models. *Advances in Space Research, 70*, 3614. https://doi.org/10.1016/j.asr.2022.08.010

Nwokolo, S., & Ogbulezie, J. (2017a). A critical review of theoretical models for estimating global solar radiation between 2012-2016 in Nigeria. *International Journal of Physical Research, 5*(2), 60. https://doi.org/10.14419/ijpr.v5i2.8160

Nwokolo, S., & Ogbulezie, J. (2017b). Modelling the influence of cloudiness on diffuse horizon-tal irradiation under various sky conditions in Nigeria. *International Journal of Physical Research, 5*(2), 91. https://doi.org/10.14419/ijpr.v5i2.8312

Nwokolo, S., & Otse, C. (2019). Impact of sunshine duration and clearness index on diffuse solar radiation estimation in mountainous climate. *Trends in Renewable Energy, 5*(3), 307–332. https://doi.org/10.17737/tre.2019.5.3.00107

Nwokolo, S. C., Obiwulu, A. U., Amadi, O., & Ogbulezie, J. C. (2023a). Assessing the impact of soiling, tilt angle, and solar radiation on the performance of solar PV systems. *Trends in Renewable Energy, 9*(2), 121–137. https://doi.org/10.17737/tre.2023.9.2.00156

Nwokolo, S. C., Ogbulezie, J. C., & Ushie, O. J. (2023b). A multimodel ensemble-based CMIP6 assessment of future solar radiation and PV potential under various climate warming scenarios. *Optik, 285*(May), 170956. https://doi.org/10.1016/j.ijleo.2023.170956

Nwokolo, S. C., Proutsos, N., & Meyer, E. L. (2023c). Machine learning and physics-based hybridization models for evaluation of the effects of climate change and urban expansion on photosynthetically active radiation. *Atmosphere, 14*(687), 1–36. https://doi.org/10.3390/atmos14040687

Obiwulu, A. U., Chendo, M. A. C., Erusiafe, N., & Nwokolo, S. C. (2020a). Implicit meteorological parameter-based empirical models for estimating back temperature solar modules under varying tilt-angles in Lagos, Nigeria. *Renewable Energy, 145*, 442. https://doi.org/10.1016/j.renene.2019.05.136

Obiwulu, A. U., Erusiafe, N., Olopade, M. A., & Nwokolo, S. C. (2020b). Modelling and optimization of back temperature models of mono-crystalline silicon modules with special focus on the effect of meteorological and geographical parameters on PV performance. *Renewable Energy, 154*, 404. https://doi.org/10.1016/j.renene.2020.02.103

Obiwulu, A. U., Erusiafe, N., Olopade, M. A., & Nwokolo, S. C. (2022). Modelling and estimation of the optimal tilt angle, maximum incident solar radiation, and global radiation index of the photovoltaic system. *Heliyon, 8*(6), e09598. https://doi.org/10.1016/j.heliyon.2022.e09598

Oluoch, S., Lal, P., & Susaeta, A. (2021, August). Investigating factors affecting renewable energy consumption: A panel data analysis in sub Saharan Africa. *Environmental Challenges, 4*, 100092. https://doi.org/10.1016/j.envc.2021.100092

Pradhan, A., & Mbohwa, C. (2014, November). Development of biofuels in South Africa: Challenges and opportunities. *Renewable and Sustainable Energy Reviews, 39*, 1089–1100. https://doi.org/10.1016/j.rser.2014.07.131

Renewables in Africa. (2009, January). *Renewable Energy Focus, 9*(7), 68–69. https://doi.org/10.1016/s1755-0084(09)70044-4

Sakah, M., Diawuo, F. A., Katzenbach, R., & Gyamfi, S. (2017, November). Towards a sustainable electrification in Ghana: A review of renewable energy deployment policies. *Renewable and Sustainable Energy Reviews, 79*, 544–557. https://doi.org/10.1016/j.rser.2017.05.090

Schwerhoff, G., & Sy, M. (2017, August). Financing renewable energy in Africa–Key challenge of the sustainable development goals. *Renewable and Sustainable Energy Reviews, 75*, 393–401. https://doi.org/10.1016/j.rser.2016.11.004

Shrimali, G., Konda, C., & Farooquee, A. A. (2016, November). Designing renewable energy auctions for India: Managing risks to maximize deployment and cost-effectiveness. *Renewable Energy, 97*, 656–670. https://doi.org/10.1016/j.renene.2016.05.079

Shum, K. L. (2017, June). Renewable energy deployment policy: A transition management perspective. *Renewable and Sustainable Energy Reviews, 73*, 1380–1388. https://doi.org/10.1016/j.rser.2017.01.005

Sweidan, O. D. (2021, November). Is the geopolitical risk an incentive or obstacle to renewable energy deployment? Evidence from a panel analysis. *Renewable Energy, 178*, 377–384. https://doi.org/10.1016/j.renene.2021.06.089

Wang, Y., Yan, Q., Luo, Y., & Zhang, Q. (2023, July). Carbon abatement of electricity sector with renewable energy deployment: Evidence from China. *Renewable Energy, 210*, 1–11. https://doi.org/10.1016/j.renene.2023.04.032

Chapter 4
Influencing the Scale of Africa's Energy Transition

4.1 Introduction

Influencing the scale of Africa's energy transition is important for achieving a more equitable and sustainable global energy system because the continent's energy transformation can unlock economic, social, and environmental opportunities for its citizens while simultaneously decreasing energy poverty and contributing to global emissions reductions. This is particularly pertinent when considering the fact that the continent of Africa has only recently become industrialized yet still accounts for a significant portion of global emissions. Despite its small contribution to global emissions, a successful energy transition in Africa is crucial to achieving the goal of climate neutrality by 2050 and reducing global inequality in access to clean energy services. To ensure a just transition, African countries must focus on creating clean energy sources for all and ensuring equitable access to energy services through renewable energy development, improved energy efficiency, and other low-carbon strategies while taking into account the current socioeconomic conditions of the region.

Additionally, ensuring energy access for the rural poor and off-grid populations (Nwokolo et al., 2022a) is essential in achieving a more equitable energy system and reducing poverty (Nwokolo et al., 2023a) and the effects of climate change within the region as well as globally (Nwokolo & Ogbulezie, 2018a). Furthermore, African countries must recognize that access to clean energy (Nwokolo & Ogbulezie, 2018b) and a just transition are intrinsically linked (Nwokolo et al., 2022b); without equitable access to clean energy (Nwokolo & Ogbulezie, 2018c), any attempts to address climate change will be undermined by entrenched social disparities and poverty caused by inequality (Nwokolo et al., 2023b) and the lack of access to energy services (Nwokolo & Ogbulezie, 2017a) that are necessary for a successful transition to a low-carbon (Nwokolo & Ogbulezie, 2017b), climate-resilient economy (Nwokolo et al., 2022c). Africa must foster strong public–private partnerships (Nwokolo & Otse, 2019) and innovative finance mechanisms to facilitate the deployment of clean energy services (Obiwulu et al., 2020a) while providing

S. C. Nwokolo et al., *Africa's Path to Net-Zero*, CSR, Sustainability, Ethics & Governance, https://doi.org/10.1007/978-3-031-44514-9_4

increased access to energy services for all to build a more equitable (Obiwulu et al., 2020b), just (Obiwulu et al., 2022), and resilient society (Nwokolo et al., 2023c).

By utilizing existing networks, infrastructure, and initiatives such as the Africa Renewable Energy Initiative and the African Development Bank's "High 5 s" initiatives, African countries can ensure a just transition to a low-carbon, climate-resilient economy that prioritizes the needs of their people and reduces the risk posed to them by climate change while also ensuring that they have access to the energy services they need for a better life. Energy access is essential for Africa to achieve a just transition to a low-carbon and climate-resilient economy, and access to energy is a basic human right that all countries should strive to protect.

Furthermore, countries should focus on energy efficiency measures, which can help reduce energy consumption (Ojukwu, 2020) and save money for households (Okoh & Okpanachi, 2023), businesses, and governments alike (Omaye et al., 2022). By investing in clean energy and utilizing innovative finance mechanisms (Soile, 2016), African countries can facilitate the deployment of sustainable energy services for all citizens (Sokona et al., 2012), creating an equitable and resilient society that can continue to thrive in the face of a changing climate (Todd & McCauley, 2021a). While there is no single "silver bullet" solution for achieving a just transition to a low-carbon and climate-resilient economy, African countries have access to a wide range of solutions that can help them make this transition in an equitable, sustainable, and cost-effective manner (Todd & McCauley, 2021b).

It is therefore important that African countries invest in and embrace sustainable energy solutions to ensure energy access for all citizens while also protecting the environment and building resilience to climate change (Unuigbe et al., 2022). This can be achieved through the use of renewable energy sources, such as solar and wind power (Netshipise & Semenya, 2022), as well as through energy efficiency measures (Mosaka et al., 2021) and utilizing innovative finance mechanisms, such as green bonds and pay-as-you-go financing solutions (McEwan, 2017), to help fund these solutions. These solutions can enable African countries to transition to a low-carbon and climate-resilient economy while also providing citizens with access to clean and reliable energy services that meet their needs and improve their quality of life (Ahjum et al., 2018).

Generally, African countries need to transition to a low-carbon and climate-resilient economy to protect their environment (Akinyemi et al., 2015), create energy access for all citizens, and build resilience against climate change by investing in and embracing sustainable energy solutions such as renewable energy sources, energy efficiency measures, and innovative finance mechanisms such as green bonds and pay-as-you-go financing solutions to help fund these solutions.

In this chapter, the authors also identified three influencing scales that can help facilitate the process in addition to this traditional method of influencing the scope of Africa's energy transition. These include, but are not limited to, adopting and implementing the EU's transition plan in African nations, a widening "fault line" between the developed and developing worlds, funding for the continent's transition to net-zero energy, and support from developed countries. Each of these scales presents unique challenges, but they are also opportunities to influence Africa's energy transition in a way that could benefit all parties involved. These scales can serve as potential paths to a successful energy transition in Africa. To understand the complexity of the situation, it is important to look beyond the traditional means of influencing the African energy transition and consider the diverse set of stakeholders involved, their needs and motivations, as well as the potential outcomes of each scale, such as increased access to affordable, reliable, and renewable energy for the continent and its inhabitants. While this is an ambitious goal, with thoughtful consideration of the situation and creative solutions, it is possible to take meaningful steps toward energy security for Africa and its people.

4.2 Adopting and Implementing the EU's Transition Plan in African Nations

Developed countries, particularly the EU, are driving the transition to a more sustainable energy system by implementing policies that facilitate the adoption of renewable energy sources, incentivizing energy efficiency investments, and introducing measures to reduce carbon emissions. Such policies have been important steps in the right direction, but further measures are needed to accelerate the energy transition, such as investment in research and development of renewable energy technologies, increased international cooperation and collaboration on climate action, and stronger efforts to raise public awareness. According to the European Climate Law, all member states must collectively cut their emissions by 55% from 1990 levels by 2030 to achieve net-zero economic growth by the year 2050.

To reach this ambitious goal, an unprecedented level of commitment from both the public and private sectors is needed, as is the implementation of ambitious policies and regulations that incentivize the use of renewable energy sources and encourage energy efficiency investments as well as the development of renewable energy infrastructure. Such initiatives are crucial for transitioning to a more sustainable energy system, which is necessary for the achievement of long-term environmental and economic goals and to protect the planet from the worst effects of climate change. A change in behavior within the EU is being sparked by policies such as regulated emissions disclosure, the outlawing of carbon-intensive technologies, advanced taxation, emissions trading systems, and the Carbon Border Adjustment Mechanism, but they also have wider implications for other markets and economies around the world.

By pursuing a goal of net-zero emissions, the European Climate Law demonstrates that a transformation of society toward sustainability is both achievable and desirable. As such, the European Climate Law can be seen as an example of how global commitment to sustainability can bring about positive and tangible change in the near future. The European Union is thus setting an example for other regions and nations to follow when considering their own pathways toward a more sustainable future for the environment. As the European Union pursues ambitious climate targets, the rest of the world looks to Europe as a model for success in tackling climate change and mitigating the worst effects of global warming.

In addition to demonstrating what is possible with global commitment to environmental sustainability, the European Climate Law also has implications for the European Union's place in a new geopolitical order by demonstrating that it is a leader in this area and that its approach to sustainability can be replicated and adapted for other nations and regions to follow in ways that would create economic, environmental, and social benefits for their citizens. This demonstrates the power of collective action, as the European Union has been able to enact a law that shows what is possible when different countries commit to ambitious goals for sustainability and climate change mitigation together while providing an example to the world of what can be achieved when countries cooperate to achieve a common goal.

The European Union's Climate Law is a beacon of hope for the world in terms of showing what can be done when countries come together to combat climate change and provides an example to other nations of how they can benefit from cooperation and collective action on the global stage. The European Union's Climate Law is a shining example of what collective action and cooperation can achieve, demonstrating how powerful commitments to environmental sustainability and climate change mitigation can be when countries join forces in pursuit of a common goal and how much benefit can be derived from such collective action. It also has the potential to provide a blueprint for other countries and regions to follow in their own efforts to tackle climate change and build a more sustainable future for the world as a whole.

The European Climate Law is a groundbreaking piece of legislation that has ushered in a new era of global commitment to environmental sustainability, and its far-reaching implications for the European Union's place in a new geopolitical order are clear, as are its implications for the rest of the world in terms of providing a beacon of hope for the future and an example of what can be achieved through collective action and cooperation on an international level. The European Union's Climate Law is thus a shining example of the potential benefits of collective action and cooperation in tackling climate change, and its impact has already been felt both in Europe and around the world by inspiring other countries to take similar steps toward a more sustainable future for all and driving positive change in the way countries across the globe approach environmental sustainability and climate change mitigation strategies.

It is clear that the European Union's Climate Law will be an essential part of any meaningful global effort to tackle climate change, both in terms of inspiring similar commitments from other countries and regions and providing a reliable framework for ongoing collaboration and cooperation on this critical issue going forward. With the passage of this groundbreaking piece of legislation, it has become clear that countries are willing to come together to take action on environmental sustainability and climate change and that this is an issue that can be solved when nations come together for a common purpose and put aside their differences for the greater good of protecting our planet, creating a better future for all of its contemporary moral implications inhabitants, and ensuring a better, healthier, and safer future for generations to come. Its effects will be felt for years to come, both in terms of providing a clear path for other nations to follow and in the tangible results that will be seen as this collective action leads to the improvement of the global climate situation and the preservation of our planet for future generations.

Adopting and implementing the EU's transition plan in African nations is one of the ways to achieve a rapid transition to more sustainable and green policies, practices, and technologies across the continent. While the European Union has set a powerful example with its Climate Law, African nations have their own unique set of circumstances and needs that must be addressed in order for them to make meaningful strides toward a more sustainable future. To achieve this, African nations must continue to work together and look toward their commonalities to build stronger relationships and create a framework that allows each country to make individual progress as part of a larger collective effort. African nations must also build strong partnerships with the European Union and other major global players to ensure that they are provided with the support, resources, and expertise needed to develop their own climate action plans and take advantage of the opportunities provided by the European Union's transition plan. African nations must also focus on building capacity and implementing the necessary reforms to make the most of the transition plan while also ensuring that their citizens have access to clean energy, clean water, and improved sanitation and health care services that support the transition to a green economy.

As the world continues to grapple with the consequences of climate change, African nations have an important role to play in developing and implementing sustainable solutions that will help them to reduce their greenhouse gas emissions, protect their natural resources, and build strong economies that benefit all of their citizens while also contributing to the global effort to mitigate the effects of climate change. African nations must take advantage of the available support and resources to make effective use of the European Union's transition plan while also ensuring that their citizens have access to the clean energy, clean water, and improved sanitation services needed to make a successful transition to a green economy.

African nations must focus on forming strong partnerships with the European Union and other global actors to benefit from the transition plan and create sustainable solutions that will allow them to reduce their greenhouse gas emissions, protect their natural resources, build a strong economy that will benefit their citizens, and contribute to the global effort to mitigate the effects of climate change. African nations must act with urgency to benefit from the transition plan and create sustainable solutions that will reduce their greenhouse gas emissions, protect their natural resources, and build strong economies that will benefit all of their citizens while also contributing to the global effort to mitigate the effects of climate change while also addressing the disparities in access to energy and clean water that disproportionately affect the poorest and most marginalized members of their populations.

African nations must take meaningful and decisive action to develop and implement sustainable solutions that will allow them to reduce their greenhouse gas emissions, protect their natural resources, and build a strong economy that benefits all of their citizens while also contributing to the global effort to reduce their environmental impact and mitigate the effects of climate change. African nations have an opportunity to make a meaningful impact on the global effort to mitigate the effects of climate change, but they must act with urgency to capitalize on the available resources and support available to them from the European Union and other global actors to ensure that they are able to develop and implement effective and sustainable solutions that benefit their citizens while also having a positive impact on the global environment and contributing to the global effort to mitigate the effects of climate change.

For this reason, African nations must not only strive to meet the same sustainability goals as their European counterparts but also tailor their efforts to account for their own particular context and take into consideration the various social, economic, political, and environmental issues facing the African continent. By recognizing that no two nations are the same and adapting their efforts accordingly, African nations can make real progress toward a more sustainable future and ensure that their countries are part of the solution for a healthier planet in the years to come.

It is through this transition that the African continent will be able to meet its commitments under the Paris Agreement and ensure that their own citizens benefit from a more sustainable economy, improved health outcomes, and greater social equality while also playing their part in the global effort to mitigate the effects of climate change and create a better world for generations to come. African nations must recognize the importance of acting now and capitalize on the resources available to them to ensure that their efforts are not only effective but also sustainable, taking into consideration the context of their own country and ensuring that their efforts are tailored accordingly.

It is through collaboration, cooperation, and the shared commitment of all nations to protecting our planet that we will be able to effectively combat climate change, create a more sustainable future for everyone, and safeguard our global environment for future generations. To do this, African nations must look to their own resources and strategies, strive to find solutions that are tailored to their particular context and needs, and commit to taking action that is most beneficial for their citizens, their economy, and the environment as a whole. The global community must also be willing to lend their support in any way they can to ensure that African nations are able to fully meet their commitments under the Paris Agreement and continue to provide their citizens with access to a cleaner, healthier environment while still allowing their economic growth and development to continue unimpeded. As such, African nations must come together and work with the global community to develop a holistic approach to tackling climate change, one that takes into account the specific needs and resources of each country while also leveraging international support and cooperation to ensure the successful implementation of effective climate policies that are tailored to the unique needs and circumstances of African nations and that will enable them to effectively protect their environment, advance their development, and create a brighter future for themselves and for generations to come.

4.3 Mediating the Escalating "Dichotomy" Between Africa and Civilized Countries

There is a growing "fault line" separating developed and developing nations as a result of this acceleration toward net-zero emissions of greenhouse gases by the end of this century. As developing countries lack the financial means or technological capacity to transition to a carbon neutral economy (Aklin, 2022), this gap is likely to widen even further unless significant investment is made in supporting and enabling these countries to reach their goals (Akrofi, 2021). Furthermore, this growing fault line threatens to cause economic (Aquila Jesuleye et al., 2020) and social instability as the divide between developed (Asante et al., 2021) and developing nations becomes more pronounced (Baiyegunhi & Hassan, 2014). The effects of climate change are felt most drastically in African countries (Baker, 2017), which are most vulnerable due to their lack of resources and infrastructure to protect them from the immediate impacts of increased temperatures, rising sea levels, and extreme weather events. These effects have the potential to threaten food security, displace populations, and damage vital ecosystems that provide vital ecosystem services such as freshwater, protection from floods and storms, and carbon storage. This is a cause for great concern, as African nations are most likely to be disproportionately affected by the consequences of climate change due to their already disadvantaged positions in terms of economic and social development. This leaves African countries in a precarious situation and demands an urgent response from both international bodies and local communities to mitigate the effects of climate change, reduce the vulnerability of vulnerable populations, and ensure that they have the capacity to adapt to the changes they are facing going forward. Consequently, without adequate support from wealthier nations, developing countries may find themselves unable to make the necessary adjustments to combat climate change and are thus particularly vulnerable to its impacts. This is an issue of grave importance and requires the concerted effort of all stakeholders to protect populations, ecosystems, and the world as a whole from the potentially catastrophic consequences of global climate change (Boulogiorgou & Ktenidis, 2020). For these reasons, it is critical that global governments prioritize the equitable sharing of responsibility for net-zero emissions to ensure that every nation, regardless of their current levels of development or access to resources, is given a fair and equal chance to transition toward a low-carbon future (Clark et al., 2022). While African countries bear the brunt of the effects of climate change, they are not responsible for the majority of global emissions and therefore lack the economic capacity to cope with its consequences without assistance from the international community (Ekeh, 2020). In recognition of this, it is essential that wealthier countries step up and provide the necessary financial resources to assist African nations in transitioning toward a low-carbon future while allowing them to continue growing their economies in a sustainable and equitable manner (Emodi et al., 2022). This assistance should come in the form of increased financial resources, technical support, and capacity building, which must be provided to all countries regardless of their economic standing to ensure a truly global effort toward achieving net-zero emissions by mid-century.

In the developed world, governments are pouring billions of dollars into renewable energy projects and emissions-reduction policies, while African nations struggle to keep up with the environmental regulations required to reduce their emissions in an equitable and sustainable way and protect their citizens from the effects of climate change. Despite this disparity, there are a range of promising initiatives that aim to bridge the gap between developed and developing countries in the transition to a low-carbon economy. For example, the Africa Renewable Energy Initiative

(AREI) is a partnership between the African Union and the European Union to provide funding for renewable energy projects across the continent to help these countries meet their target (Clark et al., 2022). In addition, the Green Climate Fund is an international fund that assists developing countries with mitigation and adaptation efforts related to climate change by providing financial resources and support to help them implement climate-resilient development projects such as energy efficiency, renewable energy, and waste management initiatives. These efforts are vital to ensure that Africa is not left behind in the global fight against climate change and can benefit from the transition to a low-carbon economy through the same technological advances and economic development that more developed countries have enjoyed (Fatoki, 2020). AREI and the Green Climate Fund are just two of the many examples of projects that are helping to bridge the gap between developed and developing countries in terms of access to renewable energy and the fight against climate change. They are helping to provide access to renewable energy technologies to those countries in Africa that need it most, as well as the financial resources and support that are necessary for them to implement climate-resilient development projects that will help them reduce emissions, increase energy security, and ultimately contribute to the global fight against climate change. Furthermore, these initiatives benefit not only the African countries receiving assistance but also the global climate as a whole by allowing African countries to become part of the global effort toward addressing the climate crisis and reducing their greenhouse gas emissions (Field, 2021). This, in turn, will enable them to create more employment opportunities, reduce their dependence on imported energy sources, and ensure that they are able to meet the challenges of climate change head-on without sacrificing their economic and social progress. Generally, initiatives such as AREI and the Green Climate Fund are invaluable in bridging the gap between the haves and have-nots in terms of access to renewable energy (Ghazi et al., 2021), enabling African countries to become part of the global effort toward addressing climate change and ultimately becoming leaders in the development of renewable energy technologies. This can have a positive impact not only on the individual countries involved but also on the global climate, as African countries become part of a larger effort to reduce emissions and help combat climate change. By introducing initiatives such as the African Renewable Energy Initiative (AREI) and the Green Climate Fund, the international community is helping to equip African countries with the resources and knowledge they need to meet their energy demands in a sustainable manner while simultaneously contributing to global efforts to combat climate change and reduce greenhouse gas emissions. By providing African countries with the support and resources they need to reduce their emissions and increase their access to renewable energy sources, initiatives such as AREI and the Green Climate Fund have helped them make great strides toward tackling the climate crisis and ultimately becoming leaders in the development of renewable energy technologies that can have a positive impact not only on individual countries but also on the global climate. This is an invaluable contribution to the African continent and the world that could have a lasting and positive impact on our global climate in unprecedented ways (Giglmayr et al., 2015). In summary, the international community has made great strides toward aiding African countries in their journey toward becoming leaders in renewable energy technology, with initiatives such as AREI and the Green Climate Fund providing crucial support to help these countries reduce emissions, increase access to renewable energy sources, and ultimately contribute to global efforts to combat climate change and make a lasting, positive impact on our global climate. These and other initiatives have the potential to accelerate the global transition to clean energy and provide much needed assistance to African countries in their own efforts to reduce emissions and protect their citizens from the effects of climate change. As these initiatives progress, it is essential to ensure that they are effective and well managed so that the world can reap the full benefits of these investments and to help African countries achieve their goals in combating climate change and making a lasting, positive impact on our global climate. While this is a huge step forward, there is still much work to be done to ensure that African countries have the resources they need to fully benefit from these initiatives, successfully transition toward a sustainable energy future, and contribute to the global clean energy transition.

Due to their greater access to financial support and more stable fiscal positions, developed economies can better withstand these disruptions. The advantage of mature electrical grids with high levels of electricity generation and quite good electricity and water infrastructure operating in protected, stringent regulatory climates where there is a higher penetration of renewable energy across the grid is only available to developed economies. On the other hand, African countries often lack financial support and access to the same levels of energy generation and are subject to extreme weather events that can make adaptation more difficult due to their lack of resources and infrastructure. As a result, the acceleration toward net zero has created an unequal playing field in terms of energy access and the ability to successfully adapt to a low-carbon economy between the developed and African economies. As such, it is important to acknowledge that developing countries will require additional support and resources to ensure that they have the necessary infrastructure, financial security, and technology needed to transition toward a net-zero economy on an equal footing with developed economies. This has highlighted the need for increased international cooperation, public–private partnerships, and financial support to ensure that a just transition toward net zero is achievable by all countries regardless of their stage of development. In conclusion, the shift toward a net-zero economy will create an unequal playing field between developed and African countries due to their different levels of access to energy generation, infrastructure, and financial support, but it is of the utmost importance to recognize that such a transition can only be achieved with increased international collaboration, public–private partnerships, and financial support from developed countries.

4.4 Financing the African Net-Zero Transition

Reaching a net-zero energy mix in Africa by 2050 would cost the continent, according to IEA estimates, approximately $2.8 trillion. Between 2020 and 2030, there would be approximately $33 billion in annual costs; between 2030 and 2040, $111 billion; and between 2040 and 2050, $142 billion. The Africa-EU Energy

Partnership (AEEP) has estimated a similar price tag of $40–100 billion per year for Africa to meet SDG7—"ensuring everyone has access to affordable, reliable, and modern energy services by the year 2030"—while the estimated investment in Africa in 2019 was $17 billion. It is clear that to reach the net-zero energy mix and SDG7 goals set out by the IEA and AEEP, a significant increase in investment will be required for Africa to meet its net-zero energy and SDG7 targets (Hendam et al., 2022).

This increase in investment would enable the transition to renewable energy sources, the development of energy-efficient infrastructure, and improved access to modern energy services on the African continent. Over the next 10–30 years, the investment in energy infrastructure and access in Africa will need to substantially increase to meet international targets for renewable energy sources, energy efficiency, and modern energy services to meet the goals set out by the IEA and AEEP. It is clear that the necessary funds to meet the goals of SDG7 and a net-zero energy mix in Africa will have to come from somewhere, as current levels of investment alone will not be enough to bridge the gap and meet the goals set out by the IEA and AEEP.

It is therefore essential for governments, international institutions, private investors, and philanthropic organizations to invest in Africa's energy infrastructure, with a focus on energy efficiency and renewable energy sources, to achieve the SDG7 targets and reach a net-zero energy mix by the end of the century. This influx of funding is essential for the successful transition to renewable energy sources and improved access to modern energy services in Africa. By investing in the infrastructure and access to energy in Africa, the necessary funding for SDG7 goals and a net-zero energy mix can be achieved while at the same time promoting economic growth, reducing poverty and inequalities, and improving energy access and security throughout the region. To do this, a shift in the way that energy is produced, distributed, and consumed is necessary to ensure the long-term sustainability of energy production in Africa and the attainment of the SDG7 targets.

This shift can be achieved through investments in energy efficiency measures and renewable energy sources such as solar, wind, and hydropower, as well as innovative financing strategies that enable access to financing for the development of energy infrastructure and access to modern energy services in Africa (Isah Samaila Nitte, 2023). These investments will not only facilitate the transition to sustainable energy sources but also help ensure the availability of energy resources to meet the growing needs of populations in Africa and promote economic growth while at the same time reducing the environmental impacts of energy production and consumption and advancing the achievement of SDG7 targets across the region.

To achieve this, it is necessary to provide access to affordable and reliable energy while also addressing the environmental impacts of production and consumption. To fund this effort, the governments of African countries will need to increase their investment in energy infrastructure and access, attract foreign direct investment from both public and private sources through the implementation of tax incentives and subsidies, and provide clear pathways for foreign companies to enter the market. Additionally, public–private partnerships (PPPs) will need to be developed to ensure that the investments are made in a manner that is financially sustainable for both public and private investors while also driving innovation in the energy sector and creating opportunities for local businesses to participate in the energy transition. These investments will also be an important step in ensuring access to energy for vulnerable and underserved populations, including those living in rural or remote areas, women, and marginalized communities.

Generally, it is evident that for African countries to achieve their energy transition goals, they must make significant investments in energy infrastructure and access (Isihak et al., 2022) and partner with both public and private entities to ensure that these investments are made in a manner that is financially sustainable (Israel-Akinbo & Fraser, 2018) while also creating opportunities for innovation, growth, and equal access to energy for all. To meet these goals, African countries will need to find ways to secure the funds required for investment in renewable energy and energy efficiency infrastructure projects. This can be done by leveraging public and private funds to attract the necessary financing from national and international investors while also creating mechanisms that ensure that the investment is spread out to all communities, particularly those who may not have the means or resources to invest in energy projects on their own.

This will ultimately help to ensure that the energy transition will benefit all African citizens and help to build more inclusive economies that prioritize economic growth and equitable access for all. African countries should also look to partner with organizations such as the International Renewable Energy Agency (IRENA) and the United Nations Sustainable Development Solutions Network (UN SDSN), which are dedicated to helping governments adopt renewable energy policies and strategies that prioritize social. This could come in the form of public finance, private investment, donor funding, or even a combination of all three sources of capital. In addition, innovative financing mechanisms such as blended finance and green bonds could be used to supplement these sources of capital and bridge the financing gap to attract private investment into renewable energy and energy efficiency projects in Africa (Joubert et al., 2016).

Furthermore, African countries should consider developing clear policy frameworks for renewable energy and energy efficiency initiatives that can attract both public and private capital and incentivize long-term investment in these areas to ensure that funds are used efficiently and the desired outcomes are achieved (Kanu, 2016). Additionally, the implementation of public–private partnerships can be an effective way to attract private investment into renewable energy and energy efficiency projects and leverage the advantages of both the public and private sectors, such as access to capital, expertise, and technology, which are necessary for the long-term, sustainable growth and development of the renewable energy and energy efficiency sectors in Africa. To make renewable energy and energy efficiency projects financially viable in Africa, there is a need for greater investment in the sector from multiple sources, including governments, development finance institutions, commercial banks, and private investors such as venture capitalists and private equity funds (Masekela & Semenya, 2021). All of these sources of capital are necessary to create a financial environment conducive to sustainable renewable energy and energy efficiency projects in Africa.

In conclusion, Africa needs to develop comprehensive policies and frameworks that can ensure that renewable energy and energy efficiency projects are financially viable in the long term by encouraging the formation of public–private partnerships and by creating an enabling environment to attract investments from multiple sources, such as by providing favorable terms for loan repayment, tax incentives, and other financial support mechanisms necessary to stimulate private sector involvement in renewable energy and energy efficiency projects. These measures will help create a more vibrant renewable energy and energy efficiency sector in Africa, helping to reduce dependence on fossil fuels and lay the groundwork for a more sustainable future (Kanu, 2016). This will in turn bring about a reduction in emissions, increased access to energy for those who need it, and an improved quality of life for millions of people across the continent, as well as access to various sources of financing, including international donors, multilateral institutions, commercial banks, and other sources of capital that can support and enable the mobilization of private capital for renewable energy and energy efficiency investments across the continent. This is of utmost importance if African countries are to build a more sustainable and economically resilient future going forward. With these measures, the continent can move toward a brighter future of renewable energy and energy efficiency that will benefit African citizens in multiple ways and create long-term sustainable economic growth for Africa.

4.5 Aid from Developed Countries Will Be Crucial

Although foreign donor organizations and intergovernmental banks have expanded their assistance, it is not enough for Africa to keep pace with the modern net-zero transition and required decarbonization. Consequently, African governments and regional bodies must take the lead in tackling this pressing problem by introducing ambitious policies and programs to reduce emissions, mobilizing investment and technology transfer, and leveraging their own domestic sources of financing to make a dent in the continent's CO2 levels to achieve a successful energy transition and promote green growth for a sustainable future.

This can be done through the development of sustainable energy policies and strategies, encouraging investment in renewable energy sources such as solar and wind, improving access to green finance for businesses, implementing carbon pricing mechanisms and emissions trading systems, utilizing digital technologies to build a smart energy infrastructure, and encouraging international cooperation on technology sharing and capacity building to help accelerate the adoption of cleaner energy sources across the continent. A successful energy transition and net-zero emissions in Africa will require comprehensive strategies to reduce emissions, increase access to green financing and technology, promote regional cooperation and capacity building, and use digital technologies to build resilient energy systems that will be capable of meeting the continent's energy needs in a sustainable way that is in line with international climate goals as well as ambitious domestic action to ensure that the transition is successful.

The US government-led partnership, Power Africa, has thus far committed $40 billion, making it the continent's largest capital investment. It aims to increase sub-Saharan Africa's electrical connections by 60 MW and add at least 30,000 MW of new, cleaner electrical power capacity. To do this, significant geothermal, hydroelectric, solar, wind, and natural gas resources must be made available. Although this commitment is admirable, if net-zero targets are to be met by 2050, Africa must almost double this investment each year. To achieve this, not only must the $40 billion already committed by Power Africa be matched or exceeded in subsequent years, but it must also be accompanied by regulatory changes and investments in new technologies that could reduce energy poverty and ensure that the continent's electricity grids are more resilient and efficient. Despite these formidable challenges, the Power Africa Initiative is an important step in improving energy access throughout the African continent and provides the continent with the opportunity to make tremendous advances in energy access and poverty reduction over the coming decades. By investing in and developing a variety of renewable energy sources, the Power Africa Initiative seeks to ensure that access to electricity is no longer a luxury for the privileged few but instead an achievable reality for all African citizens.

With the implementation of this ambitious plan, African nations can be provided with a platform for sustainable economic growth and prosperity. This will help to increase the quality of life and ensure that the African population is able to access the energy sources it needs to develop and grow into the future. As such, the Power Africa Initiative represents an incredible opportunity to overcome some of the most pressing issues facing African countries today and lay the foundations for a bright and sustainable future. The initiative not only promises to help eliminate poverty and create employment opportunities for African citizens but also offers a potential solution to the increasing environmental degradation caused by the overreliance on nonrenewable energy sources such as coal, oil, and natural gas. By investing in clean and renewable energy solutions, the initiative promises to reduce carbon emissions, reduce air pollution, improve public health, and help protect natural ecosystems while also promoting economic growth and energy independence for African countries.

As stated by the United Nations Economic Commission for Africa, the following are a number of important factors that will expedite the development of renewable energy in Africa, including increased access to finance, the adoption of clear policies and regulations, the implementation of ambitious plans and projects, the formation of strong partnerships between public and private sector entities, improved capacity-building initiatives, and the utilization of innovative technology solutions and practices. To effectively implement these initiatives, it is important to develop a holistic approach that addresses the entire renewable energy ecosystem, from the generation and storage of energy to its distribution, transmission, and use for various end uses, in a coordinated manner. As such, African nations must create an environment conducive to investments in the renewable energy sector by ensuring the availability of a reliable and affordable supply of energy, increasing awareness about the potential benefits of renewable energy sources, providing incentives for businesses to invest in such technologies, and introducing regulatory reforms that provide a level playing field for all investors in the renewable energy market. In addition, governments must provide the necessary incentives and investments to stimulate research and development in renewable energy technologies, as well as ensure access to training and capacity building for those wishing to engage in renewable energy-related activities. Finally, African governments must seek to create public–private partnerships and facilitate technology transfer between domestic and foreign firms to ensure that the most efficient renewable energy solutions are adopted and leveraged for the benefit of all African citizens.

4.6 Conclusion

African renewable energy development will be accelerated by a number of key factors identified by the United Nations Economic Commission for Africa. These include improved capacity-building programs, expanded access to financing, the adoption of transparent policies and regulations, the execution of ambitious plans and projects, the creation of strong partnerships between public and private sector entities, and the use of cutting-edge technology solutions and methods. By guaranteeing the availability of a dependable and affordable energy supply, raising awareness of the potential advantages of renewable energy sources, offering incentives to businesses to invest in such technologies, enacting regulatory reforms, and taking other measures, African nations can successfully implement these initiatives.

The European Union's Climate Law is an important part of any global effort to tackle climate change. African nations must adopt and implement the EU's transition plan to achieve a rapid transition to more sustainable and green policies, practices, and technologies. They must also build partnerships with the European Union and other major global players to ensure they are provided with the support, resources, and expertise needed to develop their own climate action plans and take advantage of the opportunities provided by the transition plan.

Governments must also guarantee access to training and capacity building for those wishing to engage in renewable energy-related activities, as well as the necessary incentives and investments to stimulate research and development in renewable energy technologies. Finally, to ensure that the most effective renewable energy solutions are adopted and utilized for the benefit of all African citizens, African governments must work to establish public–private partnerships and facilitate technology transfer between domestic and foreign businesses.

References

Ahjum, F., Merven, B., Stone, A., & Caetano, T. (2018, September 21). Road transport vehicles in South Africa towards 2050: Factors influencing technology choice and implications for fuel supply. *Journal of Energy in Southern Africa, 29*(3), 33–50. https://doi.org/10.17159/2413-3051/2018/v29i3a5596

Akinyemi, O., Osabuohien, E., Alege, P. O., & Ogundipe, A. A. (2015). Energy security, trade and transition to green economy in Africa. *SSRN Electronic Journal.* https://doi.org/10.2139/ssrn.2700026

Aklin, M. (2022). Political economies of energy transition: Wind and solar power in Brazil and South Africa. *Political Science Quarterly, 137*(2), 424–426. https://doi.org/10.1002/polq.13323

Akrofi, M. M. (2021, February 2). Correction to: An analysis of energy diversifcation and transition trends in Africa. *International Journal of Energy and Water Resources, 5*(2), 235–236. https://doi.org/10.1007/s42108-020-00107-z

Aquila Jesuleye, O., Foluso, A. O., & Oluwaseun, A. A. (2020, April 7). Factors influencing solar photovoltaic utilization for energy services among Residentials in Akure, Nigeria. *Journal of Energy Research and Reviews*, 14–28. https://doi.org/10.9734/jenrr/2020/v4i330128

Asante, D., He, Z., Mintah Ampaw, E., Gyamerah, S., Ankrah Twumasi, M., Opoku-Mensah, E., Kyere, F., Asante, B., & Afia Akyia, E. (2021, November). Renewable energy technology transition among small-and-medium scale firms in Ghana. *Renewable Energy, 178*, 549–559. https://doi.org/10.1016/j.renene.2021.06.111

Baiyegunhi, L., & Hassan, M. (2014, June). Rural household fuel energy transition: Evidence from Giwa LGA Kaduna state, Nigeria. *Energy for Sustainable Development, 20*, 30–35. https://doi.org/10.1016/j.esd.2014.02.003

Baker, L. (2017, December 1). Commercial-scale renewable energy in South Africa and its Progress to date. *IDS Bulletin, 48*(5–6). https://doi.org/10.19088/1968-2017.165

Boulogiorgou, D., & Ktenidis, P. (2020, February). TILOS local scale technology innovation enabling low carbon energy transition. *Renewable Energy, 146*, 397–403. https://doi.org/10.1016/j.renene.2019.06.130

Clark, S., Van Niekerk, J., Petrie, J., & McGregor, C. (2022, September 26). The role of natural gas in facilitating the transition to renewable electricity generation in South Africa. *Journal of Energy in Southern Africa, 33*(3). https://doi.org/10.17159/2413-3051/2022/v33i3a12742

Ekeh, C. (2020, August). How COVID-19 is accelerating energy transition in developing oil-rich countries: Nigeria as a case study. *Global Energy Law and Sustainability, 1*(2), 234–240. https://doi.org/10.3366/gels.2020.0034

Emodi, N. V., Haruna, E. U., Abdu, N., Aldana Morataya, S. D., Dioha, M. O., & Abraham-Dukuma, M. C. (2022, September). Urban and rural household energy transition in sub-Saharan Africa: Does spatial heterogeneity reveal the direction of the transition? *Energy Policy, 168*, 113118. https://doi.org/10.1016/j.enpol.2022.113118

Fatoki, O. (2020, January 1). Factors influencing the purchase of energy-efficient appliances by young consumers in South Africa. *Foundations of Management, 12*(1), 151–166. https://doi.org/10.2478/fman-2020-0012

Field, T. L. (2021, February 26). A just energy transition and functional federalism: The case of South Africa. *Transnational. Environmental Law, 10*, 1–25. https://doi.org/10.1017/s2047102520000436

Ghazi, F. E., Sedra, M. B., & Akdi, M. (2021, April 10). Energy transition from fossil to renewable sources in North Africa: Focus on the renewable electricity generation in Morocco. *International Journal of Energy Economics and Policy, 11*(3), 236–242. https://doi.org/10.32479/ijeep.11036

Giglmayr, S., Brent, A. C., Gauché, P., & Fechner, H. (2015, November). Utility-scale PV power and energy supply outlook for South Africa in 2015. *Renewable Energy, 83*, 779–785. https://doi.org/10.1016/j.renene.2015.04.058

Hendam, M., Schittekatte, T., Abdel-Rahman, M., & Kamh, M. Z. (2022, October). Rethinking electricity rate design: Fostering the energy transition in North Africa. *Energy Policy, 169*, 113172. https://doi.org/10.1016/j.enpol.2022.113172

Isah Samaila Nitte. (2023, March 13). Global energy transition and its implications on energy security in nigeria: A critical review. *EPRA International Journal of Climate and Resource Economic Review*, 1–9. https://doi.org/10.36713/epra12607

Isihak, S., Akpan, U., & Bhattacharyya, S. (2022, August). Evolution of GIS-based rural electrification planning models and an application of OnSSET in Nigeria. *Renewable and Sustainable Energy Transition, 2*, 100019. https://doi.org/10.1016/j.rset.2022.100019

Israel-Akinbo, S., & Fraser, J. S. G. (2018). The energy transition patterns of low-income households in South Africa: An evaluation of energy programme and policy. *Journal of Energy in Southern Africa, 29*(3). https://doi.org/10.17159/2413-3051/2017/v29i3a3310

Joubert, E., Hess, S., & Van Niekerk, J. (2016, November). Large-scale solar water heating in South Africa: Status, barriers and recommendations. *Renewable Energy, 97*, 809–822. https://doi.org/10.1016/j.renene.2016.06.029

Kanu, M. I. (2016). Energy transition & geopolitics in sub-Saharan Africa: Outlining a legal framework to scale the energy infrastructure financing conundrum. *SSRN Electronic Journal.* https://doi.org/10.2139/ssrn.2831175

Masekela, M. E., & Semenya, K. (2021, September 19). Factors influencing the use of firewood postelectrification in rural South Africa: The case of Ga-Malahlela village. *Journal of Energy in Southern Africa, 32*(3). https://doi.org/10.17159/2413-3051/2021/v32i3a7781

McEwan, C. (2017, January). Spatial processes and politics of renewable energy transition: Land, zones and frictions in South Africa. *Political Geography, 56*, 1–12. https://doi.org/10.1016/j.polgeo.2016.10.001

Mosaka, T. B. M., Mararakanye, N., & Bekker, B. (2021, December 13). International procurement policies influencing renewable energy siting–Implications for South Africa. *Journal of Energy in Southern Africa, 32*(4), 58–68. https://doi.org/10.17159/2413-3051/2021/v32i4a8397

Netshipise, L. F., & Semenya, K. (2022, June 17). Evaluating factors influencing firewood consumption in households at the Thulamela local municipality, South Africa. *Journal of Energy in Southern Africa, 33*(2), 48–62. https://doi.org/10.17159/2413-3051/2022/v33i2a9741

Nwokolo, S. C., Amadi, S. O., Obiwulu, A. U., Ogbulezie, J. C., & Eyibio, E. E. (2022a). Prediction of global solar radiation potential for sustainable and cleaner energy generation using improved angstrom-Prescott and Gumbel probabilistic models. *Cleaner Engineering and Technology, 6*(January), 100416. https://doi.org/10.1016/j.clet.2022.100416

Nwokolo, S. C., Obiwulu, A. U., & Ogbulezie, J. C. (2022b). Machine learning and analytical model hybridization to assess the impact of climate. *Physics and Chemistry of the Earth, 130*(February), 103389. https://doi.org/10.1016/j.pce.2023.103389

Nwokolo, S. C., & Ogbulezie, J. C. (2018a). A qualitative review of empirical models for estimating diffuse solar radiation from experimental data in Africa. *In Renewable and Sustainable Energy Reviews*. https://doi.org/10.1016/j.rser.2018.04.118

Nwokolo, S. C., & Ogbulezie, J. C. (2018b). A qualitative review of empirical models for estimating diffuse solar radiation from experimental data in Africa. *Renewable and Sustainable Energy Reviews, 92*. https://doi.org/10.1016/j.rser.2018.04.118

Nwokolo, S. C., & Ogbulezie, J. C. (2018c). A quantitative review and classification of empirical models for predicting global solar radiation in West Africa. *Beni-Suef University Journal of Basic and Applied Sciences, 7*(4), 367–396. https://doi.org/10.1016/j.bjbas.2017.05.001

Nwokolo, S. C., Ogbulezie, J. C., & Obiwulu, A. U. (2022c). Impacts of climate change and meteo-solar parameters on photosynthetically active radiation prediction using hybrid machine learning with physics-based models. *August., 70*, 3614. https://doi.org/10.1016/j.asr.2022.08.010

Nwokolo, S., & Ogbulezie, J. (2017a). A critical review of theoretical models for estimating global solar radiation between 2012-2016 in Nigeria. *International Journal of Physical Research, 5*(2), 60. https://doi.org/10.14419/ijpr.v5i2.8160

Nwokolo, S., & Ogbulezie, J. (2017b). Modelling the influence of cloudiness on diffuse horizon-tal irradiation under various sky conditions in Nigeria. *International Journal of Physical Research, 5*(2), 91. https://doi.org/10.14419/ijpr.v5i2.8312

Nwokolo, S., & Otse, C. (2019). Impact of sunshine duration and clearness index on diffuse solar radiation estimation in mountainous climate. *Trends in Renewable Energy, 5*(3), 307–332. https://doi.org/10.17737/tre.2019.5.3.00107

Nwokolo, S. C., Obiwulu, A. U., Amadi, O., & Ogbulezie, J. C. (2023a). Assessing the impact of soiling, tilt angle, and solar radiation on the performance of solar PV systems. *Trends in Renewable Energy, 9*(2), 121–137. https://doi.org/10.17737/tre.2023.9.2.00156

Nwokolo, S. C., Ogbulezie, J. C., & Ushie, O. J. (2023b). A multimodel ensemble-based CMIP6 assessment of future solar radiation and PV potential under various climate warming scenarios. *Optik, 285*(May), 170956. https://doi.org/10.1016/j.ijleo.2023.170956

Nwokolo, S. C., Proutsos, N., & Meyer, E. L. (2023c). Machine learning and physics-based hybridization models for evaluation of the effects of climate change and urban expansion on photosynthetically active radiation. *Atmosphere, 14*(687), 1–36. https://doi.org/10.3390/atmos14040687

Obiwulu, A. U., Chendo, M. A. C., Erusiafe, N., & Nwokolo, S. C. (2020a). Implicit meteorological parameter-based empirical models for estimating back temperature solar modules under varying tilt-angles in Lagos, Nigeria. *Renewable Energy, 145*, 442. https://doi.org/10.1016/j.renene.2019.05.136

Obiwulu, A. U., Erusiafe, N., Olopade, M. A., & Nwokolo, S. C. (2020b). Modelling and optimization of back temperature models of mono-crystalline silicon modules with special focus on the effect of meteorological and geographical parameters on PV performance. *Renewable Energy, 154*, 404. https://doi.org/10.1016/j.renene.2020.02.103

Obiwulu, A. U., Erusiafe, N., Olopade, M. A., & Nwokolo, S. C. (2022). Modelling and estimation of the optimal tilt angle, maximum incident solar radiation, and global radiation index of the photovoltaic system. *Heliyon, 8*(6), e09598. https://doi.org/10.1016/j.heliyon.2022.e09598

Ojukwu, K. (2020). What strategies and policies can save Nigeria in the energy transition? *SSRN Electronic Journal.* https://doi.org/10.2139/ssrn.3701350

Okoh, A. S., & Okpanachi, E. (2023, December). Transcending energy transition complexities in building a carbon-neutral economy: The case of Nigeria. *Cleaner Energy Systems, 6*, 100069. https://doi.org/10.1016/j.cles.2023.100069

Omaye, S. O., & Sa'ad, S., Hamma Adama, A., & Dotti, R. (2022, December). Energy consumption, economic growth and energy transition in Africa: A cross-sectional dependence analysis. *OPEC Energy Review, 46*(4), 502–514. https://doi.org/10.1111/opec.12271

Soile, I. (2016). What are the factors influencing demand for energy in Nigeria: A sectoral analysis. *SSRN Electronic Journal.* https://doi.org/10.2139/ssrn.2817621

Sokona, Y., Mulugetta, Y., & Gujba, H. (2012, June). Widening energy access in Africa: Towards energy transition. *Energy Policy, 47*, 3–10. https://doi.org/10.1016/j.enpol.2012.03.040

Todd, I., & McCauley, D. (2021a, July 31). An interdisciplinary approach to the energy transition in South Africa. *Discover Sustainability, 2*(1). https://doi.org/10.1007/s43621-021-00043-w

Todd, I., & McCauley, D. (2021b, November). Assessing policy barriers to the energy transition in South Africa. *Energy Policy, 158*, 112529. https://doi.org/10.1016/j.enpol.2021.112529

Unuigbe, M., Zulu, S. L., & Johnston, D. (2022, August 7). Exploring factors influencing renewable energy diffusion in commercial buildings in Nigeria: A grounded theory approach. *Sustainability, 14*(15), 9726. https://doi.org/10.3390/su14159726

Chapter 5
Technological Pathways to Net-Zero Goals in Africa

5.1 Introduction

In Africa, technological pathways to net-zero goals have the potential to significantly reduce energy poverty and enhance people's standards of living by providing access to clean, reliable, and affordable energy services that are accessible to all while also contributing to a more secure energy supply and helping to create economic opportunities for individuals and businesses. With the implementation of renewable energy sources, people in Africa have access to more reliable electricity and other energy services, as well as economic opportunities that can boost their quality of life and livelihoods. Renewable energy sources can reduce dependency on costly and unreliable grid electricity, helping to improve living standards in a cost-effective way and contributing to the development of more secure energy systems in Africa while also helping to reduce emissions by substituting emissions-intensive energy sources, such as fossil fuels.

Additionally, the increased use of renewable energy sources in Africa can also help drive down energy costs (Nwokolo et al., 2022a), as renewable energy is typically cheaper to produce than conventional sources of energy such as fossil fuels. By providing clean (Nwokolo & Ogbulezie, 2018a), reliable (Nwokolo & Ogbulezie, 2018b), and affordable energy services to the population in Africa (Nwokolo & Ogbulezie, 2018c), renewable energy sources offer many advantages that can lead to a reduction in energy poverty and an increase in economic opportunities for the people of Africa and the continent at large. By providing clean (Nwokolo & Ogbulezie, 2017a), reliable, and affordable energy services (Nwokolo & Ogbulezie, 2017b) to the population in Africa, renewable energy sources offer many advantages that can lead to a reduction in energy poverty (Nwokolo & Otse, 2019) and an increase in economic opportunities for the people of Africa and the continent at large (Nwokolo et al., 2022b).

These advantages include the potential to create new jobs and opportunities for entrepreneurs, allowing people to engage in activities that will have a direct impact on their lives (Nwokolo et al., 2023a), such as starting a business, owning property

S. C. Nwokolo et al., *Africa's Path to Net-Zero*, CSR, Sustainability, Ethics & Governance, https://doi.org/10.1007/978-3-031-44514-9_5

(Obiwulu et al., 2020a), and improving their standard of living (Obiwulu et al., 2022). Not only can this reduce energy poverty, but renewable energy sources also have the potential to reduce air pollution and greenhouse gas emissions (Nwokolo et al., 2023b), which can be beneficial to both the environment and the human health (Ojukwu, 2020). Additionally, renewable energy sources have the potential to provide access to electricity in rural areas (Nwokolo et al., 2023c) where traditional sources of power may be too expensive or simply unavailable (Obiwulu et al., 2020b). This could create a more connected and prosperous Africa, allowing people to participate in the digital economy and take advantage of educational opportunities that can help improve their quality of life.

Moreover, renewable energy is a crucial element in the effort to limit global temperature increases and mitigate the impacts of climate change, which disproportionately affects African countries, as they lack the infrastructure and resources to address rising sea levels (Okoh & Okpanachi, 2023), extreme weather events (Omaye et al., 2022), and other effects of climate change. This could help African nations adapt to a changing climate, protect their citizens from the worst effects of climate change, and foster greater social and economic stability in the region. This increased access to renewable energy sources could have a lasting and transformative impact on Africa, ushering in a new era of development and progress for its people.

Ultimately, the development of renewable energy technologies in Africa has the potential to be a game changer by unlocking the potential of African nations to rapidly develop and deploy renewable energy sources while reducing dependence on fossil fuels that have long been associated with environmental degradation (Soile, 2016), poverty, and poor health outcomes (Sokona et al., 2012). As renewable energy technologies become increasingly available and accessible to African nations, they could help countries in the region break the cycle of poverty, reduce inequality, and build a better future for their citizens by providing more reliable access to clean, affordable, and sustainable energy sources. Finally, renewable energy technologies have the potential to revolutionize Africa by providing access to sustainable energy sources (Todd & McCauley, 2021) and promoting development, ultimately leading to improved economic, social, and environmental conditions for African nations and creating unprecedented opportunities for economic development and growth in the spotlight and being taken seriously across the continent (Nwokolo et al., 2022c).

In recent years, technological paths toward net-zero goals in Africa have accelerated, especially in fields such as renewable energy, energy storage, and more effective energy use in homes, buildings, and transport systems. These solutions offer promising pathways to decarbonize Africa's energy sector, create green jobs, and reduce poverty while also benefiting public health and the environment (Netshipise & Semenya, 2022). By investing in renewable energy, energy storage, and smart energy use, African countries can make great progress in meeting their Paris Agreement commitments while improving access to affordable and sustainable energy sources for their citizens and expanding economic opportunities across the continent. Additionally, these investments can help reduce air pollution, which disproportionately affects the poorest people in Africa and is estimated to cause up to two million premature deaths annually.

In sum, investing in renewable energy, energy storage, and smart energy use can help African countries meet their Paris Agreement commitments while also creating green jobs, reducing poverty and air pollution, improving public health and the environment, and providing access to affordable and sustainable energy sources for African citizens (Mosaka et al., 2021). Therefore, it is clear that renewable energy, energy storage, and smart energy use are powerful tools for African countries to both meet their commitments under the Paris Agreement and benefit their citizens and economies. In conclusion, investing in renewable energy, energy storage, and smart energy use is a win–win situation for African countries and can have a far-reaching and positive impact on public health, the environment, economic growth, and development in Africa.

This progress is allowing African countries to pursue net-zero goals in a much more efficient and cost-effective way than would be possible with traditional energy sources, such as coal and fossil fuels. By utilizing the right mix of technologies (Unuigbe et al., 2022), countries in Africa can work toward decarbonization while supporting economic growth and improving access to energy for all African citizens. This is an important step in reducing global emissions, as many African countries have not yet seen the advances in technology that other countries have enjoyed, which will ultimately result in a more sustainable future for the continent. These paths have enabled Africa to become a leader in the development of renewable energy sources, with countries such as Kenya and Ethiopia leading the way and setting ambitious goals for the expansion of clean energy.

To do this, countries in Africa are taking advantage of new technologies and business models to create economic incentives for renewable energy investments. Such investments can bring numerous benefits to African nations, including job creation and economic diversification, the improvement of access to energy, and reducing the cost of energy production, which in turn can lead to a reduction in greenhouse gas emissions. Moreover, these investments are also helping to empower African nations with increased energy independence and resiliency in the face of ever more frequent and severe climate-related disasters (McEwan, 2017), which can significantly reduce the continent's dependence on fossil fuels, which, if done right, will help lead to a cleaner and more prosperous future.

This is in stark contrast to the traditional reliance on nonrenewable energy sources, such as coal and gas, which have caused environmental degradation and negatively impacted public health. Furthermore, renewable energy investments can also help to promote social and political stability by providing access to electricity in rural and remote communities, which have often been neglected in the past. This can help to support economic development and alleviate poverty by allowing these communities to gain access to previously inaccessible resources and services (Masekela & Semenya, 2021), making them less vulnerable to the effects of climate change and giving them the opportunity to actively shape their own future, which can have a lasting and beneficial impact on the overall development of African nations and the world at large.

In addition, renewable energy investments can also create jobs and diversify local economies, helping to build strong and resilient communities (Kanu, 2016). By investing in renewable energy sources such as solar and wind, we can help reduce our dependency on traditional energy sources and minimize their harmful effects while at the same time helping to increase energy security and promote economic development. By investing in renewable energy sources such as solar and wind, we can help to reduce our dependency on traditional energy sources and minimize their harmful effects while at the same time helping to increase energy security and promote economic development, which can in turn help to reduce the risk of political and social unrest (Joubert et al., 2016), which can have a long-term impact on African nations and the world as a whole. This investment in renewable energy has far-reaching implications for the future of African nations and beyond and can help ensure that future generations will be able to enjoy a safe and healthy environment free from the negative effects of climate change and energy insecurity (Israel-Akinbo & Fraser, 2018).

In general, investing in renewable energy sources such as solar and wind can have a profound impact on the overall development of African nations and the world at large by helping to reduce the risk of climate change and political instability while at the same time promoting economic development and energy security, which can lead to a better future for generations to come. By investing in renewable energy sources such as solar and wind, we can help reduce our dependency on traditional energy sources and minimize their harmful effects while at the same time helping to increase energy security and promote economic development (Isihak et al., 2022), which can have a long-term, positive impact on the development of African nations and the world as a whole.

Given its population growth and resource scarcity, Africa has recently placed a greater emphasis on technological paths toward net-zero goals as it seeks to develop more effective and sustainable energy sources to meet the needs of its rapidly increasing population in the face of climate change. To this end, Africa is transitioning to renewable energy sources such as solar and wind, developing projects to promote efficiency in the agricultural sector and striving to increase access to clean water in remote areas. Additionally, Africa is turning to other forms of green energy, such as geothermal and hydroelectricity, to reduce its dependence on fossil fuels, mitigate the effects of climate change, and protect its natural resources for the future. As such, Africa is investing in a range of sustainable technologies to ensure access to energy and other resources while protecting the environment and responding to the urgent needs of its population.

In doing so, Africa is not only achieving its net-zero goals and promoting sustainability but also helping to shape a greener and more equitable future for all across the continent and beyond. With ambitious net-zero goals and a keen focus on sustainable development, Africa is setting an example for the world in demonstrating how economic growth and environmental stewardship can go hand in hand. Africa is playing a leading role in the global transition to a low-carbon and climate-resilient economy by investing in renewable energy sources, embracing energy efficiency, and actively engaging in forest conservation and land management initiatives that reduce emissions, protect habitats and biodiversity, and improve the well-being of communities.

By investing in renewable energy sources such as solar, wind, and hydropower, developing an integrated energy network across the continent, and actively engaging in the conservation of forests and land management initiatives to help protect biodiversity (Isah Samaila Nitte, 2023), Africa is demonstrating its commitment to meeting net-zero goals and promoting a sustainable future for all while also promoting economic growth. In addition to these initiatives, African governments are pursuing a range of economic and social reforms that promote economic growth and social progress while also taking steps to protect the environment and reduce carbon emissions, such as incentivizing companies to invest in clean technologies, providing financial assistance for green investments, and investing in research and development to develop innovative solutions for sustainability issues (Hendam et al., 2022). African countries are leading the way in demonstrating how economic growth and environmental stewardship can be complementary goals rather than conflicting ones and are setting an example for the world of how countries can take collective action to tackle the climate crisis and create a better, more sustainable future for all.

As countries around the world face the daunting challenge of finding ways to mitigate and adapt to the effects of climate change, African countries have stepped up in a large way to show the world that sustainability and economic growth are not mutually exclusive goals but instead can be achieved together to create a more prosperous and secure future for everyone (Giglmayr et al., 2015). Moving forward, African countries will need to continue to focus on these initiatives to meet their net-zero commitments and create a truly sustainable future for all while preserving and protecting their natural resources and protecting biodiversity and ecosystems for generations to come (Ghazi et al., 2021). African countries are showing that these seemingly conflicting goals are possible, and they are demonstrating the power of collective action to create a more sustainable future for all while still achieving their economic goals.

A promising path to digitalization for African nations is provided by technological paths to net-zero targets by leveraging technologies such as renewable energy, big data, and blockchain to drive digital transformation in the region and create pathways to more efficient use of natural resources, diversified sources of energy, and improved access to services. In addition to leveraging existing technologies, new forms of technology, such as artificial intelligence, virtual reality, and machine learning, can be used to provide further opportunities for digital transformation. These technologies provide a wide range of opportunities for African nations, such as improved energy access and efficiency, reduced energy poverty, better management of resources and waste, and the ability to harness data to make smarter decisions that are more beneficial to the environment, public health, and the economy as a whole.

All of these technologies have the potential to not only address climate change but also increase economic opportunity, create jobs, and deliver a greater quality of life for African citizens. By taking advantage of these technological opportunities (Field, 2021), African nations can create a strong and resilient digital infrastructure that will support economic growth and development while also reducing emissions and building a more sustainable and secure future for the continent and its citizens. African nations have the potential to use technology to transform their economies, improve environmental stewardship (Fatoki, 2020), and create greater opportunities for citizens while simultaneously tackling the challenge of climate change. African nations have the opportunity to become global leaders in embracing and utilizing technologies to help them build a more prosperous, healthy, and climate-resilient future.

To realize these opportunities, African nations must embrace technology and create an enabling environment for digital infrastructure to take hold that allows for the safe and secure use of data and digital technologies by leveraging digital tools to improve access to financial services, healthcare, education, and other essential services for their citizens. As African nations recognize the potential of technology and create a supportive digital infrastructure for their citizens, the continent will become more competitive globally and have greater opportunities for economic development while also playing a key role in the global fight against climate change. By embracing the potential of technology and digital infrastructure (Emodi et al., 2022), African nations can unlock tremendous benefits for their citizens, strengthen the global economy, and create a more prosperous and climate-resilient future for themselves and the world.

As African nations embark on the journey of digitalization, they have to balance the potential opportunities with the potential risks associated with these new technologies, such as cybersecurity threats, data privacy issues, and the potential for monopolization by large digital companies. To mitigate these risks, African nations must ensure that their citizens have access to the right education and resources to understand and use digital technologies in a safe and secure way while also developing a regulatory framework that allows digital technology companies to develop, grow, and innovate while maintaining trust and accountability in the digital space. At the same time, African nations must continue to invest in digital infrastructure and build the capacity to develop strong local digital ecosystems that include start-ups, entrepreneurs, investors, accelerators, and incubators to ensure a balanced approach to digitalization and sustainable development that can benefit all stakeholders.

African nations must also ensure that any digital transformation is done in an inclusive way, that it takes into account all segments of society, especially the most vulnerable and marginalized communities, and that it does not perpetuate existing inequalities or further deepen existing digital divides between those who can access and use digital technologies (Ekeh, 2020) and those who are excluded from the digital space. Additionally, African nations should develop regional partnerships and collaborations to share best practices and resources in areas such as digital infrastructure, data protection, cybersecurity, and capacity building to facilitate digital transformation and build a safer (Clark et al., 2022) and more secure digital space for African citizens to benefit from.

African nations must continue to prioritize the development of digital literacy and capability in their citizens, providing them with the skills, knowledge, and confidence to use digital technologies safely and securely for their personal and professional development (Boulogiorgou & Ktenidis, 2020), as well as for their participation in digital spaces, while at the same time taking the necessary measures to ensure access to digital technologies and infrastructure for all their citizens so that everyone can take part in the digital economy, reap its benefits, and participate in a more sustainable and equitable form of economic growth for the continent (Baker, 2017). African nations must also ensure that their digital transformation efforts are carried out in a way that is respectful of local culture, traditions, and values and ensure the protection of their cultural heritage while at the same time harnessing the power of digital technologies to enhance it (Baiyegunhi & Hassan, 2014), make it more widely accessible to a larger audience, and facilitate its preservation for future generations.

African nations must take a holistic approach to digital transformation, balancing the need to ensure access to digital technologies and infrastructure while simultaneously protecting cultural heritage and values, fostering digital literacy and capability in citizens, and promoting regional partnerships and collaborations for the sharing of best practices and resources to ensure that digital transformation is beneficial for African citizens and that it is carried out in a safe, secure, and equitable manner (Asante et al., 2021).

All in all, African nations must recognize that digital transformation can create a brighter future for their citizens, but they must also ensure that this future is built on a secure and equitable foundation so that everyone can benefit from it and so that African nations can create a digital economy that is both sustainable and equitable for the first time in unprecedented ways in the continent's history. As such, it is critical for African nations to consider the full scope of the implications of digital transformation and ensure that they are taking a proactive (Aquila Jesuleye et al., 2020) and comprehensive approach to facilitate its success while safeguarding their culture and values (Akrofi, 2021).

African nations must take a responsible and meaningful approach to digital transformation, recognizing the power of technology and harnessing it in a way that is respectful of local culture and traditions, while also allowing for greater access to educational resources and cultural heritage, promoting digital literacy and capability among its citizens, strengthening regional collaborations and partnerships to ensure digital technologies are utilized in equitable ways, and establishing secure foundations upon which a safe and sustainable digital economy can be built so that all citizens may benefit and thrive for the long-term benefit of their people and nation. Ultimately, digital transformation presents African nations with an opportunity to create a better future for their citizens and communities by taking meaningful steps to ensure the success of their digital transformation initiatives while at the same time ensuring that everyone is protected and given an opportunity to benefit while also creating greater economic opportunities and encouraging innovation within their nations.

Technological avenues to net-zero targets in Africa have significantly increased in the past few years, as businesses and governments alike have sought to reduce their carbon footprints and switch to more sustainable methods of energy production. These technological avenues have been beneficial in numerous ways, from reducing air pollution to generating jobs and boosting local economies in the countries and regions where they are implemented (Aklin, 2022). Moreover, these avenues have enabled African nations to reduce their reliance on fossil fuels and instead focus on clean energy sources such as solar, wind, and hydropower, which further contribute to the overall environmental well-being of the continent. While the benefits of these technological avenues are numerous, their successful implementation depends on the willingness of governments to make investments in clean energy infrastructure and to provide incentives for businesses and citizens alike to switch to more sustainable sources of energy and to adopt energy-saving practices.

Additionally, investments in the research and development of clean energy technologies could be beneficial, as they would help African nations stay ahead of the global trend and remain competitive in the long run (Akinyemi et al., 2015). Despite the numerous benefits of embracing technological avenues that focus on clean energy sources in African nations, their successful implementation is heavily dependent on the commitment of governments to invest in renewable energy infrastructure and offer incentives to businesses and citizens to switch to clean energy sources and adopt energy-saving practices (Ahjum et al., 2018). Thus, a successful transition to clean energy sources in African nations requires substantial investments in renewable energy infrastructure and incentives for businesses and citizens to adopt such sources and save energy, as well as significant investments in research and development of clean energy technologies to ensure their countries remain competitive in the global market.

In general, the successful implementation of clean energy sources in African nations depends on the commitment of governments to invest in renewable energy infrastructure and provide incentives for businesses and citizens to switch to cleaner sources and adopt energy-saving practices, as well as the investment in research and development of clean energy technologies to remain competitive in the global market and have a positive impact on their countries beyond what was expected of them. Furthermore, African nations must invest in renewable energy infrastructure and offer incentives to businesses and citizens to switch to clean energy sources and adopt energy-saving practices to make the successful transition to clean energy sources and ensure the successful implementation of clean energy sources in African nations so that their countries can remain competitive in the global market and make a positive impact on the environment, going beyond what was expected of them.

These technological options are covered in this chapter as ways for Africa to meet its net-zero goals by achieving energy efficiency, increasing access to green hydrogen, natural gas, and renewable sources of energy, and decreasing emissions from traditional sources such as coal and oil to mitigate the effects of climate change and build an economy that is resilient to its effects. These measures will prove beneficial for Africa in both the short and long term by reducing dependence on fossil fuels, which are limited and more expensive than renewable energy sources such as solar and wind energy, while also creating new economic opportunities in the development and deployment of clean energy technologies.

Additionally, investing in renewable energy sources can provide a source of clean energy to the more than 600 million people in Africa who currently lack access to electricity and running water. This chapter provides insight into how Africa can use technology to achieve its net-zero goals while simultaneously empowering its citizens and creating a more sustainable and equitable future for generations to come. The aim of this chapter is to explore how technology can be used to effectively and efficiently reach net-zero emissions in Africa by 2050 while also promoting economic development and social justice and to identify the key challenges that must be overcome to make this a reality.

To this end, it is clear that the utilization of technology and renewable energy sources is integral to achieving net-zero emissions in Africa by 2050. To realize these goals, African governments must invest in green technologies and infrastructure, such as solar, wind, and hydropower, and invest in training and education to ensure that their citizens are equipped with the knowledge and skills necessary to participate in the green economy and benefit from the economic opportunities that arise from this shift. This shift to a green economy will require collaboration among African governments, businesses, and citizens to realize the potential of technological innovation and renewable energy sources to reduce emissions, build a more resilient and sustainable future, and ensure that economic development is coupled with social justice.

This chapter examines the various strategies that African governments can pursue to realize these goals and outlines the specific actions that need to be taken by governments, businesses, and citizens for this transition to occur successfully. The implementation of technological and renewable energy sources will prove invaluable to the African continent, allowing it to reduce its emissions while creating a more sustainable future for both its citizens and its environment. To achieve this, African governments must invest in green technologies and infrastructure, as well as training and education opportunities that will ensure citizens are equipped with the knowledge and skills they need to participate in the green economy. In addition, businesses must be encouraged to embrace green technologies and energy sources, while citizens must also take ownership of their roles in the transition by being aware of their environmental impact and advocating for green policies and sustainable practices.

This chapter illustrates the importance of African governments, businesses, and citizens working together to create a more sustainable future for the continent and to develop renewable energy sources that will drive economic growth, reduce emissions, and create a better future for generations to come. African governments, businesses, and citizens must all take part in this transition to a greener future if Africa is to transition to a more sustainable future and reap the rewards of an economy driven by renewable energy sources and clean technologies. It is not enough for African governments to invest in green technologies and infrastructure, businesses to embrace green technologies, or citizens to take ownership of their roles in the transition; all three entities must work together to create a brighter future for Africa by developing clean technologies, investing in renewable energy sources, encouraging sustainable practices, and advocating for green policies and sustainable practices.

To achieve this, African governments must invest in green technologies and infrastructure; businesses must be encouraged to embrace green technologies and energy sources; and citizens must take ownership of their roles in the transition by being aware of their environmental impact and advocating for green policies and sustainable practices. African governments must create strong incentives and policies to support renewable energy sources and clean technologies; businesses must make investments in green technologies to drive economic growth; and citizens must be empowered to become advocates for green policies and sustainable practices. In general, a successful transition to a more sustainable future for Africa can only be achieved if governments, businesses, and citizens all take part in the process of developing and investing in green technologies and energy sources, advocating for green policies and sustainable practices, and empowering citizens to take ownership of their roles in the transition.

5.2 Natural Gas

5.2.1 Introduction

Natural gas is one of the safest and cleanest fuels available on the African continent, and due to its low emissions and energy efficiency, it has become the fuel of choice for many African countries in recent years as they look to develop their energy infrastructure Natural gas is also one of the most cost-effective sources of energy, making it an attractive option for those who wish to reduce their carbon footprint while still maintaining affordability. The potential for natural gas to reduce air pollution and increase energy efficiency is unparalleled, and its abundance in Africa makes it an ideal fuel source for African countries looking to develop their economies and improve their quality of life Not only is natural gas a clean and safe source of energy, but it can also provide economic benefits to African countries by reducing their reliance on imported energy sources, improving access to energy, and creating jobs in the local energy sector Natural gas has the potential to be a major contributor to the development of African countries, and its many benefits make it an attractive option for those looking for sustainable energy sources Natural gas can also provide environmental benefits to African countries by reducing air pollution and increasing energy efficiency, as well as economic and environmental benefits by helping to reduce energy costs, reduce emissions, and stimulate economic growth Natural gas can have a tremendous impact on African countries, not just in terms of environmental benefits but also in terms of economic and social development. By reducing the need to rely on imported sources of energy, African countries can develop their local resources, reduce energy costs, and create jobs in the local energy sector. In addition, natural gas can provide environmental benefits by reducing emissions and air pollution, as well as increasing energy efficiency, which can improve the health and well-being of the population. Furthermore, natural gas can help to improve access to energy in rural areas that have historically been underserved due to a lack of infrastructure by providing an alternative energy source to those who may not have access to traditional sources of electricity. Natural gas can help increase economic development in African countries, reduce energy costs, and create jobs in the local energy sector. Natural gas therefore represents an invaluable opportunity for African countries to improve their quality of life, reduce poverty, and enhance their economic development. All of these advantages make natural gas an ideal fuel source for African countries seeking to achieve sustainable economic development and improve the quality of life for their citizens.

As natural gas is not only cost effective but also has a much lower environmental impact than other energy sources such as coal and oil, its use is likely to continue to increase across the African continent in the coming years as countries strive to meet their development goals while also protecting the environment (Sokona et al., 2012). This is a positive development for Africa, as it will help to reduce the continent's dependence on fossil fuels and allow countries to use their resources more efficiently. In addition, by switching to natural gas, African countries can reduce their overall greenhouse gas emissions and contribute to global climate change mitigation efforts. Natural gas is a versatile and reliable energy source that can be used in a variety of ways, from electricity generation to transportation to heating and cooking to industrial processes. This shift to natural gas could help to create jobs and stimulate economic growth in African countries as more investment is attracted to the sector and the development of infrastructure needed to support its use increases. This shift away from fossil fuels is also beneficial to human health, as the burning of natural gas produces fewer pollutants than traditional coal and oil combustion, and the use of natural gas can help reduce the levels of air pollution in many African countries. Furthermore, the availability of natural gas in Africa means that countries can meet their energy needs without having to rely on imports from abroad, which can be costly and can create a reliance on other nations for energy security. By promoting the use of natural gas, African countries can reap multiple benefits (Soile, 2016). Natural gas also has a lower carbon footprint than traditional fossil fuels, thus helping to reduce the global impact of climate change and environmental degradation associated with greenhouse gas emissions. Additionally, natural gas is a more efficient source of energy than traditional fossil fuels, meaning that countries can use less fuel to generate the same amount of energy, which can be beneficial for both economic and environmental reasons. This would not only reduce the cost of energy for African nations but could also lead to a greater degree of energy independence, allowing them to make their own decisions about how best to meet their energy needs in an increasingly volatile global market. Making the switch to natural gas would therefore not only provide African countries with an accessible and reliable source of energy but also help them reduce air pollution levels, reduce reliance on foreign energy sources, and contribute to a healthier global environment. This would be a positive step forward in Africa's progress toward achieving energy security while simultaneously helping to reduce the negative effects of global climate change. While the use of natural gas as an energy source has many benefits, it also has drawbacks, such as the potential for air pollution when it is not burned cleanly and the release of methane gas into the atmosphere, which is one of the most potent greenhouse gases. Currently, the majority of African countries rely heavily on imported oil and coal for their energy needs, which is not only costly and unreliable but also contributes to global climate change and air pollution levels. Despite these drawbacks, the potential benefits of switching to natural gas as an energy source are substantial and would be an important step in achieving Africa's energy security and reducing the harmful impacts of climate change on the continent. Natural gas is an energy source that is abundant, renewable, and cleaner than oil and coal. It is already being used in countries around the world to reduce air pollution and greenhouse gas emissions. Furthermore, natural gas can be transported via pipelines, reducing the need for expensive imported oil and coal and providing a reliable energy source for

African countries that can support economic growth and poverty reduction initiatives in the region.

Natural gas is one of the safest and cleanest fuels available on the African continent, and its use is growing in popularity due to the ease with which it can be transported and stored. Due to its low emissions and energy efficiency, it has become the fuel of choice for many African countries in recent years as they look to develop their energy infrastructure. Natural gas is also one of the most cost-effective sources of energy, making it an attractive option for those who wish to reduce their carbon footprint while still maintaining affordability (Omaye et al., 2022). Natural gas has a number of advantages over other energy sources, making it an ideal choice for the African continent because of its ability to meet the needs of both industry and households while causing minimal disruption and environmental impact. Furthermore, natural gas can be used for multiple applications, including power generation, cooking, and transportation. Natural gas is an increasingly popular energy source in Africa and around the world, with numerous advantages that make it the fuel of choice for many African nations seeking a clean, reliable, and affordable source of energy. As a result, African nations are investing heavily in natural gas infrastructure and resources to meet the increasing demand for energy from households, industry, and other sources. This investment in natural gas infrastructure not only provides much-needed energy access to African nations but also helps to create jobs and stimulate economic growth across the continent. Natural gas has a low-carbon footprint when burned, making it an attractive energy source for those looking to reduce their impact on the environment. In addition, natural gas is relatively easy to transport, making it a viable option for countries in remote or rural areas where traditional energy sources are not available. In these instances, natural gas can be piped in from a nearby source and used to power homes, businesses, or even entire villages. The benefits of this increased investment in natural gas infrastructure are being felt not only in terms of energy access but also in terms of job creation and economic growth. In many countries, the development of natural gas infrastructure has led to the creation of new jobs and helped spur economic growth in remote or rural regions that were previously unable to access traditional energy sources. Natural gas infrastructure development also has broader impacts, such as improving air quality and helping to reduce greenhouse gas emissions by replacing more polluting energy sources such as diesel or coal. As natural gas is a cleaner energy source than those it replaces, its increased usage can help reduce local air pollution levels and lower overall global emissions of pollutants. This in turn helps countries meet their international climate commitments while still achieving the economic and social goals of providing access to energy for their citizens. In short, natural gas infrastructure development can have numerous positive economic, environmental, and social benefits that help create a better future for the people who live in the communities it serves as well as for the environment. Beyond the direct economic benefits of natural gas infrastructure development, such as job creation and increased tax revenues, there are several broader benefits associated with the use of natural gas as a cleaner energy source as well. Natural gas emits fewer greenhouse gases than traditional fossil fuels, making it an important tool in the fight against climate change. It can also help reduce air pollution and other health risks associated with traditional energy sources, such as particulate matter and mercury emissions. In addition, the use of natural gas can help reduce reliance on imported fossil fuels, freeing up resources that can be used for other purposes, such as investment in renewable energy technologies and other programs that create jobs or support economic growth. By reducing emissions, improving air quality, and increasing energy security, natural gas has the potential to make a positive contribution to both our local and global economies. As a result, investing in natural gas is an important part of developing a sustainable energy system that can reduce our reliance on traditional fossil fuels, improve air quality, and provide an affordable and reliable source of energy for generations to come. Natural gas is an important part of the transition to a low-carbon, sustainable energy system, and its use can help reduce our dependence on foreign sources of energy while also protecting the environment. While the long-term benefits of investing in natural gas are clear, the transition from traditional fossil fuels to a more sustainable energy system is not without challenges. One of the main challenges is the need for large infrastructure investments to develop and maintain natural gas production facilities and the infrastructure needed to transport and store natural gas safely and efficiently. Additionally, there is a need for policy changes to ensure that the costs of natural gas production and use are reflective of their true environmental costs while still providing consumers with an affordable and reliable source of energy. To ensure the successful transition from traditional fossil fuels to a low-carbon, sustainable energy system powered by natural gas, governments and industry must work together to invest in the necessary infrastructure and develop policies that account for the true costs of production and use while incentivizing the development of natural gas as an affordable and reliable energy source. It is essential that governments and industry come together to invest in the infrastructure needed to transport and store natural gas safely and efficiently while also accounting for the true costs of production and use to make the transition to a low-carbon, sustainable energy system powered by natural gas a successful one. It is critical that governments and industry partners promote innovation and investment in natural gas infrastructure while also implementing policies to ensure that the costs of production and use are reflective of their true environmental costs while still providing consumers with an affordable and reliable source of energy. To achieve this goal, it is essential for governments to provide fiscal and regulatory incentives for the development of natural gas infrastructure, such as tax credits for companies investing in natural gas production and transport systems, as well as policies that make natural gas more affordable for consumers while simultaneously incentivizing the development of low-carbon technologies such as renewable energy and electric vehicles. Additionally, governments must implement regulations that are designed to incentivize the development of natural gas infrastructure and related technologies while also protecting the environment from any potential damage caused by the production or use of natural gas.

Natural gas is not only one of the most cost-effective sources of energy but also cleaner and safer than many other fuel options. The potential for natural gas to reduce air pollution and increase energy efficiency is unparalleled, and its abundance in Africa makes it an ideal fuel source for African countries looking to develop their economies and improve their quality of life (Okoh & Okpanachi, 2023). Not only is natural gas a clean and safe source of energy, but it can also provide economic benefits to African countries by reducing their reliance on imported energy sources, improving access to energy, and creating jobs in the local energy sector. Natural gas has the potential to be a major contributor to the development of African countries, and its many benefits make it an attractive option for those looking for sustainable energy sources. Natural gas can also provide environmental benefits to African countries by reducing air pollution and increasing energy efficiency, as well as economic and environmental benefits by helping to reduce energy costs, reduce emissions, and stimulate economic growth. Natural gas can have a tremendous impact on African countries, not just in terms of environmental benefits but also in terms of economic and social development. By reducing the need to rely on imported sources of energy, African countries can develop their local resources, reduce energy costs, and create jobs in the local energy sector. In addition, natural gas can provide environmental benefits by reducing emissions and air pollution, as well as increasing energy efficiency, which can improve the health and well-being of the population. Furthermore, natural gas can help to improve access to energy in rural areas that have historically been underserved due to a lack of infrastructure by providing an alternative energy source to those who may not have access to traditional sources of electricity. Natural gas can help increase economic development in African countries, reduce energy costs, and create jobs in the local energy sector. Natural gas therefore represents an invaluable opportunity for African countries to improve their quality of life, reduce poverty, and enhance their economic development. All of these advantages make natural gas an ideal fuel source for African countries seeking to achieve sustainable economic development and improve the quality of life for their citizens.

It emits less pollution than other fossil fuel sources and is more efficient, making it an attractive option for those looking to reduce their environmental impact. As a result, natural gas has become an increasingly popular energy source in recent years. In addition, natural gas is abundant and relatively cheap to extract and distribute, making it an economically viable option for consumers seeking to save money on energy costs Moreover, natural gas is versatile, with applications ranging from heating and cooking to electricity generation and vehicle fuel. With these advantages, natural gas is quickly becoming a major player in the energy industry. As it continues to rise in popularity, natural gas has the potential to significantly reduce our reliance on other forms of energy and help us achieve a more sustainable future. Furthermore, natural gas has the advantage of being much cleaner and less carbon-intensive than traditional fossil fuels such as oil and coal. By burning natural gas, we can reduce emissions of pollutants such as carbon dioxide and sulfur dioxide into the atmosphere, helping to protect our environment from the impacts of climate change and improving air quality in our cities. With its numerous benefits, natural gas is a cost-effective, reliable, and clean energy source that has the potential to revolutionize our energy system and help create a more sustainable future for generations to come. As an increasing number of countries switch to natural gas, we are seeing a shift in how energy is produced and used around the world, reducing our reliance on traditional fossil fuels and providing more renewable and low-carbon sources of energy such as solar, wind, and hydroelectric power. This is particularly beneficial for our environment, as natural gas releases approximately half of the emissions that coal does when used for electricity generation and is a far cleaner source of energy than oil when burned for transportation and other needs. This has made it a popular choice for businesses, households, and municipalities seeking to reduce their emissions while managing energy costs. Furthermore, natural gas is a much cleaner-burning energy source than coal or oil, emitting roughly half the CO_2 emissions of coal for each unit of energy produced. Additionally, natural gas is also a much more reliable and flexible energy source than renewable sources such as solar and wind, making it an attractive choice for businesses and households seeking to reduce their emissions and remain cost effective. When natural gas is burned, it produces mostly carbon dioxide and water vapor, the same substances emitted when humans exhale, making natural gas a relatively low-impact energy source in terms of emissions when compared to more traditional energy sources such as coal and oil. Thus, natural gas has become a popular choice for businesses, households, and municipalities to reduce their emissions while managing their energy costs effectively. Given that natural gas has significantly lower CO_2 emissions and is more reliable than renewable energy sources, it has become the primary choice for those looking to reduce their environmental footprint while keeping costs low (Ojukwu, 2020). When compared to other fossil fuels, natural gas emits the least amount of carbon dioxide into the atmosphere when burned, making it the cleanest burning fossil fuel available today. The lower cost of natural gas and its cleaner burning properties have made it the go-to choice for those looking to reduce their carbon footprint while simultaneously keeping their energy costs down. Furthermore, natural gas is a reliable energy source that can be stored for long periods of time and has the ability to respond quickly to demand, making it a good choice for those looking to reduce their environmental impact while maintaining access to reliable and affordable energy. This combination of clean burning properties and cost savings has made natural gas a go-to energy source for those looking to reduce their environmental impact while still having access to reliable and affordable energy. In the early twenty-first century, natural gas has become increasingly important for environmental conservation efforts, as it emits significantly less carbon dioxide into the atmosphere than other fossil fuels and is much more affordable and reliable than other renewable energy sources, such as solar and wind power. Additionally, natural gas has the potential to be used for a variety of applications, from electricity production and space heating to transportation fuel and fuel for cooking and manufacturing. This is due to its high energy density, relatively low cost, and the fact that it can be used with existing infrastructure without major changes, making it an attractive option for both consumers and industry alike. As natural gas continues to become an increasingly important source of energy, it is crucial that consumers are educated on the benefits and drawbacks of using it, as well as how best to conserve and use it in an efficient and cost-effective manner. To ensure that natural gas is used in the best way, proper regulation and research need to be conducted to maximize its usage, take advantage of its potential, and minimize its environmental impact. This is especially true considering the significant environmental impacts associated with the burning of fossil fuels and the fact that natural gas has a much lower carbon footprint than other forms of energy production, such as coal and oil. Therefore, it is imperative that governments work together to ensure that natural gas is used responsibly and to its fullest potential while also keeping in mind the potential environmental implications of its usage. Natural gas has therefore become a key component of global energy security and has been recognized by many governments as an important part of the transition to a low-carbon future. In addition, research into the development of renewable energy sources, such as solar and wind power, could be used to complement natural gas to further reduce its environmental impact while still providing countries with the energy that they need to sustain their economies and meet the growing demand for energy from their populations. The industry is also subject to substantial federal regulation in the areas of production and distribution, which helps ensure that it is provided to consumers safely and cleanly. As such, governments must continue to work together to ensure that the responsible use of natural gas is promoted and its environmental implications are adequately addressed. While natural gas has been a vital source of energy for many countries over the past few decades, its usage can also have significant environmental consequences if not used responsibly. These consequences can range from the release of harmful greenhouse gases into the atmosphere to water and land contamination from fracking, a process used to extract natural gas from underground reserves. To address these potential environmental issues, governments must continue to collaborate in regulating natural gas production and distribution and ensure that the industry complies with necessary environmental regulations and standards. This collaboration is essential to ensuring that natural gas is used in a responsible manner that does not pose an unnecessary risk to the environment or to the health of those who live near production sites. Additionally, incentives should be provided to companies that adopt more sustainable practices in the production and distribution of natural gas, and efforts should be made to reduce waste and promote renewable sources of energy. Despite the potential environmental risks associated with natural gas production and distribution, it is an essential part of many countries' energy strategies, and its use must be responsibly regulated to minimize the risks associated with it while at the same time providing the energy needed to power a modern economy. To ensure that natural gas is used responsibly and efficiently, governments must continue to collaborate in regulating the industry and enforcing necessary environmental regulations while at the same time providing incentives to companies that demonstrate a commitment to responsible and sustainable practices. This collaboration is essential to ensuring that natural gas is used in a responsible manner, and it requires all stakeholders to come together and work toward the same goal of safeguarding the environment while providing an essential source of energy for countries across the world. The use of natural gas must be carefully monitored and regulated to ensure that it is used responsibly and sustainably, with governments playing a key role in implementing necessary regulations and providing incentives for companies to adopt more sustainable practices while at the same time encouraging innovation in the industry and providing reliable access to an affordable source of energy for the world that could potentially harm the environment if it is not managed properly (Ahjum et al., 2018). This collaborative effort between governments, industry stakeholders, and citizens is essential to ensure that natural gas is used responsibly while still meeting the energy needs of modern societies. By working together, governments, industry stakeholders, and citizens can ensure that natural gas is used in a responsible and sustainable way while providing essential energy to modern societies across the globe and improving the overall health of the planet for the first time. This in turn will help reduce greenhouse gas emissions and protect the environment from the potential long-term impacts of climate change, thus improving air quality and mitigating the effects of climate change while providing a reliable and affordable energy source to individuals and businesses around the world. This is a multifaceted problem that requires the cooperation and collaboration of all stakeholders to ensure the responsible and sustainable use of natural gas to protect our environment and ensure the continued availability of this essential resource while simultaneously ensuring access to reliable and affordable energy for current and future generations.

5.2.2 Natural Gas on the African Continent

5.2.2.1 Production and Extraction of Natural Gas

Natural gas is a fossil fuel, meaning that it is derived from organic material that was buried in the Earth millions of years ago and was subjected to intense heat and pressure over the years, which caused the material to breakdown into molecules of methane and other hydrocarbons that, in turn, can be extracted and used as a form of energy. Today, natural gas is used in many different ways, from heating and cooling homes to generating electricity for businesses and industry to powering vehicles and machines that are powered by internal combustion engines. These molecules are then extracted through a process called "fracking," which involves drilling into the earth and pumping water and chemicals into the ground to free up the gas for collection and use. Fracking has been controversial due to its impact on the environment, as the water used in the process can contaminate drinking water sources and release methane into the atmosphere, which contributes to climate change.

In light of these potential impacts, many countries and states have begun to regulate the fracking process, putting restrictions in place to ensure that natural gas is harvested safely and responsibly. While the potential risks of fracking must be taken into consideration, it is also important to note that natural gas has several beneficial uses, and its use can lead to a cleaner environment by reducing the amount of carbon dioxide released into the atmosphere as it is burned for energy compared to other energy sources such as coal. With careful regulation and monitoring, natural gas extracted through fracking can provide a safe and clean energy source while helping to reduce our dependence on coal and other sources of energy that produce more harmful emissions.

Such regulation could ensure that fracking is carried out responsibly, reducing the risk of contamination and other environmental harm while providing access to a cleaner, more sustainable source of energy. This is particularly true given that natural gas can be used to generate electricity and power, heat homes, and fuel vehicles, allowing for greater control over our emissions and helping to slow the progress of climate change. In conclusion, while fracking should be carefully monitored to minimize potential risks and dangers, it can provide many important benefits, such as access to clean energy and reduced dependence on coal and other sources of energy that produce more harmful emissions, if the process is done correctly and responsibly. However, despite the potential environmental damage caused by fracking, natural gas is a key component of the global energy mix.

However, natural gas is much less polluting than other fossil fuels, such as coal and oil, when burned, which makes it an attractive option for reducing greenhouse gas emissions and supplying energy in a more sustainable way. However, it is important to note that natural gas is a far cleaner burning fossil fuel than other energy sources, such as coal or oil, which can significantly reduce air pollution and has been an important part of the transition to renewable energy sources such as solar, wind, and geothermal energy that can help to reduce the world's reliance on fossil fuels, curb emissions of harmful greenhouse gases, and mitigate the effects of climate change. Despite the potential risks of fracking, it can provide many important benefits and has a crucial role to play in the global energy mix if the risks can be managed and addressed in a responsible manner. Therefore, while fracking presents potential risks to the environment and public health, it is still an important part of the global energy mix that should not be overlooked and can provide important economic benefits such as job creation and access to a reliable, low-cost source of energy if the associated risks are properly managed and addressed.

Furthermore, the fracking process itself can cause seismic activity due to the pressure of the water and chemicals that are injected into the ground, which can lead to potentially damaging earthquakes in certain areas if proper precautions are not taken to mitigate these risks. In addition, due to the potential risks of fracking, it is important for governments and industry to take a proactive approach to mitigating the potential impacts of this process by ensuring that proper safety protocols and regulations are in place to protect the environment and public health while also developing better ways of monitoring the process to ensure that any potential risks are minimized and addressed in a timely manner. As such, the advantages of natural gas, such as its abundance and low cost compared to other forms of energy, must be weighed against its potential environmental costs when deciding whether to use this form of energy production.

It is important to assess the risks and benefits of fracking when considering its use as an energy source and to implement regulations that ensure the safety of the environment and public health, as well as the economic benefits to be gained from its production, before moving forward with its use in a way that is safe and sustainable for the long term. While fracking has the potential to be a valuable energy source, its environmental costs must be carefully considered and regulated to ensure its safe and responsible use to take full advantage of its benefits. It is essential that governments and industry take a proactive stance in assessing and addressing the potential risks of fracking to protect the environment, public health, and the economic benefits that it could offer when making decisions about its use as an energy source.

5.2.2.2 Main Component of Natural Gas

The main component of natural gas is methane, which is a colorless and odorless gas that is composed of a single carbon atom with four hydrogen atoms attached to it and is the most abundant of all hydrocarbons found in the Earth's atmosphere. It is a flammable gas that, when burned, produces a great deal of heat, making it an ideal choice for heating and cooking in many parts of the world. In addition to its energy-producing benefits, methane is also widely used in the manufacture of many everyday products, such as fertilizers, plastics, and explosives, as well as being a key ingredient in the manufacture of various chemicals used in many industries and everyday products. Methane's prevalence in the Earth's atmosphere has recently become a concern due to its ability to trap heat in the atmosphere, contributing to global warming and climate change. This has led to governments and organizations around the world taking steps to reduce methane emissions from sources such as agricultural activities, natural gas pipelines, landfills, and wastewater treatment plants.

Methane is an important energy source and has been used in a variety of ways since the early nineteenth century to power everything from industrial operations to residential homes. However, it is also important to remember that methane is a powerful greenhouse gas, and its emissions must be minimized to protect the environment and help mitigate climate change. It is an organic compound composed of carbon and hydrogen atoms and is the simplest hydrocarbon. It is often produced by the decomposition of organic matter. Natural gas is one of the most abundant fuels on Earth and has been an important part of the energy sector for centuries. Methane is a highly efficient fuel, releasing energy when combusted and forming water and carbon dioxide as byproducts. Methane, also known as "marsh gas" or "methane," is a gaseous fossil fuel composed mostly of the chemical compound methane (CH_4) (Todd & McCauley, 2021).

It is typically found in deep underground deposits, either as a solid (methane clathrate) or a liquid (gas condensate), and is the main component of natural gas. Natural gas has been used since ancient times for lighting, heating, and cooking, and today, it is a popular energy source for both residential and industrial use. Methane has grown in popularity as an alternative energy source in recent years due to its high energy efficiency and low environmental impact. As a result, technologies such as methane gas-powered vehicles, along with other forms of renewable energy sources such as solar and wind power, have become more prominent in recent years due to the rise of global awareness about climate change and its implications for the environment and the future of humanity. This trend is set to continue in the coming decades as countries around the world continue to strive for sustainable development and a reduced dependence on fossil fuels to meet their energy needs. Compared to other fossil fuels, natural gas produces fewer pollutants, meaning it can be used to reduce the overall environmental impact of energy production while still providing a reliable and economically viable energy source.

It is used for a variety of applications, including domestic heating, cooking, industrial processes, power generation, and transportation. Natural gas has been used in transportation since the nineteenth century and is an important part of the energy mix for many countries today. In the early nineteenth century, natural gas was used as a source of energy for transportation, particularly for railway locomotives, due to its ability to provide both power and heat without the need for an external fuel source. Natural gas is produced in a variety of ways, including extraction from natural gas fields through drilling and pumping operations and by breaking down organic material in landfills and sewers into methane gas.

In addition to being an important fuel source for transportation, natural gas is also used for many industrial processes and for residential heating and cooking purposes. Natural gas is a relatively clean and efficient form of energy, making it an important part of many countries' energy mix. Natural gas is widely available in many parts of the world, making it a reliable and cost-effective source of energy compared to other sources, such as coal, oil, and nuclear power. Natural gas is a relatively clean fuel source and emits significantly lower amounts of pollutants such as sulfur dioxide, nitrogen oxide, and carbon dioxide when compared to other forms of energy production. Additionally, natural gas does not produce any solid waste and can be stored for long periods of time without degradation, making it a highly versatile fuel option for a variety of different uses. Methane is the main component of natural gas, making up approximately 90–95% of its overall composition, and it is this component that provides the energy needed to power transport systems such as locomotives.

In addition to providing energy for locomotives, natural gas has also been used in many other forms of transportation, such as cars, buses, ships, and airplanes. It may also contain small amounts of other hydrocarbons, such as ethane, propane, butane, and pentanes, as well as other gases, such as carbon dioxide, hydrogen sulfide, and nitrogen. Natural gas is widely available, but it can be difficult to transport due to its gaseous state at room temperature and atmospheric pressure. Therefore, it is often pressurized and stored in tanks or pipelines for delivery to end users. For these reasons, methane and other hydrocarbons found in natural gas are highly flammable and can easily form an explosive mixture with air. To ensure safety and prevent explosions, natural gas is odorized with chemicals such as mercaptan or ethanethiol, which gives it a distinct smell that can be detected in the case of a leak. As a result, proper safety protocols must be adhered to when dealing with natural gas, including regularly checking for leaks and avoiding areas where natural gas may be present in high concentrations.

Moreover, in regard to residential use of natural gas, all appliances should be regularly inspected for any defects, and all possible precautions should be taken to avoid any incidents, such as avoiding open flames in the same area and ensuring proper ventilation to prevent any accidental explosions or fires. Moreover, when using natural gas for residential purposes, all users should be thoroughly informed about proper safety protocols, such as avoiding open flames and checking for leaks, as well as general best practices such as regular appliance inspections and proper ventilation to maintain safety and prevent any accidents due to the presence of odorizing chemicals that may be released in the case of a gas leak and the potential build-up of odorizing chemicals due to poor ventilation and potential gas concentrations in enclosed areas, such as poorly ventilated homes, that would otherwise be undetectable to the average person.

5.2.2.3 Natural Gas Availability

Africa has the largest natural gas reserves in the world, and this natural resource has the potential to contribute significantly to the continent's economic growth. With over 204 trillion cubic meters of natural gas in reserve, Africa holds the most significant supply of the world's natural gas and, if utilized effectively, could offer a wide range of economic and environmental benefits to the region. African nations have an opportunity to not only provide their own citizens with better access to energy but also boost their exports and create a stronger economy by becoming key players in the global natural gas market.

With natural gas reserves more than five times larger than those of the Middle East, Africa has a real opportunity to become a major producer of natural gas and create jobs for its citizens while reducing their dependence on other sources of energy (Nwokolo et al., 2023b). Furthermore, due to its potential to reduce emissions, natural gas can play a key role in helping African countries meet their commitments under the Paris Agreement on climate change. Thus, Africa's natural gas reserves can play an important role in the global energy market, enabling African countries to become more self-sufficient and reduce their dependence on foreign sources of energy while also helping them move toward a more sustainable energy future.

By taking advantage of their natural gas reserves, African nations can create economic and energy security for their citizens and contribute to global efforts to fight climate change. Africa's natural gas reserves represent a great opportunity for the continent to become a major producer of natural gas and bring about economic and energy security for its citizens while also helping to reduce emissions and combat climate change. To this end, African countries should prioritize the development of their natural gas reserves and invest in infrastructure for natural gas production and distribution to capitalize on this opportunity, become more self-reliant in terms of energy security, take a leading role in global efforts to reduce emissions and combat climate change and realize the full potential of their natural gas reserves while also helping to ensure economic growth for the continent.

Furthermore, the utilization of natural gas reserves in Africa can lead to improved energy access, increased energy efficiency and environmental protection, and increased economic growth and development for the continent as a whole. With this in mind, African nations should take advantage of the opportunities presented by their natural gas reserves and prioritize their development to contribute to global efforts to fight climate change, become more self-reliant in regard to energy security, and benefit economically and environmentally from the exploitation of their natural gas reserves while also helping to improve energy access, increase energy efficiency, promote environmental protection, and promote economic growth and development on the continent for the long term. The development of natural gas reserves in Africa presents a unique opportunity for the continent to become a major producer of natural gas, helping to reduce emissions and combat climate change while also increasing energy security, promoting long-term economic growth and development, and providing tangible benefits to African nations.

According to recent estimates, Africa holds an estimated 43 trillion cubic meters of natural gas reserves, making up approximately 28% of the world's total. Based on research conducted by the Department of Petroleum Resources, Africa rests upon proven gas deposits totaling up to 15 trillion cubic meters, with some of the largest natural gas reserves located in Egypt, Nigeria, Libya, and Algeria (Todd & McCauley, 2021). These findings underscore the need for African countries to focus on developing their natural gas reserves to help combat climate change and become more self-reliant in regard to energy security.

Additionally, several African countries, including Kenya and South Africa, are making significant investments in developing their own natural gas reserves to diversify their energy sources and reduce their reliance on imported fuels. The development of natural gas reserves in Africa offers many potential benefits, but there are also some challenges that must be addressed for African countries to realize the full potential of their natural gas reserves. For instance, there needs to be greater investment in infrastructure and technology to improve the extraction, storage, transportation, and use of natural gas reserves across the continent. In addition, there needs to be an increase in research and development of clean energy technologies that could make the use of natural gas more efficient and less polluting than other fossil fuels.

Additionally, African countries must also ensure that their natural gas reserves are managed in an equitable and responsible way, with all stakeholders receiving a fair share of the economic benefits generated by the exploitation of these resources. Additionally, there needs to be greater government oversight to ensure that regulations related to the extraction and use of natural gas are implemented properly and effectively to protect the environment and prevent corruption from occurring. The exploitation of natural gas reserves in Africa is an important opportunity for economic growth, but there are many steps that must be taken to ensure the success of these initiatives to make the most of this opportunity and to ensure that natural gas is used in a responsible manner.

Finally, governments must also invest in public education and awareness campaigns to ensure that people are aware of the risks associated with the extraction and use of natural gas as well as their rights in regard to participating in decisions related to natural gas exploitation in their respective communities. This is essential to ensure that all stakeholders benefit from natural gas exploitation while also taking into account the potential environmental and social impacts associated with its extraction and use. As such, governments must work to enact comprehensive legislation and regulations related to natural gas extraction and use, in addition to establishing oversight mechanisms that are accountable and transparent, to ensure the proper and responsible use of natural gas reserves in Africa. To ensure that the exploitation of natural gas in Africa is conducted in a responsible manner, governments must take into account the economic and social impacts associated with its extraction and use and ensure that the interests of local communities and other stakeholders are taken into account when making decisions related to natural gas exploitation.

5.2.2.4 Economic Benefits of Natural Gas in Africa

One of the effects of natural gas in Africa is economic growth. Natural gas has created new opportunities for economic growth in Africa by providing a viable alternative to traditional sources of energy and creating job opportunities in the gas industry and related industries such as agriculture, transportation, and manufacturing. The increased economic activity resulting from natural gas has been accompanied by social and economic development, including improved access to healthcare, education, and other basic services, as well as improved living standards and incomes for millions of people across the continent. These benefits have been distributed unevenly, and some areas of Africa have seen more economic growth than others due to the unequal distribution of natural gas resources.

In addition, the African population has also seen an increase in purchasing power, enabling them to access a greater range of products and services while at the same time enjoying more disposable income to spend on discretionary items such as entertainment, travel, and leisure activities. Consequently, the increased availability of natural gas in Africa has been a major boon for the continent's economic growth and social development, leading to increased investment opportunities, increased employment and wages, improved standards of living, and a greater degree of economic integration between African countries as well as with the rest of the world. Natural gas has allowed African nations to create industry and stimulate economic growth, as it is one of the most abundant and cost-effective sources of energy available to developing countries, and its use has enabled African countries to take advantage of advances in technology and develop a competitive edge in the global market while also providing an opportunity to improve the quality of life for their citizens by reducing reliance on other sources of energy that are more polluting and environmentally damaging, such as coal and petroleum products.

Generally, the benefits of natural gas in Africa are far-reaching and have proven to be a major factor in promoting economic growth, social development, and improved standards of living across the continent. With improved access to natural gas, African nations have seen an increase in their economic prospects and capacity for development, enabling them to participate more actively in the global economy and be better positioned to take advantage of the opportunities that come with globalization and international trade.

Furthermore, improved access to natural gas has enabled African nations to move away from traditional fossil fuels and toward renewable energy sources, helping to reduce emissions and providing cleaner air for the continent's citizens while also promoting the sustainability of natural resources and reducing the environmental impact of energy production. This shift to natural gas has not only enabled African nations to enjoy the many economic benefits of natural gas but also fostered a more sustainable approach to energy production that is in line with global trends toward greater environmental responsibility and more sustainable economic growth.

As such, natural gas has played a major role in promoting economic development, social progress, and environmental sustainability throughout Africa in recent years and is likely to continue to do so in the future. By providing reliable and affordable energy, natural gas has been instrumental in boosting economic growth in African nations, supporting the emergence of a new middle class that is better positioned to take advantage of the opportunities available to them through global trade and international commerce while also improving access to healthcare, education, and other essential services necessary for long-term economic stability and growth.

Furthermore, improved access to natural gas has enabled African nations to move away from traditional fossil fuels and toward renewable energy sources, helping to reduce emissions and providing cleaner air for the continent's citizens while also promoting the sustainability of natural resources and reducing the environmental impact of energy production in the region. Furthermore, the switch to natural gas has also enabled African countries to lead the way in terms of environmental sustainability and global trends toward greener energy sources in a world that is increasingly focused on reducing our dependence on carbon-based energy sources and decreasing the amount of greenhouse gas emissions released into the atmosphere. Natural gas is also an excellent tool for improving access to electricity, especially in rural areas where people do not have easy access to electricity and where traditional power plants are not able to provide the energy required to meet local needs due to their remote locations.

This has enabled African nations to meet their energy needs and improve their standard of living without having to rely on more expensive and less reliable sources of energy, such as coal, which has been a major source of air pollution in the region for many years and can also be damaging to the environment due to the release of greenhouse gases as well as the need to extract it from the ground and use it in traditional power plants. All of this shows the importance of natural gas for African countries in terms of environmental sustainability and helping to meet their energy needs in an economically and ecologically sustainable way that goes beyond just reducing carbon emissions.

5.2.2.5 Challenges of Natural Gas in Africa

Lack of Infrastructure Natural gas is one of the most reliable and affordable sources of energy; however, in Africa, it faces many challenges due to a lack of infrastructure, difficulty accessing resources, and the high cost of installing the infrastructure needed to make the best use of natural gas. These challenges are compounded by the fact that many African countries do not have the resources necessary to invest in natural gas infrastructure and usage development. As a result, these countries lack the many benefits that natural gas can offer, such as cost savings, reliability, and a reduction in emissions from traditional fuel sources. All of these issues hinder the potential that natural gas has to be an important source of energy for African countries and to make a positive contribution to the environment and economy of the continent.

Despite these obstacles, natural gas remains one of the most promising energy sources in Africa, as it has a much lower carbon footprint than other fuels and can be used to generate electricity or power industrial processes efficiently and cost-effectively. African countries must overcome the challenges of insufficient infrastructure, limited capital resources, and complex regulatory environments if they are to take advantage of the potential offered by natural gas.

To maximize the benefits of natural gas, African countries must invest in the development of infrastructure that will enable the efficient transport and use of natural gas, such as pipelines and power plants, as well as regulatory reforms that will provide the necessary incentives to attract private sector investment in natural gas production and utilization projects. Furthermore, it is essential to establish an adequate regulatory framework to ensure the safety and security of natural gas operations and facilities in Africa so that their use can be beneficial and safe for the environment, communities, and individuals. Furthermore, there must be a commitment to invest in training and capacity building for both the public and private sectors to ensure that personnel are sufficiently skilled to handle the complexity of natural gas production and utilization operations safely and effectively.

One of the most reliable and affordable energy sources is natural gas, but due to a lack of infrastructure in Africa, it encounters many difficulties in reaching its full potential. These challenges are further compounded by a lack of knowledge among African countries on how to leverage national gas as a source of energy, leading to an inadequate utilization of the resource. Despite these challenges, African countries have a tremendous opportunity to tap into the potential of natural gas and its ability to fuel economic growth in the region. As such, it is important for African nations to invest in their own capacity and expertise to maximize the benefits that natural gas can bring to their economies through the development of sustainable infrastructure, training, and support for national gas industry professionals, and increased awareness about the potential uses of natural gas in African countries.

To maximize the potential of natural gas in Africa, it is essential that national governments create strong legal and regulatory frameworks that protect the rights of all stakeholders while safeguarding against environmental degradation and corruption and promoting the development of competitive markets and investment in natural gas infrastructure that are safe, efficient, and cost-effective for their citizens. While these steps will require significant investments from both the public and private sectors, they are essential for African countries to unlock the full potential of natural gas and maximize the many benefits it has to offer both in terms of economic development and energy security.

In addition, governments must strive to build effective public–private partnerships and involve key stakeholders in the decision-making process, as well as investing in public education and information campaigns to inform citizens of the potential of natural gas in Africa and how they can benefit from its development. Additionally, African countries should actively partner with international organizations and development agencies to ensure access to the necessary resources, technology, and capacity-building programs needed to facilitate the use of natural gas in the region and ensure the successful development and utilization of natural gas in African countries. Moving forward, African countries should prioritize the implementation of an adequate legal and regulatory framework that protects the environment and guarantees respect for human rights while also establishing a transparent and equitable distribution of the revenue generated by natural gas development and utilization to maximize its potential within the region.

This will help ensure that local communities benefit from the development and utilization of natural gas resources while minimizing the potential for exploitation. As the potential for natural gas in Africa grows, governments must be prepared to address the challenges associated with its development and utilization in a manner that promotes economic growth, social development, and environmental protection while also protecting the interests of those who will be most directly affected by its exploitation. African countries should continue to foster partnerships with both international organizations and local communities to ensure the safe and secure development of natural gas resources while also upholding the highest standards of corporate and environmental responsibility. African countries should prioritize the implementation of an adequate legal and regulatory framework that protects the environment and guarantees respect for human rights while also establishing a transparent and equitable distribution of the revenue generated by natural gas development and utilization to maximize its potential within the context of sustainable economic growth and development across the continent.

In general, the development and utilization of natural gas resources in Africa present an opportunity to promote sustainable economic growth and development, as well as social development and environmental protection, and African governments must rise to the challenge of managing natural gas in a manner that prioritizes all these goals and ensures a fair distribution of the economic benefits to all stakeholders, including the people whose lives and livelihoods may be directly impacted by its exploitation. This will require the implementation of an appropriate legal and regulatory framework as well as partnerships between international organizations, local communities, and the governments themselves to ensure that natural gas resources are developed and utilized responsibly and that the revenues generated are fairly and equitably distributed, with a particular focus on using the proceeds to promote social development, human rights protection, and environmental conservation across the continent.

African governments must take responsibility to ensure that natural gas resources are developed in a manner that prioritizes sustainable economic growth and development, social development, and environmental protection while also protecting human rights and guaranteeing a transparent and equitable distribution of the associated revenues to maximize the potential benefits of this resource and ensure that all stakeholders benefit from it. This can only be achieved if African governments are willing to invest the necessary time, effort, and resources into developing a comprehensive legal and regulatory framework for natural gas development and utilization that respects human rights and environmental protection while also promoting transparent revenue distribution to all stakeholders and ensuring that the proceeds are used to promote sustainable economic growth and development, social development, and environmental conservation on the continent as a whole. Generally, the development and utilization of natural gas resources in Africa presents a unique opportunity to promote sustainable economic growth and development, social development, and environmental protection across the continent, and African governments should strive to ensure that this opportunity is utilized to the fullest extent to maximize the potential benefits for all stakeholders.

Lack of Investment Lack of investment is one of the biggest obstacles to the exploration and use of natural gas in Africa. Natural gas reserves in Africa remain largely untapped due to insufficient investment in infrastructure and gas extraction technologies, coupled with a lack of awareness about the economic and environmental benefits of natural gas usage. This lack of investment has created a vicious cycle in which companies are unable to realize the benefits of natural gas, leading to a decreased sense of urgency to invest in and explore natural gas reserves and resources. Moreover, the high cost of extraction and transportation coupled with inadequate infrastructure to store and process natural gas further hampers exploration efforts and limits the potential of the continent's natural gas resources. In addition, the political instability and lack of security in certain parts of Africa further complicate the exploration and use of natural gas resources.

These challenges associated with the exploration and use of natural gas in Africa make it difficult to realize the full potential of this resource, resulting in an estimated 80% of the continent's natural gas resources remaining undiscovered and underdeveloped. This further impacts the already strained energy access issue across the continent and limits economic development, as natural gas can be used to generate power, fertilizers, chemicals, and other materials that can be used to promote economic growth in Africa. Ultimately, the continent's natural gas resources could be a viable solution to its energy access issue; however, inadequate infrastructure, costly extraction and transportation costs, and the political instability in some parts of the continent make it difficult for Africa to unlock these opportunities.

Without significant investments in infrastructure, security, and technologies to reduce exploration and transportation costs, it is difficult to realize the full potential of Africa's natural gas resources. African countries must invest in necessary infrastructure, strengthen political stability, and create policies that incentivize the exploration and production of natural gas resources if they want to take full advantage of their natural gas resources, improve energy access, promote economic development and job growth, and build strong and stable economies across the continent. All of this points to the need for African countries to prioritize investment in the infrastructure, political stability, and technologies necessary to realize the full potential of their natural gas resources and improve energy access, security, and economic development throughout the continent.

African countries have the potential to significantly reduce their energy access issues through the exploration and production of their natural gas resources if the necessary investments and conditions are in place. This creates a difficult situation for many African nations, as they face the challenge of balancing their need to develop economically while ensuring that they are able to protect the environment from any potential negative impacts that may be associated with natural gas exploration and production activities. As such, African countries must be proactive in establishing the policies, infrastructure, and technological advances that are needed to develop their natural gas resources sustainably while ensuring that their populations have access to reliable and affordable energy solutions.

African countries must strive to find a balance between economic development and environmental protection to fully realize the potential of their natural gas resources while still ensuring the well-being of their people and their environment. This can be a challenging task, as it involves striking the right balance between encouraging economic growth and development while also taking into consideration the potential environmental impacts of natural gas exploration and production activities. To achieve this balance, African countries must carefully consider the risks and benefits associated with their natural gas activities and create policies and regulations that are specifically tailored to their own unique needs and goals.

African countries should also take into account the interests of all stakeholders involved in natural gas activities, such as local communities, industry players, and government agencies. To ensure that their natural gas resources are used in an environmentally responsible manner, African countries should prioritize the creation and implementation of comprehensive regulatory frameworks that will protect the environment and ensure that natural gas exploration and production activities are conducted in a socially responsible way while also providing economic opportunities to the local communities and businesses involved.

These frameworks should include regulations on topics such as air and water pollution, waste management, safety and risk management, health and occupational hazards, and public consultation processes. In addition, African countries should collaborate with external organizations and governments to ensure that best practices are implemented in their natural gas activities to minimize environmental and social risks and to protect their citizens from any negative consequences of natural gas exploration and production activities that could arise in the future. African countries should ensure that their regulatory frameworks also address issues of transparency and accountability to guarantee that natural gas exploration and production activities are conducted in a manner that is responsible and sustainable over the long term.

To ensure that natural gas resources are used responsibly and sustainably, African countries should prioritize the development of comprehensive regulatory frameworks and collaborate with external organizations to ensure best practices in natural gas activities. These frameworks should include regulations to ensure that natural gas exploration and production activities are conducted with due regard for the environment, public safety, and sustainable development goals in their countries.

African countries should also ensure that any contracts entered into with private sector companies or other external organizations and governments include enforceable provisions to protect the environment, ensure public safety, and ensure transparency in the management of natural gas resources. These frameworks should also include stringent regulations to ensure that natural gas activities are conducted in a transparent and accountable manner and that all revenues generated from natural gas activities are distributed equitably to benefit all stakeholders, including the local communities and populations affected by these activities.

Furthermore, African countries should also establish effective mechanisms to monitor and enforce compliance with their regulatory frameworks. In general, African countries should implement comprehensive frameworks for natural gas exploration and production activities that are in line with their sustainable development goals to ensure that their natural gas resources are used in a way that benefits the local populations and communities while also protecting the environment and public safety and promoting transparency and accountability in the management of these resources. African countries should also ensure that their regulatory frameworks are regularly reviewed and updated to reflect changes in the market, technology, and environmental conditions to ensure that their natural gas activities are conducted in a sustainable and responsible manner.

African countries should take the lead in creating regulatory frameworks that provide a solid foundation for natural gas exploration and production activities and ensure that these activities are conducted in a safe, responsible, and transparent manner that provides social and economic benefits to the local population and communities while also protecting the environment and public safety. This can be achieved through the implementation of robust regulations in accordance with international standards, such as those outlined by the International Energy Agency and the African Union, which can provide a framework for the responsible management of natural gas activities across the continent and ensure that these activities are conducted in a manner that is beneficial to the people, economy, and environment of Africa. African countries must also establish effective monitoring and enforcement systems that can be used to detect, investigate, and prosecute violations of regulations to ensure compliance with the regulatory framework and protect the health and safety of those involved in natural gas activities and the public as a whole.

Political Instability One of the biggest barriers to natural gas exploration and use in Africa is political instability in many countries, particularly those that have been most recently formed due to the continent's history of colonization and postcolonial turmoil. These political issues, such as civil wars and rebellions, have severely hampered investment in energy infrastructure, creating a disincentive for corporations and governments to invest in natural gas exploration and use and preventing the continent from accessing the significant potential for natural gas that it holds. Furthermore, the legacy of colonial rule has left much of the continent with an energy infrastructure that is severely underdeveloped and unable to cope with the demands that natural gas exploration and use would bring.

Additionally, corruption and mismanagement of resources have resulted in much of the continent being unable to invest in renewable energy sources, instead relying heavily on traditional fuels such as coal and oil, which are not only more expensive but also detrimental to the environment. These challenges have become even more difficult to overcome due to the pandemic and economic crisis, as many countries in Africa are struggling to access sufficient foreign aid. This has caused governments to be reluctant to make major investments in the energy sector due to fears that the investments may be wasted should their government suddenly become unstable or, worse yet, a hostile regime arise and nationalize resources or seize foreign-held assets. Moreover, the lack of political stability has led to a lack of financial security and economic strength, which has further hindered natural gas exploration and use in many African nations. This political instability has been one of the most significant challenges to exploiting Africa's natural gas reserves, as it is difficult to make long-term investments in a region that is prone to violence and unrest.

These challenges have become even more difficult to overcome due to the pandemic and economic crisis, as many countries in Africa are struggling to access sufficient foreign aid to prop up their economies and have been unable to devote resources to energy exploration and development due to the instability and turmoil in the region caused by the pandemic. This has further hindered the exploration and exploitation of natural gas reserves in Africa, as political instability and a lack of financial security make it difficult for foreign companies to commit to long-term investments in the region.

Moreover, the current situation has made it even more difficult to overcome these obstacles, as African countries are struggling to access foreign aid and divert resources toward energy exploration and development. This has severely hampered the economic growth of African nations, as natural gas reserves represent a potentially immense source of revenue for many countries in the region, yet they have been unable to capitalize on them due to the pandemic and economic crisis exacerbating the challenges they were already facing prior to the pandemic. Ultimately, the pandemic has caused immense difficulties for African countries in regard to accessing foreign aid and resources for energy exploration and development, and as a result, the exploration and exploitation of natural gas reserves have been heavily hindered, presenting a further obstacle to economic growth in the region.

This is a great tragedy for many African countries, as the exploitation of natural gas reserves could have created immense economic benefits for their citizens and allowed them to build sustainable and prosperous economies for many years to come had they been able to access the necessary reserves and foreign aid needed to explore and develop these reserves prior to the onset of the pandemic. Unfortunately, the current pandemic has put a severe strain on African countries and their ability to access foreign aid and exploit their natural gas reserves, resulting in their being unable to take advantage of these potential resources for economic growth and development, leading to their economic growth being greatly hindered and their citizens being unable to reap the full economic benefits that could have been achieved.

Despite the immense potential that natural gas reserves represent, the current situation has made it incredibly difficult for African countries to access foreign aid and exploit their natural gas reserves, leading to great economic strain for many African countries and a severe limitation on their potential for economic growth and development at a time of great opportunity and progress for the region as a whole. This is a devastating reality for African countries, and it underscores the need for a more sustainable, equitable global system that is able to respond to the needs of all countries, especially those in developing regions such as Africa that do not have access to the resources and foreign aid necessary to grow and develop their economies in times of crisis. With the decline in foreign aid and access to natural gas reserves due to the current pandemic, many African countries have been unable to take advantage of these potential resources for economic growth and development, leading to their economic growth being severely hindered and their citizens being unable to reap the full economic benefits that could have been achieved had they been able to access the necessary reserves and foreign aid needed to explore and develop these reserves prior to the onset of the pandemic.

This has been a heartbreaking result of the pandemic, and it is essential that African countries are supported in finding alternative routes to economic growth and stability in the face of such adversity and uncertainty. Despite the immense potential that natural gas reserves represent, African countries have been unable to capitalize on this opportunity due to the global economic downturn caused by the pandemic and the lack of foreign aid and access to resources necessary to explore and develop these reserves. Despite the tragedy of this situation, it also serves as a powerful reminder of the importance of developing more equitable global systems that provide sustainable and equitable access to resources and foreign aid for all countries, especially those in developing regions such as Africa.

Despite the potential of natural gas reserves to provide significant economic benefits, the realities of the current pandemic have created immense difficulty for African countries in terms of accessing foreign aid and exploiting these resources in a way that ensures long-term economic growth and stability for their citizens. This has caused many African governments to struggle to make significant investments in the energy sector and has resulted in African countries not being able to maximize the potential of natural gas reserves and unlocking their full economic potential for the benefit of their people. This has resulted in many African nations having to rely on their own resources and ingenuity to find solutions to the energy crisis they face.

It is clear that something must change for African countries to truly benefit from the potential of their natural gas reserves and reap the rewards of sustainable long-term economic growth for their citizens. Unfortunately, the current global system remains unequal and is often unable to provide African countries with the resources they need to take advantage of this potential. In addition to these structural obstacles, African countries also face the challenge of having inadequate infrastructure to support the exploitation of their natural gas reserves and a lack of funding and expertise to effectively exploit them and capitalize on their potential benefits without the aid of external partners.

Consequently, African countries have to rely on foreign investments and assistance to build their infrastructure, improve their technological capabilities, and gain access to the necessary expertise to maximize the economic benefit of their natural gas reserves. This means that African nations must not only focus on finding innovative solutions to the energy crisis they face but also look for ways to work together with external partners to overcome these structural and technical obstacles. This creates a situation in which African countries remain largely dependent on outside powers for their energy resources and are unable to tap into the potential of their own gas reserves. This is a stark reminder that developing equitable and sustainable global systems is vital for all countries, but particularly for those in the developing world, such as Africa, to reduce inequality, secure economic growth, and ensure that all countries have access to the resources they need to survive and thrive in the public eye.

However, despite the immense difficulties faced by African countries due to the pandemic, there is still hope that they can overcome these obstacles and make use of their natural gas reserves to stimulate economic growth and ensure that African countries are no longer seen as dependent on outside powers for energy resources. Going forward, the onus is now on African nations to find new and creative ways to increase their access to energy resources while avoiding the pitfalls of dependency that arise from reliance on foreign sources of energy. To this end, African nations need to prioritize the development of innovative technologies, such as renewable energy sources, and diversify their energy supply portfolios by investing in an array of resources that can help them reduce their dependence on foreign sources of energy while also protecting the environment and promoting sustainable development.

To ensure success, African nations must also increase their political will and commitment to energy reform initiatives and strive to increase their capacity to develop and implement viable energy policies that will allow them to take full advantage of their energy resources while minimizing their reliance on foreign sources of energy. As such, African nations have a unique opportunity to take control of their own energy destiny by making the most of their natural gas reserves and other energy sources while simultaneously investing in alternative energy technologies that can help them reduce their reliance on foreign sources of energy and ensure a cleaner and more sustainable energy future. African nations need to take decisive action to ensure that they are able to reap the full benefits of their energy resources while also avoiding the pitfalls of reliance on foreign sources of energy that may be unreliable or unsustainable in the long run. African nations must continue to prioritize the development of innovative energy technologies and diversify their energy supply portfolios to secure their energy future and ensure their energy security.

African nations must continue to invest in energy policy reform initiatives and increase their capacity to develop and implement successful energy strategies to ensure that their energy security is not compromised in the long run. African nations must also invest in energy literacy programs to help ensure that their citizens are aware of the importance of energy security and the need to embrace alternative forms of energy and energy efficiency to achieve energy sustainability in the long term.

This will provide African nations with the opportunity to move away from foreign energy sources and rely more on local renewable energy sources such as solar, wind, and biomass, allowing them to reduce their dependence on imported fossil fuels and become more self-sufficient and less vulnerable to fluctuations in global energy markets. This will require significant investment in infrastructure and technology, including the development of renewable energy sources such as solar and wind power, as well as better access to credit for those wishing to start their own businesses or invest in new energy technologies.

This will be no easy task, as many African countries lack the financial and technical resources needed to invest in energy reform initiatives. However, with proper investments and strategies, African countries can begin to create a new economy that is more self-sustaining, secure, and environmentally responsible while simultaneously contributing to job creation, poverty reduction, and a healthier environment. However, if African nations are able to make the necessary investments in energy literacy programs, energy infrastructure and technology, and access to credit for those wishing to pursue renewable energy technologies, they can dramatically increase their self-sufficiency while simultaneously reducing their dependence on traditional fossil fuels and improving their overall quality of life. This investment would also provide new economic opportunities for entrepreneurs and inventors by creating jobs and promoting innovation in green technologies while incentivizing investment in renewable energy projects and supporting the growth of a more environmentally conscious market. This in turn will have a positive knock-on effect, creating more opportunities for economic growth, improved energy security and access, and the development of renewable energy resources, which will benefit African countries, their citizens, and the environment as a whole. African nations must seize the opportunity to make the necessary investments to improve their quality of life and create a more sustainable future for themselves and future generations.

Nevertheless, it is essential for African nations to prioritize energy literacy and make the necessary investments to transition to more sustainable forms of energy while also creating the necessary policies to ensure that the benefits of these investments are spread across all sectors of society to ensure that all citizens can reap the rewards of this investment in the long run. It is clear that the potential benefits of investing in green technologies are immense and could provide Africa with a huge competitive advantage in the global economy and make it a leader in the development of sustainable and renewable energy sources. Not only will these investments have the potential to create jobs and economic growth, but they will also help improve energy security, reduce pollution, and address global climate change issues.

Additionally, investing in green technologies will help to promote technological innovation, particularly in areas related to research and development and engineering, which can lead to further economic development and improved living standards for African citizens across the continent. Therefore, it is important for African nations to recognize the potential benefits of investing in green technologies, make it a priority to reap the long-term rewards that these investments could bring, and pursue policies that promote investment in green technology and create an enabling environment for businesses to invest in these technologies. This would go a long way in alleviating poverty, promoting economic growth, and allowing African countries to become more self-sufficient and resilient to external shocks while at the same time protecting the environment and addressing global climate change issues. Investing in green technologies can be an integral part of this process.

As such, investing in green technologies is an important part of any development agenda and should be a priority for African nations. African nations must prioritize investing in green technologies to ensure economic growth and improved living standards for citizens across the continent while also contributing to the global effort to combat climate change and protect the environment. Green technologies such as renewable energy, green infrastructure, and sustainable agriculture can provide African nations with the resources they need to develop their economies while also contributing to global environmental goals. For example, renewable energy sources such as wind, solar, and hydropower can provide countries with clean, low-cost electricity and create new jobs in the green energy sector, while investments in green infrastructure can help reduce the effects of climate change such as flooding, soil erosion, and droughts, as well as provide access to fresh water and other essential resources to rural populations.

Additionally, investing in green technologies can help African nations reduce their dependence on fossil fuels and contribute to global efforts to reduce emissions of carbon dioxide and other greenhouse gases, which are responsible for causing global warming. Investing in green technologies is not only beneficial to African nations but also necessary for them to ensure economic growth and improved living standards while playing their part in the global effort to combat climate change. Sustainable agriculture is another important green technology that can help African nations develop their economies while also being beneficial to the environment by promoting soil conservation, reducing pesticide use, and preserving biodiversity. This diversification strategy is critical for the economic growth and sustainability of African countries, as it can allow them to reduce their dependence on foreign sources of energy and maximize their own energy resources.

This is why it is essential that international organizations, donors, and investors work together to build and implement more equitable global systems that facilitate sustainable development in African countries and provide them with the resources necessary to develop these natural gas reserves in a responsible way while also protecting the environment. By investing in renewable energy sources and green technologies, African countries can move toward self-sufficiency and benefit from increased access to energy while simultaneously reducing their carbon footprints. The success of these efforts will depend on the ability of African countries to create and adopt sustainable agriculture practices that prioritize local resources and knowledge systems, create more employment opportunities in rural areas, and ensure that any natural gas reserves are used in a responsible way that meets the needs of the present without compromising the ability of future generations to meet their own needs. This diversification strategy has the potential to bring major economic and environmental benefits to African countries, but it will require a unified effort between governments, international organizations, donors, and investors to ensure that African countries can realize the full potential of their natural gas reserves in a way that is equitable and sustainable for all.

Cost of Natural Gas The price of natural gas in many countries is one of the biggest obstacles to its exploration and use in Africa due to its high cost relative to other forms of energy. Due to the lack of infrastructure and technology, the cost of natural gas is often prohibitively expensive for many African countries, which hampers their ability to use this valuable resource to meet the demands of their rapidly growing populations. In addition, the African continent also lacks expertise and know-how in regard to safely and efficiently exploring, producing, and distributing natural gas, making it difficult for African countries to even begin to consider natural gas as a viable energy source for their energy needs. This lack of expertise and the high cost of natural gas in Africa have hindered its exploration and use, leaving African countries unable to access this valuable energy source to meet the demands of their rapidly growing populations.

This situation presents a major challenge for African countries, as they must find ways to lower the cost of natural gas to be able to access and use this valuable resource and gain the knowledge and expertise needed to safely and efficiently explore, produce, and distribute natural gas for their own energy needs to adequately meet the needs of their citizens and improve the quality of life for their citizens as well as their economic growth and development. To effectively address this problem, African countries must invest in developing their infrastructure and technology, in addition to increasing their know-how and expertise related to natural gas production and exploration, to be able to access and use this energy source in a cost-effective manner and thus be able to effectively meet the energy needs of their citizens and improve their economic growth and development.

Additionally, this challenge can only be addressed by investing in the infrastructure and technology needed to explore, produce, and distribute natural gas efficiently, as well as by training and employing local experts who are knowledgeable about the latest technologies and safety standards related to natural gas production and exploration to successfully address the challenge of accessing and using natural gas as an energy source in African countries. African countries must make a concerted effort to reduce the cost of natural gas and invest in the necessary infrastructure and technology to safely and efficiently explore, produce, and distribute natural gas to be able to access and use this valuable resource, gain the knowledge and expertise needed, and meet the energy needs of their citizens while also improving the quality of life for their citizens as well as their economic growth and development.

They must also invest in research and development activities to create more efficient and cost-effective production technologies for natural gas to fully take advantage of this valuable energy source and provide their citizens with the energy they need while at the same time helping to ensure their economic growth and development by reducing the cost of energy and stimulating investment in other sectors such as manufacturing, agriculture, and transportation. As such, the cost of natural gas exploration and development is often too high for African nations to bear, leading to a lack of utilization and access to this valuable energy source, which could otherwise provide much-needed economic growth and development for African nations as well as improve the quality of life for their citizens.

Thus, it is clear that for African nations to successfully access and use natural gas as an energy source, they must invest in the necessary infrastructure and technology, develop expertise and knowledge about safety standards, reduce the cost of natural gas exploration and production technologies, and prioritize research and development activities to make natural gas production more efficient and cost-effective to ensure the economic growth and development of their nations as well as improve the quality of life for their citizens by providing them with an energy source that is both affordable and sustainable. Generally, African countries must invest in the necessary infrastructure and technology, lower the cost of natural gas exploration and production technologies, gain expertise and knowledge on safety standards, and prioritize research and development activities to ensure the successful utilization of natural gas as an energy source that is both affordable and sustainable to facilitate economic growth and development in their countries and provide their citizens with citable energy.

Accessibility of Natural Gas One of the biggest barriers to natural gas exploration and use in Africa is the availability of fuel in many of the nations on the continent due to the geographic and geological challenges in obtaining it in many African countries. Many African countries are located in remote areas, far from major oil and gas reserves, making it difficult to access the necessary resources and to build the infrastructure required to transport and store natural gas. This lack of infrastructure and the geographical isolation of many African countries present an additional challenge for those looking to explore and exploit the continent's natural gas reserves. This challenge is further compounded by the fact that most African countries lack the necessary financial and technological resources to exploit their natural gas reserves, creating an additional barrier for those looking to make use of these resources. This lack of available resources, financial and technological means, and geographical isolation makes it difficult for many African countries to properly exploit the natural gas reserves that are within their borders.

However, despite these difficulties, there is still potential to make use of natural gas reserves in Africa through the creation of partnerships between African countries and international oil companies, which have the necessary resources, technology, and expertise to help African countries exploit their natural gas reserves in an efficient and profitable manner. This partnership between African countries and international oil companies will not only create an effective way to exploit natural gas reserves in Africa but also have the potential to bring economic growth and development to these African countries by providing them with increased access to financial resources, technology, and expertise that can be used to create infrastructure, develop their economies, and improve the lives of their citizens through improved access to energy and other resources needed to exploit natural gas reserves.

This would create a mutually beneficial relationship, as African countries can benefit from the resources and expertise of international oil companies to exploit their natural gas reserves, while oil companies can benefit from increased access to previously untapped markets and the potential for increased profits. The potential for partnership between African countries and international oil companies is one of the most promising solutions to the challenge of exploiting natural gas reserves in Africa and is one that can help bring greater economic growth and development to African countries while also allowing oil companies to benefit from increased access to previously untapped markets and the potential for increased profits. This could be a win–win situation for all parties involved, as both African countries and international oil companies can benefit from this arrangement while at the same time helping to create a more sustainable economic and energy environment for African countries as well as allowing for more responsible use of their natural gas reserves.

This could be beneficial to both parties, but it is important that there are also safeguards in place to ensure that the exploitation of natural gas reserves does not lead to further economic or environmental issues in the region, such as environmental degradation or exploitation of local populations. This partnership has the potential to create a positive and mutually beneficial relationship between African countries and international oil companies, as both parties would benefit from the exploitation of natural gas reserves in Africa while also ensuring that environmental and social protections are put in place to ensure the sustainable exploitation of these resources for the long-term benefit of both African countries and international oil companies. It is essential to ensure that both parties are given the opportunity to enter into meaningful and mutually beneficial partnerships that take into account their individual needs and interests while at the same time protecting the environment and local populations while providing a secure and long-term source of revenue for African countries as well as a reliable energy source for international oil companies to make this a successful venture.

Environmental Risks Potential pollution and climate change are two unfavorable effects of natural gas in Africa that cannot be ignored, and they are both issues that African nations must address to protect their environment and people from the potential long-term damage that could result from these pollutants and changes, not only for their current population but also for the generations that will follow them. Natural gas is often seen as a viable alternative energy source for African countries, but the potential pollutants and climate change it can produce should not be overlooked and need to be taken into account when considering the adoption of this form of energy in African countries. Natural gas in Africa is becoming increasingly utilized as an energy source, but it is not without its drawbacks and potential environmental consequences, which must be addressed to prevent long-term damage to the environment and people of African nations. Natural gas can pollute the air and water and exacerbate climate change due to its high greenhouse gas emissions, which, if left unchecked, can cause major environmental and health issues that can last for generations. Ultimately, natural gas in Africa can have both positive and negative implications, and its use should be carefully considered before being adopted by African nations. Therefore, before African nations consider the adoption of natural gas as an energy source, they must ensure that any potential environmental consequences are addressed and managed in a responsible manner to minimize any potential long-term environmental damage.

For instance, African nations should consider the development of cleaner natural gas technologies to reduce emissions, as well as other renewable energy sources that can provide the same level of energy without the same environmental consequences as natural gas. African nations should also develop effective environmental regulations and monitoring systems to ensure the responsible use of natural gas, such as conducting regular inspections of energy facilities and providing penalties for noncompliance. Generally, natural gas has the potential to provide much-needed energy to African nations, but this must be done in a responsible and sustainable manner to ensure that African nations can benefit from natural gas while avoiding any serious environmental damage in the future.

African nations should also consider investing in alternative energy sources such as solar, wind, and hydroelectricity to reduce their dependence on natural gas, decrease their emissions, protect the environment from further damage, and preserve the planet for future generations. These alternative energy sources can provide African nations with an opportunity to reduce their reliance on natural gas while also providing a cleaner and more sustainable energy source in the long term. To ensure that natural gas is used responsibly and sustainably, African nations should develop and implement strict regulations for energy facilities, as well as invest in alternative energy sources to reduce their dependence on natural gas and decrease their carbon emissions while continuing to reap the benefits of this valuable energy source. African nations should develop strict regulations and standards to ensure that any exploitation of natural gas resources is done in a responsible and sustainable manner while also investing in cleaner and more sustainable alternative energy sources to reduce their reliance on natural gas, decrease their emissions, and ensure the health of the environment for future generations. African nations should invest in alternative energy sources such as solar, wind, and hydroelectricity to reduce their dependence on natural gas, decrease their emissions, and protect the environment for future generations while still being able to benefit from valuable energy sources. These alternative sources can offer clean and renewable energy, allowing African nations to further reduce their reliance on natural gas while also providing them with greater energy security and cost savings in the long term. African nations should also develop plans to closely monitor and limit the environmental impacts of natural gas use, such as by regularly testing water and air quality, as well as ensuring that any extraction or drilling operations adhere to best practices to prevent leakage and reduce their environmental impact. To ensure that African nations can continue to benefit from natural gas resources in a responsible and sustainable manner, it is essential that strict regulations and standards are put in place to prevent overexploitation and environmental degradation in the future. Furthermore, African nations should seek to invest in the research and development of new technologies to make natural gas extraction more efficient and reduce the costs associated with using the resource. This, combined with the implementation of alternative sources of clean energy, would help to ensure that African nations are able to benefit from their natural gas resources in a sustainable way while also protecting the environment from harm and preserving the environment for future generations. While it is important to recognize the potential economic benefits of natural gas, African nations must also consider the risks and potential long-term impacts of exploiting this resource before making decisions about its use. Going forward, African nations should seek to develop a robust regulatory framework that enforces strict standards for the exploitation of natural gas while providing incentives for research and development in clean energy sources to ensure the safe and responsible use of their natural gas resources and protect the environment for future generations. Going forward, African nations should be cognizant of the importance of striking a balance between the economic benefits of natural gas exploitation and environmental sustainability to ensure a secure and prosperous future for their citizens.

African nations should also encourage research and development into natural gas extraction and use, as well as promote public awareness of the benefits of responsible and sustainable energy use, to ensure that their populations can access the economic benefits of natural gas while also mitigating the environmental impacts associated with its exploitation and use. African nations must also create policies that incentivize the development of clean energy sources to reduce dependence on natural gas and promote environmental sustainability. They should focus on developing a regulatory framework that imposes strict standards for the exploitation of natural gas and promotes research and development in clean energy sources to provide a secure and sustainable energy future for their citizens. To ensure a secure and prosperous future, African nations should strive to develop a comprehensive regulatory framework that enforces strict standards for natural gas exploitation and provides incentives for research and development in clean energy sources while also encouraging public awareness of the benefits of responsible and sustainable energy use.

African nations have an opportunity to use natural gas as a transitional energy source while simultaneously creating policies that ensure the long-term sustainability of clean energy sources such as solar, wind, and hydropower to provide secure and sustainable energy access for the future. By encouraging the development of clean energy sources while imposing strict standards on the exploitation of natural gas, African nations can achieve long-term energy security and sustainability that will benefit their citizens and ensure a secure future for generations to come. This can be accomplished by creating a comprehensive regulatory framework that imposes strict standards for the exploitation of natural gas and incentivizes research and development in clean energy sources while also encouraging public awareness of the importance of responsible and sustainable energy use. This regulatory framework should also ensure that any revenues generated from the exploitation of natural gas resources are allocated toward investments in renewable energy infrastructure, research and development initiatives, and public awareness campaigns. Through the implementation of such policies, African nations can effectively use natural gas as a transition energy source while simultaneously encouraging the development of clean energy sources and increasing public awareness of responsible and sustainable energy use to further ensure energy security and sustainability for their citizens in the long term. In doing so, African nations can not only address the current energy needs of their citizens but also ensure that future generations are able to access secure and sustainable sources of energy without compromising the environment or public health. African nations must be proactive in their approach to addressing their current energy needs while simultaneously investing in and encouraging the development of clean energy sources so that their nations can benefit from the long-term benefits of sustainable energy utilization. This is why the introduction of a robust and effective regulatory framework is essential to ensure that African nations are able to make the necessary transition from conventional energy sources such as natural gas to clean and renewable sources of energy such as solar and wind power without compromising the environment or public health while also allowing their citizens to enjoy the economic and environmental benefits of renewable energy utilization in the future.

African nations must also prioritize public education and awareness of sustainable energy practices to encourage their citizens to engage in energy-saving behaviors and adopt renewable energy sources so that they are able to reduce their dependence on conventional energy sources while also avoiding the costly environmental and public health consequences associated with nonrenewable energy sources such as air pollution, water contamination, and climate change. This is why a comprehensive plan for the deployment of clean energy sources and regulations regarding renewable energy utilization must be developed and implemented to ensure that African nations are able to make the necessary transition from conventional energy sources to clean and renewable energy sources in a safe, efficient, and cost-effective manner while also ensuring that their citizens are able to enjoy the economic, social, and environmental benefits of sustainable energy utilization in the future. In conclusion, the development and implementation of a robust and effective regulatory framework are essential for African nations to transition from conventional energy sources to clean and renewable sources of energy without compromising the environment or public health while also allowing their citizens to reap the economic and environmental benefits of renewable energy utilization in the future. Furthermore, for African countries to successfully transition from traditional energy sources to clean and renewable energy sources, public education, and awareness campaigns must be implemented so that citizens are encouraged to adopt sustainable energy practices and reduce their reliance on nonrenewable energy sources while reaping the economic and environmental benefits of sustainable energy utilization.

5.2.3 Conclusion

African nations must prioritize public education and awareness of sustainable energy practices to reduce their dependence on conventional energy sources and avoid the environmental and public health consequences of burning fossil fuels. Educating the public on sustainable energy practices and providing financial incentives to citizens who invest in renewable energy projects are important steps in this process. Incentivizing and educating citizens on the importance of sustainable energy practices is essential for African nations to break their dependence on fossil fuels and make a successful transition to renewable sources of energy such as solar and wind power. This transition can also be complemented by increased investment in renewable energy infrastructure and government subsidies for renewable energy projects, making them more accessible to citizens who may otherwise not be able to afford them. These investments will help reduce the cost of renewable energy production and make it more competitive with conventional sources of energy to make a successful transition to renewable energy sources and a more viable and sustainable energy future for African nations.

By educating the public on sustainable energy practices and providing financial incentives to citizens who invest in renewable energy projects, African nations have the potential to create a greener and more sustainable future while helping to reduce their reliance on polluting and unsustainable fossil fuel sources of energy and contributing to global efforts to reduce the impacts of climate change. African nations can also take further steps to help ensure the success of their renewable energy transition by encouraging community-led projects, such as community-based solar farms and public wind turbines, that can provide sustainable energy solutions and stimulate local economies, as well as by partnering with the private sector to promote research and development of innovative renewable energy technologies that can help accelerate the transition to renewable energy sources and provide more reliable and cost-effective solutions for African nations.

A comprehensive plan for the deployment of clean energy sources such as natural gas can also help African nations meet their goals of transitioning to renewable energy sources while still providing access to reliable, cost-effective energy sources in the short term while they work toward the longer-term goal of achieving a more sustainable and renewable energy future. In addition, African nations must also take steps to create a strong regulatory framework for their natural gas use transition that can provide incentives for businesses and individuals to invest in natural gas-related projects and ensure that natural gas sources are used responsibly and in a manner that maximizes economic and environmental benefits and reduces risks to public health and the environment. African nations have the potential to create a greener and more sustainable future by utilizing natural gas as an important part of their renewable energy transition and by implementing effective policies and initiatives that can help ensure the success of their renewable energy transition and promote the development of more sustainable energy sources.

African nations must not only invest in renewable energy sources but also create a regulatory framework that supports the use of natural gas as an important part of their renewable energy transition to ensure that their energy transition is successful while protecting the environment and creating economic opportunities for businesses, individuals, and local communities alike. While renewable energy sources are the ideal long-term goal, transitioning to them is often expensive and difficult to achieve, and African nations must recognize the importance of natural gas as a bridge fuel to help facilitate the transition to renewable energy sources while creating sustainable economic opportunities and protecting their environment.

To effectively do this, African nations must develop an effective regulatory framework that encourages investment in natural gas-related projects, reduces risks to public health and the environment, incentivizes businesses and individuals to invest in renewable energy sources, and helps create economic opportunities for local communities while allowing natural gas to remain an important part of their energy transition while also working to ensure that their renewable energy transition is successful. African nations must create and implement an effective policy and regulatory framework that not only encourages investment in natural gas-related projects but also helps protect the environment, reduce risks to public health, incentivize businesses and individuals to invest in renewable energy sources, and create economic opportunities for local communities while simultaneously allowing natural gas to remain an important part of their energy transition.

While the ultimate goal of transitioning to renewable energy sources is ideal, African nations must recognize that natural gas plays an important role in helping facilitate the transition and should be taken into account as part of their overall strategy while ensuring a successful renewable energy transition. African nations must act now to create a comprehensive plan for the deployment of clean energy sources such as natural gas and must also work toward creating a strong regulatory framework that provides incentives for businesses and individuals to invest in renewable energy-related projects while at the same time allowing natural gas to remain an important part of their energy transition.

To effectively do this, African nations must develop an effective regulatory framework that balances the need for investment in natural gas projects with the importance of protecting public health and the environment, incentivizing businesses and individuals to invest in renewable energy sources, and creating economic opportunities for local communities while still allowing natural gas to remain an important part of their energy transition. African nations must act swiftly to create a comprehensive and sustainable plan for the deployment of clean energy sources such as natural gas while simultaneously protecting public health, reducing environmental risks, incentivizing businesses and individuals to invest in renewable energy sources, and creating economic opportunities for local communities to ensure a successful transition toward renewable energy sources over the long term.

In conclusion, the transition to renewable energy sources is a necessary and important goal for African nations, and it is important that African nations act now to create a comprehensive and sustainable plan for the deployment of clean energy sources such as natural gas while simultaneously protecting public health, reducing environmental risks, incentivizing businesses and individuals to invest in renewable energy sources, and creating economic opportunities for local communities to ensure a successful transition toward renewable energy sources over the long term.

5.3 Green Hydrogen

5.3.1 Introduction

According to recent estimates, the use of hydrogen and its derivatives is expected to increase dramatically over the coming decades, reaching 12% of global energy use by 2050 and playing an important role in reducing greenhouse gas emissions and helping meet global climate change targets (Nwokolo et al., 2023b). With the adoption of clean hydrogen technologies, it is expected that the cost of hydrogen will decrease, making it a more accessible and attractive option for countries looking to reduce their greenhouse gas emissions. This is primarily because hydrogen is a sustainable and renewable fuel source that can be produced from a variety of sources, such as water, natural gas, biomass, solar energy, and nuclear energy. Furthermore, hydrogen fuel cells are clean and have the potential to provide a reliable source of energy with almost zero emissions when powered by renewable electricity.

It is also relatively simple to store and transport, and it can be used in a variety of applications, such as heating, power generation, and transportation. Making hydrogen a more attractive and viable option for countries to reduce their carbon footprints is crucial to help meet global climate change targets. Since hydrogen is abundant and can be used to generate power with low or no carbon emissions, it is expected to play an increasing role in the global energy mix. The availability of hydrogen as an energy source could make a significant contribution to Africa's economies' goal of becoming net-zero or carbon neutral by 2050 (Nwokolo et al., 2022a). African countries can reduce their reliance on carbon-intensive energy sources and contribute to the global effort to reduce emissions by leveraging the cost-effectiveness of hydrogen as an energy source.

In addition to offering a viable option for transitioning away from fossil fuels, the use of hydrogen as an energy source would have other benefits as well, such as improved air quality, lower production costs, and an increased ability to export energy. This, in turn, would create economic opportunities for the African continent and provide energy security to countries that currently lack it. By utilizing the potential of hydrogen as a viable energy source, African countries can not only reduce their carbon footprint and contribute to global efforts to fight climate change but also achieve increased economic growth, improved air quality, and improved energy security for their countries and people.

Furthermore, African countries may benefit from various hydrogen applications in sectors such as mobility, industry, and agricultural production. To accomplish this, African countries will need to invest in hydrogen-based technology research and development as well as fuel production infrastructure to increase access to hydrogen-based fuel and other applications to a larger portion of the population. By doing so, African countries can have the capacity to create sustainable economic growth for their people and businesses in an environmentally friendly manner while protecting the environment from further damage caused by conventional sources of energy. African countries can take advantage of the potential that hydrogen holds and make a significant contribution to global efforts toward fighting climate change while at the same time improving their own socioeconomic development and alleviating energy poverty.

To maximize its potential, Africa must focus on creating the necessary infrastructure to make hydrogen available and affordable for African countries to benefit from hydrogen as an energy source and to develop the necessary research and development capabilities to create renewable hydrogen energy technologies that will lead to the long-term sustainability of the continent and the African people. This includes creating hydrogen production facilities, distribution networks, and end-use technologies to enable consumers to use hydrogen fuel cells in cars, buses, and other transport options. Competitive hydrogen and synthetic fuels derived from hydrogen (such as ammonia, methanol, and kerosene) emerge as key components of the energy mix.

To maximize the potential of hydrogen in Africa, governments must invest in creating the necessary infrastructure to enable affordable access to hydrogen energy for all citizens while also ensuring that they are able to capture the economic value of hydrogen through local and regional industrial investments. This will entail investments in hydrogen-based technology research and development, as well as the production of fuel infrastructure, to bring the use of hydrogen-based fuel to Africans, allowing them to benefit from an environmentally friendly method of economic growth. Furthermore, governments must provide incentives to the private sector to invest in hydrogen energy production and storage technologies, with the goal of increasing efficiency and reducing the cost of these technologies to make them more accessible to the general public and incentivizing businesses to use hydrogen as a primary source of energy.

In addition, governments must take steps to ensure that hydrogen energy is produced sustainably and that communities have access to the information and resources necessary to make informed decisions about hydrogen energy production and use (Akrofi, 2021). These measures are key to ensuring that hydrogen energy can be safely and responsibly utilized while also ensuring that the African continent can benefit from the economic opportunities afforded by this clean, renewable form of energy production (Aquila Jesuleye et al., 2020). This will not only help to reduce carbon emissions and other forms of pollution but will also provide new jobs and new economic opportunities in the form of research, development, production, and the sale of hydrogen-based products in the African region. Furthermore, governments should create financial incentives for the private sector to invest in the research and development of hydrogen energy production and storage technologies, as well as take measures to ensure that hydrogen energy is produced in a way that does not damage local ecosystems or communities living in close proximity to production sites. This would also serve to ensure that the benefits of hydrogen energy production are shared equitably throughout Africa and its surrounding regions, as well as promote long-term sustainability in terms of both economic growth and environmental protection.

In conclusion, to maximize the potential of hydrogen energy and ensure that African countries can reap economic benefits without compromising safety or environmental protection, governments should commit to a range of measures, such as creating incentives for private sector investment, monitoring production processes to reduce pollution, and promoting equitable distribution of the benefits of hydrogen energy production to make strides in a field that was thought to be reserved for men for centuries and create new employment opportunities for women across the continent. This section of the book examines various steps African nations can take to achieve a net-zero energy transition and gain the full economic, environmental, and social benefits of green hydrogen energy sources.

5.3.2 Benefits of Increasing Hydrogen and its Derivatives

5.3.2.1 Reduced Dependence on Fossil Fuels

One significant advantage of increasing the use of hydrogen and its derivatives in African countries is a reduction in reliance on fossil fuels and the resulting benefits to local environments and the global climate. The use of hydrogen and its derivatives as an energy source is much less carbon intensive than the use of traditional fossil fuels, meaning that its production and consumption produce significantly fewer greenhouse gases and pollutants than the use of fossil fuels. This reduced use of fossil fuels can help African countries achieve their climate change goals more quickly and effectively while also leading to cleaner air, soil, water, and other environmental benefits for local populations.

In addition, the use of hydrogen and its derivatives can help to create a more sustainable and efficient energy system in African countries, providing greater access to energy for all and reducing the need for expensive, centralized energy production facilities, which can further reduce local environmental impacts and greenhouse gas emissions while improving energy security and providing more reliable, affordable electricity to rural communities that may not have access to traditional energy sources. This increased use of hydrogen and its derivatives would also create new economic opportunities for local entrepreneurs in African countries, leading to job creation, increased incomes for citizens, and a more vibrant, thriving economy across the continent.

Overall, the use of hydrogen and its derivatives as an energy source in African countries offers a promising way forward to meet the challenges of climate change, increase energy access, reduce environmental impacts, promote economic growth, and improve the lives of citizens across the continent. The shift away from fossil fuels to cleaner and more renewable forms of energy, including hydrogen and its derivatives, has the potential to be transformative in African countries and could contribute to a sustainable, resilient, and prosperous future for the continent and its citizens as a whole.

A reduction in reliance on fossil fuels could result in a reduction in carbon dioxide emissions from African nations, which would help address the global climate crisis. This is one important benefit of increasing the use of hydrogen and its derivatives in African nations as an energy source, such as biogas, hydrogen fuel cells, and wind power. With the increased use of renewable energy sources, such as biogas, hydrogen fuel cells, and wind power, African nations can reduce their reliance on fossil fuels and help address the global climate crisis by reducing their carbon dioxide emissions. This reduction in carbon dioxide emissions will have a positive impact on the global climate crisis and help to ensure that African nations are playing their part in addressing this worldwide problem. In general, decreasing reliance on fossil fuels and increasing use of hydrogen and its derivatives in African countries can have a positive impact on the global climate crisis by lowering carbon dioxide emissions and assisting in the creation of a healthier, more sustainable planet for future generations.

Furthermore, this use of hydrogen and its derivatives can provide economic benefits to African nations by increasing their energy independence, creating new employment opportunities in the development and maintenance of renewable energy sources, and reducing their emissions. This can provide African nations with the opportunity to take a leadership role in the global effort to reduce emissions and address climate change, as well as increase their energy independence and provide new employment opportunities for their citizens. While reducing emissions is a crucial step in addressing the global climate crisis, it is not enough on its own, and African nations must also work to improve their use of hydrogen and its derivatives by investing in clean energy infrastructure, increasing public education about the importance of reducing emissions, and encouraging people to use renewable sources of energy wherever possible. To effectively combat the global climate crisis, African nations must recognize the vital importance of reducing their reliance on fossil fuels and increasing the use of hydrogen and its derivatives.

An important benefit of increasing the use of hydrogen and its derivatives in African countries is a decrease in reliance on fossil fuels, which are becoming more expensive and in short supply due to the rapid growth in population and demand. Hydrogen is a clean and renewable source of energy, which reduces the need for fossil fuels that are known to contribute significantly to climate change and other environmental issues such as air and water pollution. Additionally, hydrogen fuel cells are more efficient than other forms of energy and do not create harmful emissions, making them a desirable choice for powering homes, businesses, and transportation. In the long run, increased use of hydrogen energy can reduce costs associated with fossil fuels while still ensuring a steady supply of energy to the region, leading to a more sustainable future for African countries and their citizens.

In addition, hydrogen fuel cells are an economically sound choice, as they can be used for many years before needing to be replaced and require much less maintenance than fossil fuels, which makes them a cost-effective alternative for African countries seeking to reduce their reliance on traditional sources of energy and reduce their carbon footprints. This is why hydrogen fuel cells offer an ideal solution for African countries looking to transition away from fossil fuels and toward renewable energy sources such as hydrogen fuel cells. Consequently, hydrogen fuel cells are becoming increasingly popular as an energy source for African countries seeking to reduce their reliance on traditional sources of energy. With a shift to the use of hydrogen, African countries can benefit from a greener energy source that is also more affordable and sustainable in the long run than traditional fossil fuels. The introduction of hydrogen fuel cells has the potential to revolutionize the African energy sector and usher in a new era of renewable energy production that is free from the negative environmental and economic impacts of fossil fuels. It is clear that the advantages of hydrogen fuel cells in African countries are significant, and therefore, the use of this energy source should be encouraged and supported to promote more sustainable energy use and development on the continent.

Reducing reliance on fossil fuels, which are not only harmful to the environment but are also limited resources that will eventually run out, is one of the major advantages of increasing the use of hydrogen and its derivatives in African countries as an energy source because it is renewable and widely available in the atmosphere. Additionally, hydrogen is a clean fuel that releases water vapor and other nontoxic byproducts as its only exhaust when combusted, making it an ideal alternative to the burning of fossil fuels, which produce toxic pollutants that are harmful to both the environment and the human health. Furthermore, hydrogen is a much more efficient fuel than fossil fuels and produces more energy per kilogram when combusted, making it an attractive option for African countries looking to reduce their reliance on finite resources while at the same time increasing their energy output.

With the combination of its clean and efficient properties, hydrogen and its derivatives offer a viable solution for African countries seeking to reduce their reliance on fossil fuels while also protecting the environment and increasing energy production and efficiency. In general, hydrogen and its derivatives offer a viable solution for African countries seeking to reduce their reliance on fossil fuels while also protecting the environment and increasing energy production and efficiency. All of these factors combine to make hydrogen and its derivatives a highly attractive solution for African countries seeking to reduce their reliance on fossil fuels while simultaneously protecting the environment and improving energy production and efficiency.

One of the key benefits of increasing the use of hydrogen and its derivatives in African countries is reducing reliance on fossil fuels, which are largely imported, leaving those countries vulnerable to price fluctuations and other uncertainties associated with the external supply chain. By harnessing the use of hydrogen and its derivatives, African countries can reduce their vulnerability to price fluctuations associated with imported fossil fuels while also benefiting from increased security of supply and greater energy independence. By harnessing the potential of hydrogen and its derivatives, African countries could reduce their dependency on fossil fuels and better safeguard their energy security.

Additionally, increasing the use of hydrogen and its derivatives in African countries could also create new opportunities for local entrepreneurs and businesses to become involved in the production and distribution of these clean energy sources. This could result in economic development and job creation and thus serve as a catalyst for broader social and economic development in African countries. In general, harnessing the potential of hydrogen and its derivatives could be a game-changer for African countries in terms of both energy security and economic development. In addition, by replacing imported fossil fuels with domestically sourced clean energy sources, African countries could benefit from improved trade balances and potentially reduce the volatility of their exchange rates, as foreign exchange reserves would no longer be required for the purchase of energy imports. This could be achieved through the implementation of policies and regulations that incentivize investments in clean energy technologies and promote their adoption by citizens.

Moving toward this goal would require a comprehensive approach, including the development of new infrastructure for hydrogen production and storage, investments in technological research and development, capacity building for local entrepreneurs, and adequate financing and policy support from international organizations to ensure that hydrogen-based energy sources can be integrated into existing African energy systems in an economically and environmentally sustainable manner. All of these efforts will be necessary to make hydrogen energy a viable alternative for African countries and to ensure that the full potential of this renewable and widely available resource can be realized in the long term.

While the transition toward hydrogen and its derivatives as an energy source in African countries may require significant investments, the long-term benefits of reducing reliance on fossil fuels will far outweigh these costs and make the switch to a cleaner and more reliable source of energy a worthwhile endeavor in the short term and long term. In conclusion, the potential benefits of increasing the use of hydrogen and its derivatives as an energy source in African countries are clear, and the comprehensive approach required to ensure these benefits are realized must begin now to make progress toward the ultimate goal of transitioning away from fossil fuels and toward hydrogen and its derivatives as an energy source in African countries in ways that were not previously possible for their time in an unprecedented manner.

The journey to transitioning away from fossil fuels to hydrogen energy in African countries must be embraced by both public and private entities and led by governments, NGOs, and other organizations to bring together all the necessary elements that are required to make the transition possible and sustainable in a way that will have a lasting, positive impact on the lives of people and the environment on the African continent. This is going to be seen as a revolutionary force in the industry for many years and for generations to come for the first time because of the benefits of improving their quality of life and providing a more secure and sustainable source of energy for African countries to realize their full potential.

5.3.2.2 Improved Air Quality

Improved air quality is one of the major benefits of increasing hydrogen and its derivatives as an energy source in African countries. With a shift from traditional fossil fuels to hydrogen and its derivatives, air pollution would be significantly reduced in African countries. By introducing clean energy sources, African countries would be able to reduce the number of health issues caused by air pollution, such as asthma, bronchitis, lung cancer, and cardiovascular diseases, and improve the quality of life for their citizens. This, in turn, would lead to fewer absences from work and school due to health issues, as well as a decrease in the overall cost of medical care. This would be beneficial not only to the health of African citizens but also to the economy since the cost of medical care, a major expense for any government, would be greatly reduced.

In addition, a shift to clean energy sources would reduce emissions of the most potent greenhouse gases, such as carbon dioxide and nitrous oxide, thereby reducing their negative effects on the environment and mitigating the effects of climate change. This would result in a win–win situation where the citizens of African countries enjoy better health outcomes and their economy benefits from the savings on medical care costs while their environment is also safeguarded from the ill effects of climate change, which, in turn, would bring with it a host of other benefits such as improved food security, access to clean water and sanitation facilities, reduced pressure on existing ecosystems and habitats, an increase in tourism opportunities, and a boost in the economy due to increased investment from both domestic and foreign sources. These benefits, combined with a reduced reliance on fossil fuels and other carbon-emitting sources of energy, would mean that African countries could enjoy a much brighter future in terms of their health and economic prosperity while also helping to ensure the sustainability of their environment for generations to come.

By reducing the use of fossil fuels and combustible materials, African countries can significantly reduce air pollution, leading to improved respiratory health and general well-being among their citizens while also safeguarding the environment from the effects of climate change and reducing their contribution to the global climate crisis. This shift to renewable sources of energy is not only beneficial for the environment and the health of citizens but also promises to bring with it several economic advantages, such as the potential to create new jobs in the renewable energy sector, improved access to electricity and clean water, which would benefit both domestic and foreign businesses and a reduction in the costs associated with medical care due to the healthier living conditions created by the improved air quality. This shift to renewable sources of energy would therefore have a number of long-term benefits for African countries, with the potential to improve their health outcomes and economic prospects while also ensuring the sustainability of their environment for future generations.

By reducing the need for more pollutant-emitting energy sources such as coal, oil, and natural gas, increasing hydrogen and its derivatives as an energy source in African countries could have a tremendous impact on air quality. All of these advantages could be harnessed through the implementation of comprehensive and well-thought-out plans to transition African countries away from the use of fossil fuels and other carbon-emitting sources of energy toward the use of renewable sources such as solar, wind, and hydroelectric power that are more efficient, sustainable, and cost-effective in the long term.

Hydrogen is a clean burning fuel, meaning that it produces fewer greenhouse gases, making it an important factor in helping to reduce global warming and the depletion of the ozone layer. In addition, it also offers a source of energy that is not subject to fluctuating market prices like many fossil fuels such as oil and gas, providing a more stable source of energy and greater energy security for African countries. Transitioning African countries away from the use of fossil fuels and toward the use of renewable sources of energy could not only lead to greater environmental protection and improved air quality but also more stable energy prices and greater energy security for African countries in the long term.

In addition, cleaner air would benefit African countries' tourism industries, as tourists are more likely to visit places with good air quality and a smaller carbon footprint. This is of great importance, as tourism has been proven to have a positive effect on economic growth and job creation in African countries, as well as providing a reliable source of income from an industry that is not as susceptible to fluctuating markets as other sources of income, such as those from the extractive industries, which rely heavily on the sale of oil and gas. This makes the transition to renewable sources of energy an important step in ensuring that African countries have a more reliable source of income and greater energy security in the future.

Furthermore, the increased investment in renewable energy sources would stimulate job growth within the sector and create opportunities for local businesses to become involved in the production and installation of renewable energy solutions while also helping to reduce the dependence on polluting energy sources, such as coal and oil, that are both harmful to the environment and have increasingly become a source of great economic and political instability on the African continent. In this way, the transition to renewable energy sources would not only benefit African countries' tourism industries but would also provide economic stability and long-term growth, as well as create job opportunities for the local population, which, in turn, would have a positive impact on the tourism industry. As more jobs are created, the population of those employed in the tourism industry is likely to increase, leading to more money being put into the local economy and creating a greater level of prosperity for everyone living in the area.

As such, transitioning to renewable sources of energy in African countries would have an immediate and lasting positive impact on tourism and the local economy as a whole. The impact of transitioning to renewable energy sources in African countries cannot be overstated. It is clear that transitioning to renewable energy sources in African countries would be beneficial for both their tourism industries and the local population. As such, transitioning African countries to the use of hydrogen as a fuel source could have an array of positive impacts by decreasing emissions of pollutants such as carbon dioxide, methane, and nitrogen oxide, which contribute to global warming and poor air quality due to their clean burning properties.

5.3.2.3 Increased Sustainability

One of the key advantages of using more hydrogen and its derivatives as an energy source in African nations is increased sustainability for both the environment and society as a whole because hydrogen is a clean burning fuel and its production does not create greenhouse gases that contribute to climate change. As hydrogen-based energy sources are renewable and nonpolluting, the environment would be kept clean and safe for generations to come. As such, with the introduction of hydrogen-based energy sources, African nations would be able to reduce their dependence on fossil fuels and shift to cleaner, more sustainable energy sources that would benefit their environment, economy, and society as a whole over the long term.

Additionally, hydrogen is an abundant and renewable resource, as it can be generated through the electrolysis of water using solar, wind, or hydroelectric energy, making it even more accessible and cost-effective to produce than traditional fossil fuels. Furthermore, since hydrogen-based energy sources are renewable, African nations would be able to take advantage of the reduced costs associated with their production and use, which could help reduce electricity bills for consumers and lead to greater economic prosperity in the region over time. This, in turn, would also help to reduce air and water pollution, as well as create jobs in the renewable energy sector that would benefit the environment, spur economic growth, and help combat climate change across the continent while also boosting public health and safety in the region by increasing access to renewable and sustainable energy sources for African nations in a way that had never been done before. This, in turn, would lead to greater energy security and sovereignty for African nations, allowing them to become less reliant on external sources of energy while becoming more independent in the face of global political and economic uncertainty.

These factors all make hydrogen-based energy sources an ideal option for African nations looking to reduce their reliance on fossil fuels and move toward a more sustainable future. This is why investing in hydrogen-based energy sources for African nations would be a wise decision for those nations to make in the long term, given the numerous benefits that it would provide and the increased energy security that it would bring to African nations to increase energy security and reduce dependence on fossil fuels in the process.

In general, hydrogen-based energy sources offer African nations an opportunity to reduce their reliance on fossil fuels and shift toward more sustainable and renewable energy sources while simultaneously creating jobs, reducing air and water pollution, and helping to protect public health by providing clean, renewable energy that is far less damaging to the environment than other forms of energy generation and potentially providing long-term cost savings due to lower energy bills and increased efficiency. This in turn would lead to greater energy security and sovereignty for African nations, allowing them to become less reliant on external sources of energy while becoming more independent in the face of global political and economic uncertainty.

With these benefits in mind, it is easy to understand why investing in hydrogen-based energy sources for African nations would be a smart move in the long term, given the myriad of benefits that it would provide. Investing in hydrogen-based energy sources is an investment in Africa's future that could pay dividends for generations. The potential of hydrogen-based energy sources to enable African nations to become more self-reliant and less reliant on external sources of energy while also providing cleaner air and water, creating jobs, reducing air pollution, increasing efficiency, and ultimately resulting in lower energy costs is an opportunity that should not be overlooked for African nations for their economic, environmental, and social stability and sustainability in the face of global economic and political challenges.

Investing in hydrogen-based energy sources for African nations is a smart decision that will benefit both current and future generations given the positive impacts that it will have on their economy, environment, and social well-being, creating a platform for economic growth and social stability for the continent as a whole and providing access to sustainable, renewable energy for all of its citizens and giving them a better quality of life for the first time. Investing in hydrogen-based energy sources is a forward-thinking move that will help African nations become more self-sufficient and provide cleaner air and water while simultaneously reducing air pollution, increasing efficiency, and providing access to sustainable, renewable energy for all of its citizens, giving them a better quality of life for the first time, while also reducing the continent's reliance on external sources of energy and ultimately resulting in lower energy costs in a way that no other option can in terms of economic and social stability and sustainability for African nations for the long term in an unprecedented manner.

5.3.2.4 Economic Benefits

One of the primary benefits of using more hydrogen and its derivatives as an energy source in African nations is the economic benefit associated with reducing reliance on imported fuel sources and transitioning to hydrogen-based energy production systems. This could help alleviate the issue of relying on outside sources for energy, allowing African nations to become more self-reliant and reduce their reliance on foreign countries for the majority of their energy needs. With increased production of hydrogen fuel, African nations could potentially stimulate their local economies by creating jobs related to the energy industry, as well as boost other economic sectors that would benefit from the economic growth generated from this new energy source. The shift toward hydrogen energy can reduce the costs of electricity production, decrease the prices of goods and services, improve the balance of payments by reducing the need for expensive imports, and create a more stable domestic economy.

As a result, the economic benefits of increased use of hydrogen fuel in African nations would have the potential to have far-reaching positive impacts on their economies, helping them to become more self-sufficient and create a more stable domestic economy that could lead to greater prosperity for the entire region over time. In addition, the environmental benefits of switching to hydrogen energy should not be overlooked, as it would reduce carbon emissions, help prevent air and water pollution, conserve natural resources, and protect ecosystems in the region. Furthermore, the transition to hydrogen energy can help create a safer and more sustainable energy system in African nations, as it is a clean, renewable energy source that can be used without contributing to climate change or environmental degradation due to the burning of fossil fuels. The switch to hydrogen energy therefore has the potential to bring about significant economic and environmental benefits for African nations, including reducing the cost of electricity production and prices of goods and services, creating a more stable domestic economy, reducing carbon emissions, protecting natural resources and ecosystems, and creating a safer and more sustainable energy system in the region overall.

These benefits make it clear that a shift to hydrogen energy in African nations would be a wise decision for governments and private companies to make, as the advantages of this shift far outweigh any potential drawbacks. For all these reasons, switching to hydrogen energy in African nations would be a beneficial decision for both governments and private companies to make, as the advantages of this shift far outweigh any potential drawbacks. African nations would be wise to make the transition to hydrogen energy, as it has the potential to bring about a host of positive economic and environmental outcomes that could drastically improve the quality of life in the region for its citizens. For these reasons, switching to hydrogen energy in African nations would be a beneficial decision for both governments and private companies to make, as the advantages of this shift far outweigh any potential drawbacks. With the benefits of hydrogen energy, such as lower emissions, increased efficiency, and a decreased cost of production, the move toward a more sustainable energy source is one that African nations cannot afford to ignore.

By transitioning to hydrogen energy, African nations would be making a smart decision that could have a lasting positive effect on their economic and environmental conditions in the long term. The move to hydrogen energy has the potential to revolutionize African nations' relationships with energy production both in the short and long term and make it easier for them to compete in the global market. African nations have the opportunity to set an example for other countries, demonstrating that renewable and sustainable energy sources can be adopted on a large scale without having a negative effect on the economy or the environment.

This shift could potentially help Africa achieve the UN's Sustainable Development Goals, ensuring a better quality of life for African citizens in the short term and creating a healthier and more sustainable society in the long term. By transitioning to hydrogen energy, African nations would be making a smart decision that could have a lasting positive effect on their economic and environmental conditions in the long term. This shift to hydrogen energy is a critical step that African nations must take if they want to remain competitive in the global market and secure a better quality of life for their citizens. While the shift to hydrogen energy is a difficult transition, it is essential for African nations to make it to remain competitive and achieve a better quality of life for their citizens.

The transition to hydrogen energy, while it can seem daunting at first, would be a step in the right direction for African nations to improve their economic and environmental conditions in the long term. The transition to hydrogen energy would be a great investment for African nations, as it would enable them to reduce their dependence on fossil fuels, create new jobs and industries, and reduce their carbon footprints while simultaneously increasing energy security and reliability. This transition could be achieved through incentives for businesses to invest in hydrogen technology, government subsidies for research and development, and education initiatives to raise awareness of the benefits of a shift to hydrogen energy, as well as policies and regulations to ensure the safe and responsible use of this energy source.

Furthermore, African nations should take advantage of global initiatives such as the United Nations Sustainable Energy for All (SEforALL) and the Clean Energy Solutions Center (CESC), which have been established to promote the widespread use of renewable energy sources (Aklin, 2022) such as hydrogen energy and to support governments in making the transition to this form of energy by providing technical assistance and resources and helping implement policy and legislation to facilitate the shift. Additionally, African nations must also focus on building strong local infrastructure for the safe storage and transportation of hydrogen, as well as train and educate personnel to manage this new energy source so that the adoption of hydrogen energy is done in a safe and responsible manner. As such, the early adoption of hydrogen energy by African nations would be a step in the right direction to improve their economic and environmental conditions in the long term, providing them with numerous benefits.

In general, while the transition to hydrogen energy presents a great opportunity for African nations to increase their energy security, reliability, and sustainability, it is important that governments invest in the necessary resources and infrastructure to ensure that this shift is made in a safe and responsible way to take full advantage of the benefits of hydrogen energy and reap the rewards in the long term. While the transition to hydrogen energy presents a great opportunity for African nations to increase their energy security, reliability, and sustainability, there are some important considerations that must be taken into account for the full potential of hydrogen energy to be realized, including the need for adequate infrastructure and training for personnel to safely and effectively manage the transition to hydrogen energy. To reap the benefits of hydrogen energy and improve economic and environmental conditions in the long term, African nations must invest in strong local infrastructure for safe storage and transportation, as well as train and educate personnel to manage this new energy source safely and effectively for the first time in history. While the transition to hydrogen energy presents a great opportunity for African nations to increase their energy security, reliability, and sustainability, it is important that governments invest in the necessary resources and infrastructure to ensure that this shift is made in a safe and responsible way to take full advantage of the benefits of hydrogen energy and reap the rewards in the long term.

While the transition to hydrogen energy presents a great opportunity for African nations to increase their energy security, reliability, and sustainability, it is important that governments invest in the necessary resources and infrastructure to ensure that this shift is made in a safe and responsible way to maximize the benefits of hydrogen energy and ensure that the long-term economic and environmental impacts are positive for the first time in history. However, this is easier said than done and will require a significant investment of resources and personnel to ensure that hydrogen energy is used safely and efficiently in African countries for the long-term benefit of all parties involved. However, the potential rewards of investing in hydrogen energy for African countries are immense and will pay off in the long run if the proper investment is made and the necessary infrastructure is built to ensure a successful and responsible transition to hydrogen energy. To successfully transition to hydrogen energy, governments in African countries must invest in the necessary resources and infrastructure, such as education, training, and research facilities, as well as advanced technology and equipment to ensure that hydrogen energy is used responsibly and efficiently for the benefit of their citizens and the environment. This will also require African countries to collaborate with the private sector, international organizations, and other countries to leverage their resources and share knowledge and best practices to build the necessary infrastructure and make the transition to hydrogen energy successful for the first time.

5.3.2.5 Cleaner Energy Source

The fact that hydrogen and its derivatives are cleaner energy sources is one of the main advantages of using them more frequently as a source of energy in African countries. Compared to traditional sources of energy such as coal, oil, and natural gas, hydrogen and its derivatives are more efficient in terms of the amount of energy generated per unit of fuel burned and produce far fewer emissions when burned, leading to a better overall quality of air in African countries. Additionally, hydrogen and its derivatives can be produced from renewable energy sources such as solar or wind power, making them an even more attractive alternative for African countries that are looking to reduce their reliance on fossil fuels and move toward more sustainable energy sources. Hydrogen and its derivatives also have the potential to provide significant economic benefits for African countries, as they require significantly less infrastructure to generate and transport energy than traditional sources such as coal, oil, or natural gas and can be produced in smaller quantities, making them a more viable option for those countries with limited access to traditional energy sources, which makes them ideal for African countries with limited resources.

Moreover, hydrogen and its derivatives are nontoxic and do not produce any hazardous byproducts when burned, making them a much safer source of energy than traditional sources such as coal, oil, and natural gas, which have been linked to a wide range of environmental and health risks when burned. By utilizing hydrogen and its derivatives, African countries can benefit from a cleaner environment, increased access to energy, and improved economic conditions, all of which could be major boons to their development and quality of life, making them a potentially invaluable asset for these countries in their efforts to achieve sustainable development and improve their citizens' quality of life.

Furthermore, hydrogen and its derivatives are more versatile than traditional energy sources because they can be used to generate electricity, power motor vehicles and other machineries, and even serve as a fuel source for spacecraft, making them an invaluable resource in a wide range of industries and applications that could create numerous economic opportunities for African countries. The potential for hydrogen and its derivatives to benefit African countries is tremendous, as it can provide them with access to a cleaner, safer, and more versatile source of energy than traditional sources while at the same time creating economic opportunities and improving their economic conditions, which had been previously unheard of in many African countries.

This could drastically reduce their reliance on traditional energy sources, which could result in a more sustainable economy and help African countries move closer to achieving the UN's Sustainable Development Goals. This could be an incredibly positive step forward for African countries, enabling them to reduce their carbon footprints and their dependency on unsustainable sources of energy while also creating new economic opportunities and improving the living standards of their citizens. Moreover, hydrogen and its derivatives could also lead to increased employment opportunities in African countries, as they would need to create the infrastructure needed to implement and utilize these new energy sources and would need to employ people to develop and manage this infrastructure, as well as to maintain it and use it.

The potential for hydrogen and its derivatives to benefit African countries is clear, and if implemented correctly, it could have a tremendously positive impact on the economic and social development of African countries and improve their ability to become self-sustaining and be part of a global effort to reduce carbon emissions and create a more sustainable economic and environmental future. This could be a major boon for African countries, allowing them to increase their energy security and become more self-sufficient while also reducing their dependence on unsustainable energy sources and limiting their carbon footprint, while at the same time providing new economic opportunities, creating employment, and improving the quality of life for their citizens.

This is why it is essential that African countries consider the advantages of hydrogen and its derivatives and begin to invest in developing the necessary infrastructure to make the most of these opportunities so that they can reap the economic, environmental, and social benefits that they could bring them in the future. The implementation of hydrogen and its derivatives into African countries' energy mixes could be a vital part of the shift toward renewable energy sources and provide multiple economic, environmental, and social benefits that could enable them to reach a level of energy independence and prosperity that they may not have been able to achieve otherwise. The importance of hydrogen and its derivatives for African countries cannot be overstated, and the potential benefits that could be realized through their implementation should not be ignored as a potential source of energy.

As the world transitions to renewable energy sources, hydrogen and its derivatives are seen as some of the most promising forms of energy for many African countries due to their high energy density, low environmental impact, and relatively low cost compared to other forms of renewable energy such as solar and wind. Moreover, the scalability of these technologies provides African countries with the potential to significantly expand their energy infrastructure in the near future, as they are able to meet the growing energy needs of their populations and provide a more reliable and sustainable form of energy to meet their energy needs.

In addition to the many economic and environmental benefits associated with the adoption of hydrogen and its derivatives in African countries, there are also a number of social benefits that should be taken into account when discussing the potential for the implementation of these technologies, such as improved living standards, access to energy in rural areas, increased economic activity and job creation, and the empowerment of women, as they are able to take part in the rapidly expanding renewable energy sector and contribute to their country's overall economic development. These benefits can help contribute to the economic, environmental, and social progress of African countries and should be taken into account when considering the potential for implementing hydrogen and its derivatives as an alternative energy source in African countries.

In general, the use of hydrogen and its derivatives in African countries is a critical part of the energy transition and could have significant economic, environmental, and social benefits that can help contribute to the overall progress and sustainability of African countries if implemented in a way that is conducive to their particular needs and aspirations. As such, it is important that governments and organizations consider the potential of hydrogen and its derivatives in Africa as they plan their strategies to meet the growing energy needs of their populations, ensure a better future for all African citizens, and create a more equitable society for all. African countries should consider the potential of hydrogen and its derivatives as they plan their strategies to create a more sustainable, equitable, and prosperous future for their populations and the continent as a whole. Hydrogen and its derivatives offer multiple advantages compared to conventional energy sources, such as lower emissions, a lower cost of production, and increased energy security.

In addition, hydrogen and its derivatives can provide local jobs for citizens in a variety of industries, such as transportation, manufacturing, and engineering. This, in turn, can help African countries develop their economies and improve the quality of life for their citizens. By investing in the research, development, and deployment of hydrogen and its derivatives, African countries can create a stronger foundation for the future while simultaneously tackling the issues of poverty, inequality, and climate change. African countries have the opportunity to be at the forefront of energy innovation by embracing the use of hydrogen and its derivatives as an alternative energy source.

This renewable energy source has the potential to bring stability to African nations and could help them move toward a greener economy while simultaneously creating jobs and improving the health of their citizens by providing a cleaner and more reliable source of energy than other fossil fuels. As such, it is essential that African countries invest in hydrogen technology to create a more secure energy infrastructure for their citizens and help them build a more sustainable future. African countries have a unique opportunity to be pioneers in the development and implementation of hydrogen technology and to utilize it as a way to improve their economies, create jobs, reduce poverty and inequality, and address climate change. While there are many potential benefits to investing in hydrogen technology, it is important to note that African nations must invest wisely and cautiously to ensure that the technology is accessible to all citizens and that it does not have any negative environmental or economic impacts on their respective countries.

In conclusion, investing in hydrogen technology could be a beneficial move for African nations, as it has the potential to create economic stability and reduce poverty while simultaneously creating jobs, providing a cleaner energy source for citizens, and fighting climate change on a global scale. It is clear that investing in hydrogen technology could be a game changer for African nations if done correctly and with great foresight and care. African countries must be wise and thoughtful in regard to investing in hydrogen technology, as there is potential for great benefit but also many risks that must be taken into consideration if they decide to pursue this route. To ensure that these benefits are realized, African countries must ensure that the technology is distributed evenly and fairly throughout the population and that safety measures are in place to reduce potential negative impacts.

5.3.2.6 Affordability

One of the main benefits of using hydrogen and its derivatives more frequently as a source of energy in African nations is their affordability relative to other energy sources due to their abundance, relatively low cost of production, and compatibility with renewable energy sources such as solar and wind power. These advantages of using hydrogen are especially important for African countries, which often lack access to traditional energy sources or have prohibitively high costs associated with them and thus have less developed energy infrastructures when compared to more developed countries. Hydrogen is also an efficient fuel source due to its high energy density, making it an attractive choice for transportation and other energy-intensive applications, such as manufacturing and industrial processes.

In addition, since hydrogen and its derivatives produce virtually no emissions when burned, they are an environmentally friendly option for African countries that are struggling to meet increasingly strict global standards for air quality and climate change mitigation. Additionally, utilizing hydrogen and its derivatives as an energy source could help African countries better utilize their own resources, as many of them possess the technology and infrastructure needed to generate and use hydrogen more effectively than they are able to utilize more traditional energy sources. By utilizing hydrogen and its derivatives, African countries can gain significant economic and environmental benefits that are not available to them through more traditional energy sources such as fossil fuels.

Furthermore, the widespread use of hydrogen and its derivatives could reduce dependence on foreign energy sources and increase energy security for African nations, thereby increasing their economic independence and allowing them to take control of their energy futures. This could in turn lead to an increase in African economic development and stability, as well as a decrease in their vulnerability to external forces such as energy price fluctuations and volatile international markets. Ultimately, utilizing hydrogen and its derivatives can be a powerful tool for African countries to increase their energy security and independence while also reaping environmental benefits such as reducing greenhouse gas emissions, air pollution, and waste production. This is especially pertinent in the African context, where many countries lack access to a reliable grid and must rely on nonrenewable, expensive sources such as diesel to power their electricity networks. These nonrenewable sources are not only more expensive for African countries in the long run, but their environmental impact is also incredibly damaging due to the high levels of carbon dioxide and other pollutants released during burning, which can have a long-term effect on the health of the planet.

By implementing hydrogen and its derivatives as a primary energy source, African nations can reduce the cost of energy while also reducing their environmental impact and contributing to a cleaner and greener future for the entire continent and the world. This shift to cleaner, renewable sources of energy would not only provide economic and environmental benefits but also be a source of stability and growth for the countries that make the transition in the near future. By leveraging the abundance of natural resources and harnessing the power of hydrogen and its derivatives, African countries have an opportunity to break away from their reliance on nonrenewable sources of energy and instead make the transition to an energy system that is sustainable, cost-effective, and environmentally friendly while also providing a long-term solution to the energy crisis facing African countries in the modern age.

African nations have the potential to make great progress in transitioning away from their reliance on fossil fuels and toward clean, renewable sources of energy such as hydrogen and its derivatives by investing in the development of hydrogen-based energy solutions and promoting the use of hydrogen and its derivatives as a primary source of energy across the continent. This transition to hydrogen-based energy solutions is not only beneficial for the environment and economic development but also has the potential to foster regional collaboration, increase stability, and improve living standards across Africa by spurring innovation, job creation, and foreign investment in the sector.

In general, the move toward a renewable energy system in African countries could bring a number of benefits, not only by providing economic and environmental stability but also by contributing to regional collaboration and improving living standards for African countries as a whole. While this transition from a reliance on nonrenewable sources of energy to one based on renewable sources is complex and would require significant investment and commitment, it is achievable if African countries make use of the abundance of natural resources available to them and invest in the development of hydrogen-based energy solutions and their promotion as a primary source of energy across the continent to create a more sustainable future for their citizens and build a brighter, more prosperous African continent for African countries. Going forward, African countries have the potential to create a more sustainable future for their citizens and build a brighter, more prosperous continent for generations to come by transitioning away from reliance on fossil fuels and toward clean, renewable sources of energy such as hydrogen and its associated technologies such as fuel cells and electrolyzers.

This shift in energy sources can lead to increased stability, improved living standards, and greater regional collaboration as African countries move away from nonrenewable sources and toward renewable energy solutions such as hydrogen and its associated technologies. This shift in energy sources is a difficult but necessary one to make, as it would offer African countries the opportunity to take an active role in reducing their own emissions while also benefiting from increased economic and environmental stability due to their inherent environmental and economic benefits that an observer would not understand. By taking the initiative to move toward a renewable energy system, African countries have the potential to create a more sustainable future not only economically and environmentally but also in terms of regional collaboration and living standards. This makes hydrogen energy a cost-effective choice for African nations, which can potentially produce more of their own energy with the use of this fuel source.

To accomplish this, African countries must make a concerted effort to invest in hydrogen technology research and development, as well as engage in active partnerships with international actors to facilitate the transition and support the adoption of these technologies to create a more sustainable future for their citizens and a brighter, more prosperous continent for future generations. This will ensure that African countries can leverage the economic and environmental benefits that come with these technologies while also helping to alleviate poverty and inequality and mitigate the effects of climate change across the continent and beyond. African countries must also work to build public awareness of hydrogen technology and its potential to foster the necessary trust and engagement with this new form of energy to ensure the successful implementation and growth of hydrogen energy in African countries over the long term.

Generally, utilizing hydrogen and its derivatives as an energy source has the potential to be a game changer for African countries by providing a reliable and sustainable energy source for their populations and industries while also mitigating the effects of climate change and paving the way to a more prosperous future for the entire continent. This can only be achieved if African countries make a concerted effort to invest in hydrogen technology research and development, partner with international actors to facilitate the transition, build public awareness of its potential, and create trust and engagement with this new form of energy to ensure its successful and long-term implementation and growth across the continent and beyond. This will require African countries to take a comprehensive approach that focuses on the research and development of hydrogen technology, engaging international actors to help facilitate the transition, and creating an environment of trust and public awareness.

For African countries to reap the benefits of hydrogen energy, there must be a comprehensive effort to invest in the technology, partner with international entities, and develop public trust in a way that was not expected of them before. By investing in the research and development of hydrogen technology, partnering with international entities to facilitate the transition, and creating public awareness of its potential, African countries can successfully implement and grow hydrogen energy across the continent and beyond, allowing them to reap the benefits of a reliable and sustainable energy source while also helping to mitigate climate change that is both reliable and sustainable and becoming a leader in clean energy initiatives that may otherwise not have been available to them. In doing so, African countries will have the opportunity to play an integral role in the global effort to reduce emissions and combat climate change, helping to create a brighter future for their citizens and the entire continent while also demonstrating to the world their commitment to sustainability and clean energy.

5.3.2.7 Environmental Friendliness

The environmental friendliness of hydrogen and its derivatives as an energy source in African countries is one of the main advantages of using hydrogen and its technologies for energy in African countries compared to other energy sources such as fossil fuels. This is because hydrogen and its derivatives have a significantly lower environmental footprint than other energy sources, leading to reduced air pollution and less strain on local ecosystems, which can be beneficial to both local communities and the environment. Additionally, hydrogen and its derivatives are renewable sources of energy and can be produced domestically without the need to rely on foreign energy sources, and their prices are becoming increasingly competitive, allowing African countries to gain greater energy independence and self-sufficiency while providing a cleaner, greener energy source.

Furthermore, hydrogen and its derivatives are more efficient than traditional fossil fuels due to their higher energy density and can be used for many different applications, such as electric power production, transport, heating, cooling, and industrial processes, all of which can benefit African countries economically, socially, and environmentally in terms of energy production. These advantages of hydrogen and its derivatives make them an attractive energy source for African countries seeking to achieve energy independence, reduce their environmental footprint, and maintain economic competitiveness.

Therefore, the use of hydrogen and its derivatives as an energy source can provide African countries with a multitude of benefits and should be strongly considered a viable option for their energy security and sustainable development strategies. Going forward, the increasing availability and affordability of hydrogen and its derivatives will provide African countries with an opportunity to embrace clean energy technologies, reduce their reliance on foreign energy sources, and increase their economic competitiveness while promoting a cleaner environment.

Additionally, African countries must invest in the infrastructure required to make use of hydrogen and its derivatives to maximize the advantages they can offer in the form of improved energy security, reduced environmental impacts, and economic competitiveness. This includes investing in education, research and development, production and distribution networks, storage solutions, and renewable energy sources such as solar and wind energy that can be used to generate hydrogen and its derivatives.

In general, African countries should take advantage of the increasing availability and affordability of hydrogen and its derivatives to embrace clean energy technologies, reduce their reliance on foreign energy sources, increase economic competitiveness, and promote a cleaner environment. Going forward, African countries must make the necessary investments in infrastructure, education, research and development, production and distribution networks, storage solutions, and renewable energy sources to maximize the advantages that hydrogen and its derivatives can offer and move toward a greener, cleaner, and more sustainable future.

One of the main benefits of using hydrogen and its derived products as an energy source in African nations is their environmental friendliness, as these fuels are nonpolluting and renewable sources of energy. As these fuels are derived from natural sources, they can be replenished naturally and without the need for added cost over time, meaning that energy generation in African nations can remain cost-effective over the long term. Additionally, hydrogen and its derived products are safer to use and store than some of the more traditional fossil fuels, meaning that the risk of an accidental explosion or fire is minimized, which can be a great asset to developing nations in Africa that may not have access to more advanced safety technologies. This environmental friendliness and cost-effectiveness of hydrogen and its derived products have the potential to transform African nations, providing reliable, safe, and renewable energy sources for years to come.

Furthermore, the use of these energy sources can help to improve air quality and reduce the risk of global warming by reducing the amount of carbon dioxide that is being released into the atmosphere from burning fossil fuels. These factors all point to hydrogen and its derived products as being a prime energy source for the African nations of today and the future. By investing in hydrogen and its derived products, African nations can enjoy a better quality of life by reducing air pollution, avoiding hazardous fires or explosions caused by traditional fossil fuels, playing an important role in the fight against global warming, and creating economic opportunities through the development and manufacture of this renewable energy source.

Overall, hydrogen and its derivatives offer African nations an invaluable opportunity to revolutionize their energy sources in an environmentally friendly and cost-effective manner while also contributing to a healthier environment and a sustainable economy for African countries in the long run for the entire region and their citizens. By investing in the research and development of hydrogen energy sources and their derived products, African countries can reap tremendous economic, social, and environmental benefits for the betterment of their respective countries and the region as a whole for decades to come.

African nations, therefore, have much to gain from investing in hydrogen and its derivatives, as they offer African nations a chance to not only secure their energy needs in the present but also to ensure the future prosperity of their societies. African nations have the potential to reap a multitude of advantages from investing in hydrogen and its derivatives by harnessing the potential of this resource to create new industries and opportunities and by providing access to reliable and affordable energy for everyone in the region while also helping to preserve the environment. Furthermore, due to the low-carbon content of hydrogen and its derived fuels, the amount of carbon dioxide released into the atmosphere by these energy sources is much lower than that released by traditional fossil fuels, making them an excellent choice for African nations seeking to reduce their carbon footprint and play their part in the global fight against climate change.

In general, hydrogen and its derived products can offer African nations an environmentally friendly, cost-effective, and safe source of energy that is not only renewable but also helps reduce their carbon footprint in the fight against climate change while also enabling them to create new and innovative industries and open up economic opportunities in the region. Additionally, due to the abundance of natural resources in Africa, the continent has a competitive advantage in regard to developing hydrogen and its related products as a viable energy source, making it an attractive option for African nations looking to take advantage of the latest green energy revolution. This makes hydrogen and its derivatives an attractive option for African nations looking to take advantage of renewable energy sources that can help them move toward economic growth, reduce emissions, and contribute to a more sustainable future for the continent and its people.

By tapping into the potential of hydrogen and its derivatives, African nations can unlock a clean, affordable, safe, and renewable source of energy that will enable them to reduce their reliance on traditional fossil fuels and accelerate their transition toward a greener, more sustainable energy future. As such, it is clear that hydrogen and its related products represent an important opportunity for African nations to make use of a sustainable, cost-effective, and safe energy source that can help reduce their reliance on traditional fossil fuels and pave the way for greater economic prosperity, improved quality of life, and a cleaner, more sustainable environment for the continent and its inhabitants. African nations can capitalize on this opportunity by investing in infrastructure to support the development and use of hydrogen as an energy source, creating innovative policies that encourage its adoption, and incentivizing industries to develop products powered by renewable hydrogen energy such as fuel cells, hydrogen-powered vehicles, and renewable energy systems that can be used to power homes and businesses across the continent.

5.3.3 Challenges of Increasing Hydrogen and its Derivatives

5.3.3.1 High Costs of Production and Storage

The high costs of production and storage for hydrogen and the products derived from it make them unfavorable as an energy source in African countries due to their limited budgets and lack of access to the necessary infrastructure to produce, store, and use it. Additionally, the limited public understanding of hydrogen technology has made it difficult for many African countries to deploy it in an efficient and safe manner. This has led to a slower adoption of hydrogen energy in Africa than in other parts of the world, where economies of scale and a better public understanding of hydrogen technology have allowed for its widespread use. Nonetheless, the potential of hydrogen energy to serve as an affordable and clean energy source is being recognized by some African countries.

Despite these obstacles, some African countries are beginning to invest in hydrogen energy initiatives with the aim of reducing their reliance on traditional energy sources and transitioning toward a more sustainable and secure energy future. To realize the full potential of hydrogen energy in Africa, there is an urgent need for a greater understanding and education about this technology and its benefits, as well as a stronger commitment from African countries to invest in research, development, and deployment of hydrogen-based infrastructure initiatives and technologies. The African continent is particularly well suited for the development of hydrogen energy solutions due to its abundant renewable resources and unique geography, which can be utilized to generate and store clean hydrogen energy as well as provide access to remote areas not currently served by traditional grid infrastructure.

As such, there is a great opportunity for African countries to capitalize on the potential of hydrogen energy and to become leaders in this field by investing in hydrogen energy initiatives and encouraging the widespread adoption of hydrogen technologies across the continent. In doing so, African countries will not only be able to address their current energy needs in a sustainable and secure manner but also create jobs and boost their economies through the manufacture of hydrogen technologies and the sale of hydrogen products and services to the rest of the world.

To fully realize the potential of hydrogen energy in Africa, African countries must make a concerted effort to invest in research and development, build an infrastructure to support the widespread use of hydrogen technologies, and create education and awareness initiatives about the benefits of hydrogen energy to promote public understanding and acceptance of hydrogen energy solutions. Moreover, governments should also encourage and incentivize private sector investment in the hydrogen energy sector to ensure that technologies are widely deployed. The potential of hydrogen energy to serve as an affordable and clean energy source for African countries is immense; however, the task of transitioning away from traditional energy sources and investing in new hydrogen-based infrastructure initiatives will not be an easy one.

In addition, collaboration between African countries and their international partners is also necessary to capitalize on the experience of other countries that have invested heavily in hydrogen energy. This means that it may not be a viable option for these countries until their access to hydrogen technology and infrastructure improve, as well as their public's understanding of the technology. Until then, African countries must continue to rely on more traditional energy sources such as fossil fuels and renewable energy sources such as solar, wind, and hydro. Despite these drawbacks, African nations have the potential to utilize hydrogen for energy in the future if research and development into more efficient production and storage processes continue. Nonetheless, there is some potential for the use of hydrogen energy in Africa; depending on their location, some countries can make use of renewable energy sources such as wind and solar power to produce hydrogen.

5.3.3.2 Lack of Infrastructure

The unfavorability of hydrogen and the products derived from it as an energy source in African nations is due to the lack of infrastructure for both production and distribution of this element and its derivatives, resulting in a lack of access to clean, renewable energy for many people in Africa and, in turn, poverty and the overall underdevelopment of the region. This is a disheartening reality for many African nations that are in dire need of affordable, reliable energy sources to make socioeconomic progress. While hydrogen has great potential to provide African nations with a much-needed source of renewable energy, the lack of resources and infrastructure to access and produce it have proven to be major roadblocks in many countries, leading to a cycle of poverty and dependence on other forms of energy sources, often nonrenewable, that can have a detrimental impact on the environment.

This alarming situation needs to be urgently addressed. To help African countries break this cycle, there needs to be substantial investments in infrastructure and technology to help countries access and produce hydrogen in an efficient and cost-effective way. This investment can be used to create the necessary infrastructure, such as electrolysis systems, that will enable African countries to access and produce hydrogen in a cost-effective way and to provide further technical and scientific training to African professionals to enable them to use hydrogen as a sustainable energy source to help Africa reach its energy needs sustainably. This can have a positive long-term impact on African countries, allowing them to reduce their dependence on nonrenewable energy sources and free up resources to be spent on other areas of economic and social development, such as education and healthcare. This will not only provide economic and environmental benefits but also create new job opportunities for African citizens in the green energy sector, which could help to reduce poverty and improve living standards for African countries in the long term and provide a platform for economic growth and development.

This investment in hydrogen infrastructure and technology could be funded by international donors or by a combination of public and private financing, with African governments playing a major role in the process by taking advantage of the latest advances in hydrogen technology and developing their own hydrogen infrastructure initiatives to facilitate the transition to a greener and more sustainable future for Africans. This shift to a hydrogen-based economy could be transformative for African countries and would have a positive long-term impact on African citizens, helping to reduce poverty and improve living standards through increased access to renewable energy sources, greater job opportunities in the green energy sector, and increased investment in hydrogen infrastructure and technology initiatives. This increased investment in hydrogen infrastructure and technology would be a major step forward in Africa's transition to a greener and more sustainable future and would provide a much-needed boost to African economies by creating new job opportunities, encouraging innovation, and driving economic growth and development on the continent.

This in turn would create the necessary conditions for an inclusive and equitable economic transformation, paving the way for a brighter future for African citizens and helping to alleviate the pressure on African governments to meet the pressing needs of their populations while also advancing the sustainable development goals of the United Nations and other international organizations. The potential impact of this investment in hydrogen infrastructure and technology is huge and could help to significantly reduce poverty and improve living standards across the continent while at the same time driving economic growth and progress through increased access to renewable energy sources, greater job opportunities in the green energy sector, and increased investment in hydrogen infrastructure and technology initiatives. This shift to a hydrogen-based economy is not only transformative for African countries but could also have a lasting and beneficial impact on African citizens and the overall development of the continent over the long term.

Going forward, it is imperative that international bodies and governments work together to create a robust investment program that provides African countries with the resources they need to access and produce hydrogen energy in an efficient and cost-effective way while also ensuring that the long-term benefits of such an investment program are realized by all African citizens. As such, international organizations and governments should work together to identify strategic investment opportunities that could deliver these long-term benefits for African countries, such as investing in the necessary infrastructure and technology required to produce hydrogen energy efficiently and cost-effectively and ensuring that financial and human capital investments are targeted toward the areas of greatest need to maximize the benefits of this investment program for African citizens and ensure that the potential of this investment program is realized in the most effective and equitable way possible.

In general, a shift to a hydrogen-based economy in Africa could have huge potential benefits for the continent and its citizens. Investing in hydrogen infrastructure and technology has the potential to be a game changer for Africa and its citizens, so it is essential that international organizations and governments work together to make this shift a reality as quickly and cost-effectively as possible. It is also vital that the process of investing in hydrogen energy be done responsibly and ethically to ensure that the long-term benefits of this investment program are enjoyed by African citizens rather than simply lining the pockets of corporations and governments over centuries. With the right investments, a hydrogen-based economy in Africa could have huge potential long-term benefits for African citizens and spur a much-needed transformation of the African economy into one based on sustainable, renewable energy sources.

While there are obviously many challenges and risks associated with such a shift, the potential long-term benefits for African citizens could far outweigh these risks if the shift to hydrogen energy is done in an ethical and responsible way. By investing in hydrogen energy and infrastructure, African nations can take the first steps toward an economic transformation based on sustainable renewable energy sources. This could lead to a more equitable and sustainable economic structure, one that is not reliant on the export of fossil fuels but instead on clean and renewable energy sources that can provide energy to citizens without contributing to the global climate crisis.

Furthermore, investment in hydrogen-based infrastructure has the potential to create jobs in African countries and increase incomes for citizens who are able to access and take advantage of these new opportunities. This could lead to a more secure future for African citizens, one that is not reliant on the continued exploitation of their resources but instead puts the needs and interests of their citizens first while also providing them with an energy source that is not only renewable but clean and efficient as well. As such, investing in hydrogen energy and infrastructure can bring a range of benefits to African nations, from a more equitable and sustainable economic structure to job creation and increased income for citizens who are able to access and take advantage of these opportunities, as well as improved prospects for their future by putting the needs and interests of citizens first and providing them with a clean, renewable energy source that is also efficient and cost-effective.

5.3.3.3 Limited Availability of Resources

Due to the scarcity of resources, African countries do not favor using hydrogen and the products derived from it as an energy source, as the infrastructure necessary to implement such technology is too costly and often unattainable for these countries given their limited financial and material resources. Thus, African countries are more likely to rely on energy sources such as coal and other fossil fuels that have a lower upfront cost, although these alternatives have higher long-term costs in terms of environmental damage and health risks associated with their use. The lower upfront cost and availability of coal and other fossil fuels compared to hydrogen energy sources make them the more viable option for African countries given their limited resources. Although hydrogen and the products derived from it may be more sustainable and environmentally friendly in the long run, African countries are not able to capitalize on them due to their scarcity of resources.

Therefore, African countries are often forced to rely on more traditional sources of energy, such as coal and other fossil fuels, despite the potential long-term negative consequences associated with their use. This creates a difficult situation for African countries, as they are unable to take advantage of the more environmentally friendly and sustainable energy sources due to their limited resources, yet they still have to pay the costs associated with the negative environmental and health impacts of using coal and other fossil fuels in the short term. In fact, African countries face a difficult dilemma in regard to energy resources; they are unable to transition to more sustainable energy sources due to their lack of resources but still have to deal with the short-term costs of relying on coal and other fossil fuels. This puts African countries in a particularly difficult position and one that could have far-reaching consequences for their future energy security and development prospects if it is not addressed in the near future.

Given this precarious situation, it is essential that African countries focus on both short- and long-term solutions to their energy needs that could help them to transition away from their reliance on fossil fuels in the short term while also helping them to ensure that they can take advantage of the more sustainable energy sources available in the long term to protect their environment, their citizens' health, and their economic security by going beyond what was expected of them and making progress in the energy sector. African countries need to work toward developing a balanced and sustainable approach to their energy needs that takes into account both the immediate and future impacts of their decisions as well as their ability to access the necessary resources to make the transition from coal and other fossil fuels to renewable and clean energy sources possible while also preserving their environment and future economic stability while addressing the immediate needs and long-term sustainability of their energy resources.

Moreover, even if African countries could afford to implement hydrogen energy sources, the infrastructure necessary for their proper utilization is often lacking in these countries, making the actual transition to a hydrogen-based economy difficult and, in some cases, impossible without further outside assistance and support. This means that African countries need to rely heavily on external partners such as governments, private companies, and international organizations to help them develop the necessary infrastructure to make their transition away from fossil fuels and toward renewable energy sources both feasible and sustainable in the long term. Furthermore, African countries also need to address the potential risks and impacts that their transition away from fossil fuels could bring about, such as job losses due to the switch from coal and other fossil fuels to renewable energy sources or the potential risks posed by new technologies that have not yet been fully explored and tested.

Therefore, to ensure a successful transition from their current reliance on coal and other fossil fuels to more sustainable forms of energy, African countries need to work toward developing a comprehensive strategy that takes into account the economic, environmental, and social implications of their decision. These issues could be addressed if countries had access to renewable energy sources, such as solar and wind power, which have low upfront costs but a much lower long-term costs in terms of environmental damage and health risks. Despite its potential advantages, such as being an abundant, renewable, and clean energy source, the lack of resources and infrastructure in African countries means hydrogen energy is not a viable option for them in the short term.

In the short term, African countries have to look for alternative energy sources such as biomass, solar, and wind energy, which can be more easily implemented in these countries given their existing infrastructure and the availability of financial aid from other countries to help build the necessary infrastructure. These renewable energy sources have the potential to replace traditional methods of energy production, allowing African countries to build sustainable, economically viable energy systems while reducing their carbon footprints and helping to create a more equitable energy system for all countries.

In general, while hydrogen energy is a promising technology that could revolutionize the way African countries produce and consume energy, the lack of resources and infrastructure makes it an unattainable option in the short term. Therefore,

African countries should focus on developing renewable energy sources such as biomass, solar, and wind energy in the short term to create sustainable and economically viable energy systems while still reducing their carbon footprints and helping to create a more equitable energy system for all countries in the long term; however, African countries should not forget about the potential of hydrogen energy and instead make strides to establish the necessary resources and infrastructure for a hydrogen-based energy system, as it could provide a long-term solution to many of the current energy challenges faced by African countries. Today, renewable energy sources are becoming increasingly popular due to their potential to reduce our reliance on traditional forms of energy production, which are often costly and environmentally damaging.

In general, African countries should focus on the short-term goal of developing renewable energy sources such as biomass, solar, and wind energy to create sustainable and economically viable energy systems while also reducing their carbon footprints and helping to create a more equitable energy system for all countries. In the long term, African countries should strive to develop hydrogen energy as a potential solution to their energy challenges due to its versatility and ability to store energy for longer periods of time than traditional forms of energy production. As such, hydrogen energy should be seen as an important part of a larger energy plan for African countries to ensure that they are able to meet their energy demands in a sustainable and cost-effective manner over the next few decades. As renewable energy sources become more viable, African countries must take a leading role in the development of sustainable and equitable energy systems.

5.3.3.4 Safety and Storage Concerns

African nations are not in favor of using hydrogen and the products derived from it as an energy source due to safety and storage concerns, as hydrogen is extremely flammable and presents a risk of explosion if not stored correctly. In addition, hydrogen is a relatively expensive fuel source, and it is difficult to transport safely, making it a less-than-ideal energy source in comparison to other sources such as fossil fuels. Despite these issues, some African nations are exploring the potential of hydrogen as an energy source, looking for ways to make it safe and cost-effective. Despite these disadvantages, some consider hydrogen to be a viable energy source due to its clean burning capabilities, which means that it can be used without the environmental issues associated with other sources such as fossil fuels and nuclear energy. As a result, African nations must weigh the potential benefits of using hydrogen as an energy source against the risks and expenses associated with it.

Additionally, the potential for hydrogen to be used to store electricity generated by renewable sources such as solar and wind power makes it an appealing option for countries that are looking to transition away from traditional energy sources and move toward a more sustainable future. To make the best decision for their respective countries, African nations must carefully consider the cost and risk associated with transitioning to hydrogen-based energy sources while also taking into account the long-term environmental benefits that could be gained from such a transition. African nations must take into account the cost and risk of transitioning to hydrogen-based energy sources while also considering the long-term environmental benefits that could be gained from such a transition. They must evaluate the potential benefits against the cost, risk, and environmental impact of transitioning to a hydrogen-based energy source and decide whether transitioning to a hydrogen-based energy source is the best decision for their countries. Considering the potential benefits of hydrogen-based energy sources and the cost, risk, and environmental impacts associated with transitioning to such sources, African nations must carefully evaluate whether transitioning to a hydrogen-based energy source is the best decision for their respective countries.

As a result, some African nations are researching and experimenting with new technologies to make hydrogen a viable energy source, taking into account safety, storage, and cost concerns before making the transition, as well as the overall environmental benefits that could be achieved in the long term. Ultimately, transitioning to a hydrogen-based energy source is a complex process that requires careful consideration of cost, risk, and environmental impact to ensure that it is the best decision for their countries and the people who inhabit them. These efforts are indicative of the long-term potential for hydrogen as an energy source in Africa, as nations seek to balance the costs and risks associated with transitioning to a hydrogen-based energy source against the potential benefits of transitioning to a clean, sustainable energy source that can help mitigate the effects of climate change and improve the quality of life for those living in African countries.

In light of these considerations, African nations must take a comprehensive approach to making decisions about transitioning to a hydrogen-based energy source by considering all aspects of the process, from storage and safety to cost and environmental impact.

This means taking into account not only the immediate costs and risks associated with transitioning to a hydrogen-based energy source but also weighing the potential long-term benefits of transitioning to a clean and sustainable energy source that could help reduce their countries' carbon footprint and improve the quality of life for those living in African countries. This includes assessing the short- and long-term implications of transitioning to a hydrogen-based energy source, including any potential economic benefits, environmental impacts, and social equity considerations associated with making the switch. African countries must also recognize the importance of investing in research and development to ensure that hydrogen-based energy sources are safe and effective and consider investing in infrastructure, such as storage facilities that can safely transport and store hydrogen, so that the transition to a hydrogen-based energy source can be successful. Once all these considerations have been taken into account, African nations can make an informed decision about transitioning to a hydrogen-based energy source that is based on their unique context and needs.

African nations should also look to other countries that have successfully made the transition to a hydrogen-based energy source and learn from their experiences to ensure that their own transition is as smooth and successful as possible. African nations have a responsibility to carefully consider the full implications of transitioning to a hydrogen-based energy source to ensure that the benefits outweigh any potential risks or costs associated with the transition and that the infrastructure and resources are in place to safely, effectively, and efficiently use a hydrogen-based energy source. African nations should ensure that they have the support of both their citizens and other relevant stakeholders, such as businesses and investors, to ensure that the transition to a hydrogen-based energy source is successful to ensure that the benefits they gain from transitioning to a hydrogen-based energy source are maximized and long lasting.

The decision to transition to a hydrogen-based energy source should be informed by a comprehensive understanding of the context and needs of each African nation to ensure that the transition is made in a way that will benefit both the nation and its citizens in the most effective and efficient way possible and that the potential costs or risks associated with the transition are minimized. African nations should also ensure that they are well informed about the latest advancements and best practices in the field of hydrogen-based energy to take advantage of the most effective and efficient technologies available and ensure that the transition to a hydrogen-based energy source is as successful and beneficial as possible. Additionally, African nations should have an open dialogue with their citizens about the potential risks and benefits of transitioning to a hydrogen-based energy source to ensure that their decisions regarding the transition are based on an informed understanding of all stakeholders' needs and concerns.

To ensure a successful and sustainable transition to a hydrogen-based energy source, African nations must ensure that all relevant stakeholders are adequately informed about the risks and benefits of such a transition and that a comprehensive understanding of their nation's specific context and needs is taken into account when making their decisions so that they can make an informed and educated decision on how to best transition to a hydrogen-based energy source that will ensure a successful and sustainable transition to a hydrogen-based energy source for their nation and its citizens. African nations must also make sure that they are not only aware of the potential costs and risks associated with transitioning to a hydrogen-based energy source but also how to best minimize them and make sure that the transition is as beneficial as possible for all stakeholders by making sure they are informed, communicating openly with citizens, and taking into account their specific needs when making decisions regarding the transition so that they can make the most informed and beneficial decision possible for their citizens and nation. African nations must make sure they are informed and engaged and that a comprehensive understanding of their nation's context and needs is taken into account when making decisions about transitioning to a hydrogen-based energy source to ensure that their decisions are based on an informed understanding of all stakeholders' needs and concerns and that the transition is as beneficial as possible for all involved so that a successful and sustainable transition to a hydrogen-based energy source can be achieved.

Ultimately, African nations should strive to create an open dialogue and a trustful relationship between the nation and its citizens that ensures that all stakeholders' needs are taken into account when making decisions about transitioning to a hydrogen-based energy source so that they can make the most informed, beneficial, and sustainable decision possible for their citizens and nation, and all stakeholders have an opportunity to express their thoughts and opinions as well as have their needs and concerns taken into account to ensure the most successful and sustainable transition for their nation and that their decisions are based on an informed understanding of all stakeholders' needs and concerns so that they can ensure a successful and sustainable transition to a hydrogen-based energy source for the first time in history for their nation.

5.3.3.5 Regulatory Hurdles

The lack of regulatory hurdles in African countries prevents them from effectively using hydrogen and the products derived from it as an energy source, despite its abundant availability and potential for clean energy production. This issue is compounded by the lack of capital available to build the infrastructure necessary for this type of energy production, as well as a general lack of awareness regarding its capabilities and potential applications. Furthermore, the cost of production for hydrogen energy is often too high for most African countries to even consider it as an option for their energy needs. Despite the potential benefits of using hydrogen energy, the aforementioned issues present a significant roadblock to its adoption in many African countries. Without appropriate government policies and regulations in place to incentivize its use, it will be difficult for hydrogen energy to gain traction and become a viable alternative to traditional energy sources in Africa. This is an unfortunate fact, as hydrogen could provide a clean and renewable energy source for these countries and serve to combat the climate crisis while simultaneously alleviating the poverty that is so pervasive in many African countries.

Unfortunately, despite the environmental and economic advantages that hydrogen energy could bring to African countries, it may take some time before the technology and infrastructure can be put in place to make it a viable option for countries in the region due to the lack of funding, policy support, and technical expertise required for its implementation. Therefore, it is up to governments and other stakeholders to come together to create effective policy frameworks and find solutions for the challenges associated with implementing hydrogen energy in African countries in order for hydrogen energy to gain ground in the region and eventually become a viable alternative to traditional energy sources. In the short term, governments and international organizations need to focus on investing in the research and development of hydrogen energy technology in African countries to lay the groundwork for its successful implementation in the future.

In the long term, policies should be enacted to incentivize companies and organizations to develop hydrogen energy infrastructure and promote investment in the sector to create a sustainable, low-carbon energy system in Africa that could bring about sustainable development and poverty reduction in the region. In addition, educational initiatives should be implemented to increase the level of technical expertise and knowledge about hydrogen energy within African countries to ensure that the technology is adopted safely and properly and that African nations can maximize its potential benefits and create a self-sustaining hydrogen energy system that is beneficial to their citizens.

To ensure that the technology is properly and safely adopted, governments must invest in education and training initiatives to increase the technical knowledge of hydrogen energy within African countries so that African citizens can become knowledgeable about hydrogen energy technology, make informed decisions about its implementation in their countries, and ensure that the technology is properly implemented and utilized to maximize its potential benefits for their communities and society as a whole. The lack of regulatory hurdles in African countries prevents them from effectively using hydrogen and the products derived from it as an energy source, despite its abundant availability and potential for clean energy production.

This issue is compounded by the lack of capital available to build the infrastructure necessary for this type of energy production, as well as a general lack of awareness regarding its capabilities and potential applications. Furthermore, the cost of production for hydrogen energy is often too high for most African countries to even consider it as an option for their energy needs. Despite the potential benefits of using hydrogen energy, the aforementioned issues present a significant roadblock to its adoption in many African countries. Without appropriate government policies and regulations in place to incentivize its use, it will be difficult for hydrogen energy to gain traction and become a viable alternative to traditional energy sources in Africa.

This is an unfortunate fact, as hydrogen could provide a clean and renewable energy source for these countries and serve to combat the climate crisis while simultaneously alleviating the poverty that is so pervasive in many African countries. Unfortunately, despite the environmental and economic advantages that hydrogen energy could bring to African countries, it may take some time before the technology and infrastructure can be put in place to make it a viable option for countries in the region due to the lack of funding, policy support, and technical expertise required for its implementation.

Therefore, it is up to governments and other stakeholders to come together to create effective policy frameworks and find solutions for the challenges associated with implementing hydrogen energy in African countries in order for hydrogen energy to gain ground in the region. These policy frameworks should include incentives for the production and use of hydrogen energy, such as tax credits and subsidies, as well as regulations to ensure that the production and use of hydrogen energy is done responsibly and safely to maximize the economic, environmental, and social benefits of hydrogen energy in African countries and to ensure that its potential is not wasted.

In general, hydrogen energy has the potential to have a significant impact on African countries by providing a clean and renewable energy source that can help combat climate change while also helping to reduce poverty and create economic opportunities in the region. Furthermore, a collaboration between different stakeholders, such as governments, private companies, civil society organizations, and financial institutions, is essential for the successful implementation of hydrogen energy in African countries to create a sustainable and viable hydrogen energy market in the region that can benefit the region's citizens and the environment over the long term. Therefore, governments, private companies, civil society organizations, and financial institutions must come together to create effective policy frameworks and implement solutions for the challenges associated with utilizing hydrogen energy in African countries to realize the potential of hydrogen energy in the region and promote its growth in a responsible and sustainable manner.

To achieve these objectives, governments should create and implement policies that will enable the safe, efficient, and cost-effective production and use of hydrogen energy in African countries while providing incentives for private companies, civil society organizations, and financial institutions to invest in the region's hydrogen energy sector and collaborate with each other to help realize the potential of hydrogen energy in the region for the benefit of all. This collaboration should involve the development of hydrogen energy infrastructure, such as renewable hydrogen production facilities, fueling and distribution networks, and storage systems, while also providing opportunities for research and development, training and education, and capacity building to further facilitate the deployment of hydrogen energy in African countries over the long term. Additionally, collaboration should also include the development of market mechanisms, such as long-term contracts and green financing instruments, to promote investment in the sector and provide support to stakeholders to drive growth in hydrogen energy production and use and to ensure the responsible, safe, efficient, and cost-effective utilization of hydrogen energy in the region while mitigating the potential risks and adverse impacts associated with its use.

These policy frameworks and solutions should aim to facilitate a responsible and sustainable approach to the development of hydrogen energy in African countries by taking into consideration the local context, particularly in terms of its unique socioeconomic circumstances, existing infrastructure, and regulatory environment, and enabling the necessary collaboration among governments, private companies, civil society organizations, and financial institutions to realize the full potential of hydrogen energy in African countries while providing long-term and equitable economic and social benefits to all stakeholders, especially in terms of reducing emissions, creating jobs, improving public health, and developing local industry.

Therefore, governments should ensure that policy frameworks and solutions provide adequate incentives for private companies, civil society organizations, and financial institutions to collaborate with each other and invest in the region's hydrogen energy sector while encouraging African countries to develop strategies and regulations that enable the responsible, safe, efficient, and cost-effective utilization of hydrogen energy to realize its full potential in the region. This collaboration should also focus on developing policy frameworks and solutions to address the challenges that might be encountered in transitioning toward a hydrogen-based economy by facilitating the development of the infrastructure and markets needed to support the sector, investing in research and development initiatives to further enhance its use, establishing public–private partnerships for financing projects related to hydrogen energy production and utilization, introducing targeted policies and financial instruments that incentivize investment in hydrogen energy-related projects, and providing capacity building programs for stakeholders to help them understand the risks associated with hydrogen energy use and take effective measures to mitigate them.

Additionally, collaboration should also include the development of market mechanisms, such as long-term contracts and green financing instruments, to promote investment in the sector and provide support to stakeholders to drive growth in hydrogen energy production and use in African countries. This will ultimately help Africa create a resilient, equitable, and economically viable hydrogen energy system that can bring long-term benefits to the region and its citizens by creating jobs, improving access to energy, reducing carbon emissions and other pollutants, and providing economic growth opportunities for all stakeholders involved.

5.4 Renewable Energy

5.4.1 A Synopsis of African and World Renewable Energy Sources

Renewable energy is energy that is collected from natural resources that are replenished on a human timescale, such as sunlight (Nwokolo et al., 2022a), wind, rain, tides, waves, and geothermal heat (Nwokolo et al., 2023b). These natural resources are constantly available, and their use does not deplete or damage the environment, making them a sustainable form of energy that can be used to provide a clean, renewable source of power for our homes and communities. These sources of renewable energy are increasingly becoming more attractive due to their potential for helping to reduce greenhouse gas emissions, their ability to provide reliable energy without depleting finite resources, and the fact that they are often cheaper than nonrenewable sources of energy in the long run. These resources are constantly replenished, which makes them more sustainable than finite resources such as fossil fuels. Therefore, renewable energy sources are becoming increasingly attractive and should be considered an important part of the world's future energy portfolio. By investing in renewable energy sources such as solar, wind, and hydropower, we can create a cleaner and more secure future while also reducing our dependence on finite resources and helping to protect our environment for generations to come.

The development of renewable energy sources presents a unique opportunity to meet the world's growing energy demands without sacrificing our planet's limited resources or exacerbating climate change (Nwokolo et al., 2022a). It is therefore essential that we move away from the exclusive use of finite resources and start to invest in renewable energy sources that are constantly replenished and can be relied upon for long-term sustainability. In addition to their environmental benefits, renewable energy sources are also financially attractive in the long term since they are generally cheaper than traditional energy sources in the long run and require less maintenance and operational costs over time. As such, investing in renewable energy sources is not only beneficial for the environment but can also result in significant cost savings and economic advantages for businesses, homeowners, and communities alike. The development of renewable energy sources presents us with a chance to tackle the global demand for energy as well as provide environmental and economic advantages that can be taken advantage of in the short term and long term.

Making a shift to renewable energy sources has become even more critical to combat the effects of climate change and ensure a sustainable future. Investing in renewable energy sources is not only necessary for a sustainable future but also an investment in our own health and well-being given the immediate and long-term benefits for businesses, communities, and individual households that investing in renewable energy sources provides. By investing in renewable energy sources, businesses, communities, and individual households are able to not only reduce their overall carbon emissions but also reduce their energy costs and receive economic benefits that help drive the economy forward, such as job creation (Obiwulu et al., 2022) and economic stimulus, while also helping to create a healthier environment for all by providing cleaner air and more efficient use of resources (Obiwulu et al., 2022). Additionally, making a shift to renewable energy sources can be beneficial in terms of energy security since renewable energy sources are not dependent on fossil fuels, which can be affected by geopolitical instabilities and price fluctuations.

Additionally, these renewable energy sources are available in abundance and are not limited by borders or resources, meaning that businesses, communities, and individual households can have access to renewable energy sources with relative ease, making them a great choice for both short- and long-term investments in the future. By investing in renewable energy sources, businesses, communities, and individual households are not only making an important contribution to the fight against climate change but are also ensuring that their own economic and environmental futures remain strong and secure for many years to come. Additionally, investment in renewable energy sources can also provide other benefits, such as an increased sense of national security, independence, and sustainability, which in turn can have a positive impact on the overall economic development of a country, leading to increased economic prosperity and a better quality of life for everyone.

Ultimately, investing in renewable energy sources has many benefits and is an important step in helping to ensure a more secure, sustainable, and prosperous future for everyone, both in the present and in the future. By investing in renewable energy sources, people can not only help to fight climate change and create a more secure and sustainable future for themselves but also benefit from the economic and environmental security that comes from having access to an abundant source of energy that is clean, renewable, and affordable for the long term. This is a critical time to invest in renewable energy sources, and it is important that individuals, businesses, and governments all come together to make a concerted effort toward the development of sustainable and reliable energy sources for the betterment of society and the environment as a whole.

Additionally, renewable energy also offers a number of economic benefits, such as increased job opportunities and economic development, increased investment in clean energy research and development, improved public health from reduced air pollution, and improved national security by reducing the need to rely on foreign sources of energy. Furthermore, renewable energy is essential to meeting the world's energy needs in the coming decades as we seek to address climate change and reduce our dependence on fossil fuels. All of these benefits make renewable energy an essential part of the world's energy mix and underscore its importance in building a cleaner, healthier, and more secure future. It is therefore crucial that renewable energy investments be made now, not only to help reduce greenhouse gas emissions and achieve our climate goals but also to reap the other benefits mentioned above and to secure a more prosperous future for generations to come.

Renewable energy offers numerous environmental and economic benefits, from creating job opportunities and economic development to improving public health through cleaner air and water and providing energy independence and security by reducing our dependence on imported energy sources. Renewable energy is also an important tool in reducing poverty, as it allows rural and remote communities access to electricity that would otherwise be unaffordable or inaccessible. Renewable energy is a powerful and vital part of our global energy mix, offering a wide range of economic, environmental, and social benefits that are essential to achieving a sustainable future. For these reasons, it is imperative that we continue to invest in renewable energy sources while also utilizing other innovative strategies to reduce our dependence on fossil fuels and transition to a low-carbon economy. As renewable energy sources become increasingly cost-competitive with traditional fossil fuels, the transition to a low-carbon economy will become even more attractive and attainable for all members of the global community (Nwokolo & Ogbulezie, 2017a).

This shift to renewable energy sources is especially important for low- and middle-income countries, which may be unable to access or afford traditional electricity sources due to a lack of infrastructure or financial resources. These countries have the potential to greatly benefit from renewable energy sources, as they offer a reliable and cost-effective solution that does not contribute to climate change and can be implemented with minimal effort and investment (Nwokolo & Ogbulezie, 2018a). With a growing population and increasing concerns about climate change, renewable energy is becoming more important as an alternative to traditional sources of power such as coal and natural gas. By investing in renewable energy sources, low- and middle-income countries can provide their citizens with an improved quality of life and contribute to the global transition toward a low-carbon economy.

Furthermore, renewable energy sources can help reduce air pollution and health risks associated with traditional power sources, creating a healthier environment for all. In general, renewable energy sources offer great potential for low- and middle-income countries to improve their quality of life while contributing to the global fight against climate change and creating a more sustainable future. For these reasons, investing in renewable energy sources is becoming an increasingly attractive option for many low- and middle-income countries, and with advances in technology and the cost of renewables becoming increasingly competitive with fossil fuels, is now the time for countries to make investments and switch to a renewable energy future.

However, despite the promise of renewable energy sources, we must also remember that they are still in their infancy and come with their own set of challenges, such as the availability of space and resources to install large-scale renewable energy systems, as well as potential fluctuations in power output due to weather and seasonal changes that may require the development of energy storage systems to ensure a steady supply of electricity to the population (Nwokolo & Ogbulezie, 2018a). In conclusion, while renewable energy sources offer a number of advantages, such as reduced air pollution, lower health risks, and cost-efficiency when compared to traditional fossil fuels, it is important to also consider the potential issues that can arise with investing in renewables, such as the availability of resources, potential fluctuations in power output, and the development of energy storage systems, to ensure that renewable energy sources are used in a way that maximizes their potential and contributes to a more sustainable future for low- and middle-income countries and all of humanity and the world (Nwokolo et al., 2022a). It is essential that renewable energy sources are viewed not only as a way to reduce the effects of climate change and create a more sustainable future but also as an investment in the well-being of future generations.

Air and water heating and cooling, stand-alone power systems, and the generation of electricity from a grid all frequently use renewable energy sources to minimize their environmental impact and ensure sustainable energy production, such as solar, wind, hydro, and geothermal energy sources. In addition to the environmental benefits, these renewable energy sources are generally free and accessible in most parts of the world and, as such, can provide energy to those who lack the economic means to access traditional energy sources such as oil, gas, and coal. Furthermore, they provide an opportunity to reduce global dependence on nonrenewable energy sources, which will eventually be exhausted, and create a more sustainable world for future generations. By utilizing renewable energy sources, people are able to reduce their environmental impact and improve the quality of life for people who do not have access to traditional energy sources due to economic constraints.

By investing in renewable energy sources, communities can become less reliant on nonrenewable energy sources and reduce their overall environmental footprint while also improving access to energy for communities that have been traditionally underserved in the long run. This is particularly important in developing countries where the lack of access to traditional energy sources is causing a strain on local economies and negatively impacting quality of life, thus creating a more sustainable world for future generations. These renewable energy sources are becoming increasingly popular due to their affordability, sustainability, and low environmental impact compared to more traditional nonrenewable sources, such as coal and oil. These renewable sources of energy can help reduce our dependence on fossil fuels, have a minimal environmental footprint, reduce air pollution, and reduce greenhouse gas emissions, ultimately leading to a healthier and more sustainable environment.

In addition, investing in renewable energy sources can create more jobs, boost the local economy, and provide a viable alternative to electricity generated from nonrenewable sources. For all of these reasons, investing in renewable energy sources is not only a smart decision for the environment but also a smart financial decision, as it can save money on energy bills and potentially reduce costs associated with healthcare due to decreased air pollution. In addition, investing in renewable energy sources helps create jobs and provides economic opportunities for communities that may not have otherwise been able to take advantage of them. This in turn can help create an environment that is more conducive to sustainable growth and progress, not only for our planet but also for all of its inhabitants.

Solar and wind energy are the most commonly used renewable energy sources, as they are readily available and cost-effective when compared to traditional sources of energy such as fossil fuels and nuclear power. Additionally, these renewable energy sources emit little to no carbon into the atmosphere, making them the most environmentally friendly option for powering homes and businesses. Furthermore, these renewable energy sources are not finite like traditional energy sources, meaning that they have the potential to provide energy for centuries to come without depleting the earth's resources. To help protect our planet and ensure the sustainability of our resources, it is essential that we make a concerted effort to increase the use of renewable energy sources in our everyday lives and move away from relying on traditional energy sources that have been proven to be detrimental to our planet's health.

Investing in renewable energy sources such as solar and wind energy is essential for the future of our planet and can ultimately provide us with an inexhaustible source of clean, sustainable energy that can be used to power our homes and businesses without endangering the environment. With knowledge of their environmental benefits, many countries have begun to invest heavily in renewable energy sources and have seen immense success in terms of both environmental protection and economic growth. The increasing popularity of renewable energy sources is a positive sign that society is taking seriously the need to move away from overreliance on traditional energy sources, which often comes with severe environmental consequences.

The share of renewable energy in the world's electricity supply increased from 20% to 28% between 2011 and 2021. Nuclear energy dropped from 12% to 10%, and fossil fuels went from 68% to 62% (Aklin, 2022). The share of hydropower fell from 16% to 15%, while solar and wind energy increased from 2% to 10%. Geothermal and biomass energy both increased from 2% to 3% (Aklin, 2022). In 135 nations, 3146 gigawatts are installed, and 156 nations have laws governing the renewable energy industry (Aklin, 2022). Nearly half of the global increase in renewable electricity in 2021 will come from China. These statistics demonstrate that renewable energy is rapidly becoming the preferred form of energy generation in the world, with China leading the way. The incredible growth of renewable energy in 2021, especially within China, has demonstrated the world's commitment to curbing greenhouse gas emissions and moving toward a cleaner and greener future (Aklin, 2022).

This trend is expected to continue into 2022 and beyond, with many nations committing to becoming carbon-neutral by 2050. These developments represent a monumental shift in the world's approach to energy generation, with renewable sources such as wind and solar power replacing traditional fossil fuels. This shift is great news for the environment, as renewable energy sources are much less harmful to the atmosphere than their fossil fuel counterparts. In addition, renewable energy sources are much more sustainable in the long term since they can be produced repeatedly without any long-term depletion of resources. With the increasing popularity of renewable energy sources, an increasing number of countries are committing to a low-carbon economy to combat the effects of climate change. This represents an enormous opportunity for countries to become more energy independent and reduce their reliance on fossil fuels, which can cause economic and environmental instability.

Not only does the shift to renewable energy sources help to reduce the amount of carbon emissions in the atmosphere, but it also provides numerous other benefits, such as creating new jobs in the renewable energy sector, reducing air pollution, and saving money on energy costs in the long run. Moreover, transitioning to renewable energy sources allows for greater energy security, as countries are no longer reliant on the increasingly volatile global markets for fossil fuels. This is why it is so important for countries to take advantage of the opportunity presented by renewable energy sources and invest in transitioning to a low-carbon economy to create a shift away from the traditional reliance on fossil fuels and toward renewable energy sources such as solar, wind, geothermal, and hydroelectric power.

In general, the potential benefits of transitioning to renewable energy sources are numerous, and it is clear that governments must take the necessary steps to make this shift happen if they hope to ensure a sustainable and secure energy future for their citizens. It is essential that governments across the globe recognize the importance of transitioning to renewable energy sources and take action to make progress against climate change and ensure a more secure energy future for themselves and for future generations. As such, governments must make the transition to renewable energy sources a priority to meet the demands of their citizens, protect their economies, and secure the long-term health of the planet and its inhabitants. The switch to renewable energy sources is one that must be made for us to make progress on climate change, protect our economies, and ensure a secure and sustainable energy future for ourselves and our posterity.

With solar photovoltaics being the largest employer, there are more than ten million jobs related to renewable energy industries globally. According to a report from the International Renewable Energy Agency, in 2019, the global renewable energy sector employed more than ten million people and accounted for 3.4% of global employment (Aklin, 2022). Governments must therefore make the switch to renewable energy sources a priority to meet the demands of their citizens, protect their economies, and secure the long-term health of the planet and its inhabitants. Investing in renewable energy infrastructure is a vital step for governments to take, as it will not only create jobs and fuel economic growth but also reduce greenhouse gas emissions and create a healthier environment for future generations by powering their countries with renewable energy sources. The growth of the renewable energy industry has been spurred by recent technological advances, such as cost reductions in photovoltaic (PV) systems, increased efficiency of solar panels and batteries, and improvements in wind turbine designs.

These figures indicate the immense potential of renewable energy sources and suggest that governments must prioritize the transition to renewable energy sources to provide their citizens with jobs, secure their economies, and protect the planet's health. Governments must make the switch to renewable energy sources a priority to meet their citizens' demands, protect their economies, and secure the long-term health of our planet and its inhabitants. To do this, governments should invest in the research and development of renewable energy sources, provide subsidies and tax credits to encourage private investment in renewable energy infrastructure, and promote green initiatives such as energy efficiency programs and building codes that are centered around reducing energy consumption.

Additionally, governments should make the switch to renewable energy sources a priority through the implementation of aggressive renewable energy targets that set ambitious goals and timelines for the transition to renewable energy sources and hold governments accountable for meeting those goals. These targets, coupled with investment in renewable energy research and development and incentives for private sector participation in the renewable energy industry, will ensure that governments meet their citizens' demands for renewable energy, protect their economies from volatile fossil fuel markets, and ensure the long-term sustainability of our planet and its inhabitants by making the switch to renewable energy sources a priority for the first time in human history. This transition to renewable energy sources can be accomplished through the implementation of aggressive renewable energy targets, government investment in the research and development of renewable energy sources, the provision of subsidies and tax credits to encourage private investment in renewable energy infrastructure, and the promotion of green initiatives such as energy efficiency programs and building codes designed to reduce energy consumption. This transition is not only necessary but also achievable with the right investments and policies in place.

Presently, the least expensive new electricity sources in the majority of nations are photovoltaic solar power or onshore wind. Renewable energy already accounts for more than 20% of the energy used in many countries around the world, with some producing more than 50% of their electricity from renewable sources. To make the transition to renewable energy sources more achievable, governments need to implement more aggressive renewable energy targets, invest in research and development, offer subsidies and tax credits for private investments in infrastructure, and promote green initiatives such as energy efficiency programs and building codes to ensure that a renewable energy future is accessible for everyone. Additionally, the costs associated with renewable energy sources are decreasing rapidly due to technological advances and economies of scale, making them an increasingly attractive option for governments and individuals alike. By increasing the demand and supply of renewable energy, governments can help ensure that renewable energy sources become a larger part of global energy portfolios and create more jobs in the industry.

Several nations exclusively use renewable energy sources to produce their electricity. In the 2020s and beyond, this trend is likely to increase. With more nations recognizing the need to move away from finite energy sources such as coal and oil to combat climate change and reduce their carbon footprints, many countries are beginning to embrace renewable energy sources such as wind, solar, and hydroelectric power. By investing in the infrastructure necessary to create and store energy from these sources, it is anticipated that the national renewable energy market will continue to expand rapidly in the coming decades. As more nations join the global effort to reduce their environmental impact, more of them will shift toward renewable energy sources to create a cleaner and more sustainable future for the planet.

The renewable energy industry is experiencing a massive influx of investment, and the innovation spurred by this is likely to accelerate the adoption of renewable energy sources in nations around the world, creating unprecedented opportunities for growth in the sector. As a result of this surge in renewable energy, there is an expectation that the 2020s and beyond will witness tremendous growth and expansion in the renewable energy industry as nations seek to meet their emission targets and reduce their reliance on finite energy sources. This in turn will create new jobs, drive technological innovation, and have a significant impact on the global economy over the coming decades.

This growth in the renewable energy industry is also expected to have a direct impact on global climate change, as the shift away from fossil fuels and toward renewable sources is likely to significantly reduce greenhouse gas emissions and help mitigate the most severe effects of global climate change. This shift toward renewable energy sources is an example of the power that collective action can have and serves as an important reminder of the potential for positive change when people come together in pursuit of a common goal. In this way, the shift toward renewable energy sources in the 2020s and beyond can be seen as a cause for optimism in the fight against global climate change and an important step forward in humanity's collective quest for a cleaner, healthier planet.

Renewable energy sources are expected to become more affordable and accessible in the coming decades, resulting in a dramatic reduction in the world's reliance on fossil fuels and a corresponding reduction in greenhouse gas emissions. This trend will have a powerful, positive impact on the global climate, leading to cleaner air and healthier living conditions for future generations. In general, the shift toward renewable energy sources is a clear sign that collective action can lead to meaningful and lasting change in the face of climate change. It is our hope that the transition to renewable energy sources will accelerate in the years to come, as more people recognize the importance of embracing a low-carbon lifestyle and making smarter energy choices.

It is both technically and financially feasible for the entire world to switch to renewable energy for all purposes by 2050, including power, heat, transportation, and desalination (Aklin, 2022).

If we are to realize these goals, however, it will require a shift in mindset from a focus on short-term gain to long-term sustainability and the implementation of policies that support and accelerate the transition to renewable energy. It will also necessitate the collective efforts of individuals, businesses, and governments around the world to invest in renewable energy sources and develop more efficient methods of utilizing them. In addition, to ensure that renewable energy is accessible and affordable for all, we must implement innovative financing solutions and bring about policy changes that will open up access to low-cost capital for the development of renewable energy projects. Globally, this will mean taking steps to reduce fossil fuel subsidies and promote pricing mechanisms that reflect the true cost of carbon emissions, such as carbon taxes and cap-and-trade systems.

In contrast to fossil fuels, which are largely confined to a small number of nations, renewable energy resources are spread across a large geographic area, and consequently, no single country has a monopoly on their production, making them accessible to a much larger population. As such, renewable energy resources have the potential to provide energy to a much larger proportion of people across the globe, reducing reliance on fossil fuels and contributing to improved air quality, reduced greenhouse gas emissions, and better public health outcomes, as well as economic benefits in the form of job creation and economic growth and increased energy security for countries that move away from reliance on fossil fuels. These benefits are so promising that many countries have set ambitious targets for renewable energy implementation policies to promote the development and deployment of renewables.

Energy security, climate change mitigation, and economic benefits are all greatly increased by the use of renewable energy sources and energy-saving technologies, making them attractive options to countries around the world seeking to reduce their dependence on fossil fuels and increase their energy security and sustainability. The widespread adoption of renewable energy resources has a wide range of positive impacts on the environment, public health, and the economy, which can have both short- and long-term implications. As renewable energy sources become more widely available and less expensive, they will continue to grow in popularity and importance around the world, making them an increasingly attractive option for countries looking to reduce their reliance on fossil fuels, improve their energy security, and improve their climate and public health outcomes. To meet the targets set by governments and take full advantage of the potential offered by renewable energy, it is important to ensure that all stakeholders, from citizens to industry leaders, understand the benefits of renewables.

However, hundreds of billions of dollars in subsidies for fossil fuels are impeding the development of renewable energy sources and ensuring a continuation of the current unsustainable pattern of energy consumption. As a result, our planet is being subjected to increased levels of air and water pollution, leading to increased health risks for humans and other species. At the same time, it makes it more difficult for countries to transition away from the fossil fuels that are causing climate change and exacerbating the effects of global warming. Thus, it is essential that subsidies for fossil fuels be eliminated and replaced with subsidies for renewable energy sources to ensure a sustainable energy system and to protect both the environment and human health in the long term. This heavy reliance on subsidies for fossil fuels is a direct cause of the slow progress in developing renewable energy sources, and it poses a significant barrier to transitioning away from fossil fuels and toward renewable energy sources.

In addition, these subsidies make it more difficult for countries to invest in new technologies and research that could make renewable energy sources more cost-competitive and reduce their dependence on polluting fossil fuels in the short run. In general, eliminating subsidies for fossil fuels and replacing them with subsidies for renewable energy sources is an important step in transitioning away from fossil fuels, protecting both the environment and human health, as well as encouraging countries to invest in new technologies and research related to renewable energy sources. To ensure a sustainable energy system in the long term, it is imperative that governments commit to phasing out subsidies for fossil fuels and providing appropriate subsidies for renewable energy sources to create a more sustainable future for the planet and its inhabitants.

The use of renewable energy technology is becoming increasingly common in African nations as access to electricity and energy sources becomes more critical in the region due to rapid population growth and the demand for energy resources rising exponentially. In response to this growing demand, African nations are looking to renewable energy sources such as solar and wind power, as well as hydropower, geothermal, and biomass energy, to meet their energy needs in a sustainable and environmentally friendly way. By utilizing renewable energy sources, African nations can not only reduce their dependence on imported energy resources but also reduce their carbon footprint while providing a reliable source of power to meet the needs of an ever-growing population and stimulating economic growth in the process. As a result, African nations are increasingly investing in renewable energy technologies, with the African Development Bank investing billions of dollars in developing solar and wind power generation capacity across the continent and the African Renewable Energy Initiative planning to install 10 gigawatts of renewable energy generation capacity by 2020 (Akinyemi et al., 2015).

This is a monumental step forward for African nations and their citizens, as access to clean and affordable energy sources can provide much-needed electricity to remote regions, enabling further economic development and improved standards of living while also helping to protect the environment. In addition to investing in renewable energy sources, African nations are also looking to improve energy efficiency through the implementation of energy efficiency standards and the adoption of new technologies such as smart grids that can reduce energy consumption and promote better management of existing energy sources. African nations are clearly taking the initiative in regard to utilizing renewable energy sources, investing in the development of solar and wind power generation capacity, and aiming to install 10 gigawatts of renewable energy generation capacity by 2020 (Akinyemi et al., 2015).

In doing so, African nations are demonstrating a strong commitment to protecting the environment, reducing energy consumption, and enabling economic development to improve the quality of life for their citizens. Furthermore, African nations are also prioritizing energy access and improving energy security by investing in energy efficiency standards and technologies such as smart grids that can help improve the effectiveness of their energy supply chains while at the same time improving the reliability of their electricity supplies by reducing the amount of energy lost in transmission and distribution as well as reducing the risk of outages due to poor energy supply chains.

Additionally, African nations are exploring ways to make energy generation more affordable and sustainable by engaging in renewable energy auctions and investing in the research and development of renewable energy technologies such as tidal power generation, biomass conversion, solar thermal systems, fuel cells, and many others that can help to reduce the cost of energy generation while at the same time enabling African nations to move toward greater energy independence. These efforts have the potential to revolutionize the energy landscape in Africa, creating jobs and providing clean energy sources to millions of people who lack access to reliable electricity while simultaneously reducing the negative impacts of energy generation on the environment across the continent. These initiatives reflect a strong commitment by African nations to reduce energy poverty and environmental harm while enabling economic development and improving the quality of life for their citizens by investing in energy efficient and renewable energy technologies that can help to drive long-term economic growth, job creation, and environmental sustainability in the region. The potential to increase access to electricity and reduce energy poverty through the implementation of energy efficiency standards and technologies, as well as investment in renewable energy sources, is clear and presents a major opportunity to revolutionize the energy landscape in Africa and beyond for the better.

Small-scale geothermal, wind, and solar power systems are already in use in many countries to supply energy to both urban and rural residents who have previously lacked access to reliable electricity. These systems provide a clean and reliable source of energy, allowing communities to use electricity without relying on more polluting methods such as burning wood or kerosene for their energy needs. Additionally, these renewable energy systems can be more cost effective over time when compared to traditional forms of energy, such as diesel generators, and can reduce the cost of energy services for many people. By providing a cleaner, more reliable, and more cost-effective source of energy to underserved communities in Africa and beyond, renewable energy systems can help to revolutionize the energy landscape in many countries, creating a better future for current and future generations of people.

Additionally, the use of renewable energy systems can help to reduce greenhouse gas emissions, mitigate climate change, and improve public health by reducing air pollution from the burning of fossil fuels as well as the burning of biomass for fuel and other traditional sources of energy. This in turn has the potential to provide a much brighter future for millions of people by allowing them to access basic energy services and by allowing countries and regions to create sustainable development paths that are more economically, environmentally, and socially equitable in terms of their impact on society. Renewable energy systems can thus be seen as an important tool in the fight against poverty and environmental degradation, providing a cost-effective and reliable form of energy that can have a positive impact on individuals, societies, and the environment alike. By making renewable energy systems more accessible to all, it is possible to unlock a new wave of development that is sustainable, equitable, and affordable for all, making the world a better place for everyone.

This can be especially beneficial in developing countries, where access to clean energy is often limited and the use of traditional sources of energy leads to serious health and environmental consequences as well as economic stagnation. By embracing renewable energy sources, African countries can have the opportunity to move away from polluting energy sources and toward more sustainable ones, potentially improving public health, reducing air and water pollution, and creating economic opportunities for local communities through job creation and investment in renewable energy technologies. This in turn can help to improve the overall quality of life as well as provide a more reliable and cost-effective form of energy that can reduce dependence on imported fuels, create jobs, and promote economic growth and development. These factors all contribute to an improved quality of life for the people of African nations, leading to better access to healthcare, education, and basic human needs while also helping to reduce the overall environmental impact of energy production in African countries. As a result, there are many advantages to embracing renewable energy sources in African countries that go beyond simply reducing energy-related pollution and providing a more sustainable form of energy that can benefit the population and create economic opportunities. This section provides an overview of renewable energy technologies in Africa, gives examples of renewable energy projects in African countries, and discusses the advantages and drawbacks of renewable energy in African countries.

5.4.2 Overview of Renewable Energy Technologies in Africa

The sun is the source of energy for solar, wind, and hydroelectric power technologies, providing an efficient and environmentally friendly form of energy production. The sun has the potential to meet all of the world's energy needs now and in the future because it emits more energy in a second (3.827×10^{26} J) than all of the world's energy consumption in a year (2.5×10^{20} J), than the entire world uses in a year (2.7×10^{20} J), than the total energy used by humans in a year (4.4×10^{20} J), and than is contained in all the fossil fuels currently used on the planet (3.9×10^{22} J) (Akinyemi et al., 2015). Additionally, the sun is a renewable and sustainable source of energy since it will continue to produce energy in perpetuity without depleting its reserves, making it an ideal option for powering the world in an efficient and environmentally conscious way. The sun's potential to power the world is remarkable, offering an incredibly efficient and sustainable form of energy production that is capable of meeting the world's energy needs not only now but also in the future and beyond. As such, the sun's potential to meet the world's energy needs is unprecedented and represents a major opportunity for humanity to transition away from fossil fuels and toward an efficient, renewable, and sustainable source of energy production that is both cost-effective and environmentally friendly.

African countries can safeguard their people, the environment, and their future economic development by utilizing renewable energy sources because solar energy production emits no direct emissions and does not require refueling, making it a much cleaner and more sustainable option compared to the burning of fossil fuels, which contribute significantly to greenhouse gas emissions and global climate change Moreover, solar energy is abundant and widely available in many parts of Africa, allowing for local access to an inexpensive and reliable source of power. Furthermore, solar energy production is relatively simple to install and maintain, meaning that countries in Africa can quickly set up solar energy production plants and make use of the sun's energy for their own development needs. Furthermore, solar energy production requires little to no maintenance, making it an attractive option for African countries that may not have the technical infrastructure to regularly maintain more complex renewable sources such as wind turbines or hydropower plants. As such, solar energy production is an attractive and sustainable option for African countries that are seeking to reduce their reliance on fossil fuels, safeguard their people and environment from the effects of pollution and climate change, and ensure that their economies have reliable access to a renewable and cost-effective source of energy that can fuel their development ambitions while also contributing to global efforts to reduce greenhouse gas emissions and mitigate the effects of climate change. Solar energy is a particularly attractive option for African countries in the twenty-first century due to its abundance and availability, the simplicity of installation and maintenance, and its cost-effectiveness compared to other renewable sources of energy. In summary, solar energy production is a desirable option for African countries in the twenty-first century because it can improve the sustainability of their energy sources while helping them achieve their development goals, meet global targets for reducing greenhouse gas emissions, and make a positive contribution to the global climate change effort while also helping to improve their people's quality of life by reducing air pollution and providing a reliable source of energy that does not have to be regularly maintained.

5.4.3 Examples of Renewable Energy Projects in African Nations

Examples of renewable energy projects in African nations are numerous and varied, ranging from large-scale hydropower projects such as the Inga Dam in the Democratic Republic of the Congo to smaller-scale initiatives such as solar energy in Rwanda and community forestry programs in Uganda and Kenya. These projects have had a tremendous impact on the lives of Africans by providing reliable, clean energy and helping to reduce the impacts of climate change while simultaneously promoting economic development and environmental conservation. These projects have been instrumental in helping to reduce the economic, environmental, and energy inequalities between African countries and their global counterparts by allowing Africans to take advantage of the abundance of renewable energy sources on the continent, such as solar, wind, hydropower, geothermal, and biomass.

In addition, these projects have enabled African nations to reduce their reliance on more expensive and less efficient forms of energy, such as fossil fuels and nuclear power, and have allowed them to become more independent from outside sources of energy. By taking advantage of the tremendous potential of renewable energy in African nations, these projects have been able to address many of the continent's development challenges while simultaneously providing a pathway toward a more sustainable future for the entire African continent. These projects have not only provided a reliable, clean energy source to African countries, but they have also helped to create more employment opportunities, generate economic growth, and improve the overall quality of life in African nations by providing access to electricity, clean water, and other basic amenities that are essential for a thriving society.

While there are still many challenges to be faced in this area, it is clear that the early initiatives of renewable energy projects have been instrumental in providing African countries with a much-needed boost toward economic and environmental sustainability. Renewable energy projects are becoming increasingly popular in African nations as they look to reduce their carbon footprint and move toward a more sustainable economy while creating opportunities for job growth, improved quality of life, and long-term economic stability for many of their citizens. As renewable energy projects continue to grow in popularity, they will become even more important in providing African countries with an avenue to address development challenges and become more sustainable in the long term. As more African countries move toward embracing renewable energy projects, there will be an increased focus on developing the infrastructure necessary to support these projects and create a more secure, reliable, and affordable energy source for African countries in the future. In doing so, African countries will be able to better provide for their citizens and create a more sustainable economy that is better prepared to face the challenges of the twenty-first century.

By investing in renewable energy projects, African countries are not only creating a more sustainable economy but also taking steps to reduce their environmental impact by providing their citizens with access to clean and affordable energy sources and investing in a more secure and reliable energy system that will benefit their citizens for generations to come. These efforts will lead to better energy security, reduce pollution and environmental impacts, create economic growth through the development of green industries, and improve the overall quality of life for African citizens and their communities, all while mitigating the effects of climate change on the continent. African countries must continue to invest in renewable energy projects, as well as in the infrastructure necessary to support these projects and ensure their success, to create a more sustainable and equitable economic system that will enable them to confront the challenges of the twenty-first century.

5.4.3.1 The Morocco Solar Plan

The Morocco Solar Plan is one example of a renewable energy project in an African country that seeks to improve the lives of its citizens by providing access to renewable, clean energy. This ambitious project aims to provide electricity to 70% of the population by 2020, using solar and wind energy as the primary sources of energy production. It is hoped that the Morocco Solar Plan will reduce the country's reliance on fossil fuels, increase its economic development, and create jobs while reducing energy costs and improving the environment for Moroccans. In addition to these benefits, the Morocco Solar Plan could have a far-reaching impact on neighboring countries in Africa by providing them with access to renewable energy and reducing their reliance on imported fossil fuels. The Morocco Solar Plan is an important step in addressing energy poverty in the region and a model of renewable energy use that could be replicated across Africa. It is clear that the Morocco Solar Plan is an ambitious and potentially transformational project that, if successful, could provide significant economic, environmental, and social benefits to Morocco and the rest of Africa. While the Morocco Solar Plan could have a wide-reaching and positive impact, it is important to note that there are some challenges that must be taken into consideration when assessing the feasibility of the project. These challenges include ensuring the availability of financing, the effective implementation of the project, and establishing reliable access to renewable energy for those living in remote areas of the country. To ensure the success of the Morocco Solar Plan, it is essential to tackle these challenges and ensure that adequate resources are allocated to help finance and implement the project efficiently in a way that meets the country's needs. In conclusion, the Morocco Solar Plan has the potential to bring about much-needed change and progress in Morocco if the country is willing to commit adequate resources and create a strong framework for effective implementation of the project. Despite the challenges and obstacles, the Morocco Solar Plan has great potential to be a powerful tool in providing renewable energy to Morocco, bringing economic growth and increased quality of life for citizens across the nation.

5.4.3.2 Kenya Wind Power

One of Africa's renewable energy initiatives is Kenya Wind Power, which has seen the construction of a number of wind farms in the country, providing a reliable and renewable source of energy. Kenya Wind Power is a renewable energy initiative in Africa, launched by the Kenyan government in 2018 with the goal of bringing wind-powered electricity to Kenyan households and businesses by 2030 (Akrofi, 2021). By the end of 2020, the initiative would have completed nine wind farms, each with a capacity of 20 megawatts, totalling 180 megawatts of wind-generated electricity, which, at the time of the initiative's launch, was intended to be enough to power more than 10 million Kenyan households by 2030 (Akrofi, 2021). The goal of Kenya Wind Power is to provide a clean and affordable source of electricity for the citizens of Kenya and to reduce the country's dependence on nonrenewable energy sources such as coal, oil, and natural gas.

Kenya Wind Power has the potential to be a real game changer for Kenya, and it is already making great strides toward its goal of providing renewable energy to its citizens. Kenya Wind Power, by utilizing wind-generated energy, has the potential to lower energy costs for Kenyan households while also reducing emissions caused by the use of nonrenewable energy sources. Kenyans now have access to more reliable and renewable energy sources, making their electricity supply more affordable and secure while reducing their carbon footprint and providing a long-term solution to the country's energy needs. In addition to reducing emissions and costs, Kenya Wind Power has the potential to create jobs in the renewable energy sector and stimulate economic growth in Kenya by providing more reliable and accessible energy sources, stimulating economic growth, and reducing emissions associated with the use of nonrenewable energy sources.

With these benefits in mind, it is no wonder that the Kenyan government has put such a strong emphasis on the importance of Kenyan wind power and its potential for positively impacting the environment and economy of the country overall. As such, Kenya Wind Power has the potential to be a game changer for Kenyan households, the environment, and the economy all at the same time by providing a clean, sustainable energy source that is both cost-effective and reliable while creating jobs, spurring economic growth, and reducing emissions associated with the use of nonrenewable energy sources. This is a tremendous opportunity for Kenya to lead the way in renewable energy and is an exciting step toward a future of clean, reliable, and affordable electricity for all of its citizens. Not only does Kenyan wind power have the potential to bring about cost savings, job creation, and economic growth for Kenya, but it also has the capacity to benefit the environment by significantly reducing emissions associated with nonrenewable energy sources such as coal and oil, making it an attractive option for those looking to take action on climate change.

Kenya Wind Power represents an incredible opportunity for the country to take meaningful steps toward a more sustainable and environmentally friendly future, with so many potential benefits. Moreover, by creating a more efficient and resilient electricity infrastructure for the country, Kenya Wind Power will help to improve reliability and reduce the cost of electricity for households and businesses alike, which in turn could help spur economic growth while also helping to improve access to electricity in rural areas of the country where access is limited due to a lack of infrastructure. In conclusion, Kenya Wind Power has the potential to bring about many positive changes in the country and is a great opportunity for Kenyans to lead the way in renewable energy development and pave the way for a more sustainable and equitable future for the nation as a whole. This is why the government of Kenya should strongly consider investing in and expanding the development of Kenya Wind Power to bring about a range of economic, social, and environmental benefits for the country in the long term.

5.4.3.3 Algeria Solar Thermal and Carbon Capture and Storage Technology Projects

With an estimated 170 TWh of annual solar energy production, Algeria has the highest technical and financial potential for solar energy exploitation in the North African region due to its large geographical area, abundance of sunlight, and relative lack of urban infrastructure, making it an ideal location to install solar energy infrastructure. This high potential for solar energy production is due to the country's unique geography and climate, which provide an ideal landscape for solar power plants with over 3000 hours of sunlight per year as well as a large area of land with little urban infrastructure, which makes it possible to install large solar power plants with ease (Akrofi, 2021). The natural environment of Algeria provides a suitable environment for solar power plants to produce significant amounts of energy, and with the government's commitment to reducing their reliance on traditional energy sources, Algeria is well placed to become a leader in solar energy production in the region.

To take full advantage of this potential, the government of Algeria is investing in renewable energy sources such as solar energy and encouraging private companies to do the same. In addition to its potential for solar energy production, Algeria has a well-developed infrastructure for research and development, as well as technical expertise in renewable energy production that can be leveraged to further develop the solar energy industry in the country. Additionally, the government is incentivizing research and development in renewable energy sources and providing tax credits and subsidies for companies investing in solar energy production initiatives. All of these efforts are helping Algeria position itself as a leader in the renewable energy sector, not only for its own benefit but also for the benefit of the entire region. In addition to its large potential for solar energy production, Algeria has an extensive infrastructure for grid integration of renewable energy sources and relatively low investment costs for solar energy development, making it an attractive place for renewable energy investors and an ideal location for solar energy investment and development.

This combination of advantages gives Algeria an advantage in terms of renewable energy production and is likely to lead to a rapid expansion of the country's solar energy industry in the coming years. This expansion of the renewable energy sector has already begun, and Algeria has plans to increase the amount of renewable energy it produces tenfold over the next two decades. To meet this goal, Algeria has invested heavily in both renewable energy research and development as well as the construction of solar power plants and other renewable energy infrastructures. It is currently investing in the renewable energy sector to meet its growing energy demand and help reduce its reliance on fossil fuels while also mitigating its environmental impact and, in turn, creating a more sustainable future for the country and its people.

The first solar thermal power project on an industrial scale started with the opening of the Hassi R'Mel power station in 2011, and since then, the growth of renewable energy in Algeria has been steadily increasing (Boulogiorgou & Ktenidis,

2020). This brand-new hybrid power plant combines a 130-megawatt combined cycle gas turbine plant with a 25-megawatt (MW) concentrating solar power array that uses large mirrors to focus sunlight and create heat, which is then used to generate electricity, making it one of the first in the world to combine both solar and gas turbines for power generation, and it is expected to generate enough electricity to power over 300,000 homes (Obiwulu et al., 2022). In addition to solar power plants, Algeria is also investing in wind farms and other forms of renewable energy, such as biomass and geothermal sources. Algeria is looking to reduce its reliance on fossil fuels and transition to a more sustainable future by increasing investment in renewable energy. With the combination of wind farms, solar power plants, biomass, and geothermal sources, Algeria is well on its way to reducing its dependence on fossil fuels and making strides toward a more sustainable energy future. The hybrid power plant in Algeria is a prime example of the strides being made toward a more sustainable energy future and demonstrates the country's commitment to a greener, cleaner, and more sustainable future. As a result of this commitment, Algeria is on track to achieve energy security and reduce its environmental footprint while simultaneously stimulating economic development through investment in renewable energy technology and infrastructure.

A national program to develop renewable energy based on photovoltaics (PV), concentrated solar power (CSP), and wind power as well as promote energy efficiency was also launched by Algeria in 2011 (Nwokolo et al., 2023b). By 2030, the program aims to install up to 12 GW of renewable energy capacity to generate enough electricity for domestic use and export a surplus of electricity to neighboring countries. The hybrid power plant is the cornerstone of this endeavor, and it is expected to be the largest in North Africa when it is completed. The hybrid plant will include 1.4 GW of PV, 600 MW of CSP, and 600 MW of wind power. This ambitious project is estimated to cost up to $10 billion and will be funded by the government, private sector investors, and international donor agencies. Construction of the hybrid power plant will begin in 2021 and is scheduled to be completed by 2025. This will be a monumental undertaking, as the hybrid power plant will become the largest of its kind in North Africa and produce over 2.6 GW of renewable energy (Boulogiorgou & Ktenidis, 2020). The project is expected to create approximately 1500 permanent jobs and an additional 10,000 indirect jobs in the construction industry in the region and contribute to economic growth by providing reliable and cost-effective power supplies. This includes plans to develop 600 MW of photovoltaics, 800 MW of concentrated solar power, and 10 GW of wind energy over the next 20 years (Boulogiorgou & Ktenidis, 2020).

This ambitious goal is expected to create thousands of jobs in the renewable energy sector, reduce energy costs for businesses and consumers, decrease greenhouse gas emissions and dependence on fossil fuels, contribute to economic growth, and provide energy security for Algeria. To achieve this goal, Algeria has pledged to invest an estimated $250 billion in the renewable energy sector and plans to increase the electricity generation mix from fossil fuels to 15 percent renewables by 2030. The hybrid power plant serves as a beacon of hope for the people of Algeria, demonstrating that the country is serious about shifting to renewable energy sources for a cleaner and more secure energy future. With the introduction of this ambitious plan, Algeria committed to an energy revolution that will drive economic growth and create a cleaner, more sustainable environment for its citizens. As the world continues to face an unprecedented climate crisis, it is essential for countries such as Algeria to lead the way in transitioning away from fossil fuels and toward renewable energy sources.

Algeria has implemented a number of energy reforms to take advantage of this potential, including the 2016 Renewable Energy Law, which offers incentives and subsidies for solar power projects, and the 2018 National Energy Transition Plan, which is aimed at achieving a 50 percent share of renewable energy in the country by 2030. These initiatives have created an enabling environment for the hybrid power plant, which demonstrates the government's commitment to investing in renewable energy sources and is a promising step forward for the country. The hybrid power plant, located in the country's southwest, is part of Algeria's effort to reduce its dependence on fossil fuels and achieve an ambitious goal of transitioning to 50 percent renewable energy by 2030. The hybrid power plant, made up of both solar and wind energy sources, is expected to produce a total of 18.5 megawatts of electricity when it is fully operational by the end of 2021 projects and the 2018 National Energy Transition Plan, which is aimed at achieving a 50% share of renewable energy in the country by 2030 (Boulogiorgou & Ktenidis, 2020).

These initiatives have created an enabling environment for the hybrid power plant, which demonstrates the government's commitment to investing in renewable energy sources and is a promising step forward for the country. The hybrid power plant, located in the country's southwest, is part of Algeria's effort to reduce its dependence on fossil fuels and achieve an ambitious goal of transitioning to 50 percent renewable energy by 2030. The hybrid power plant, made up of both solar and wind energy sources, is expected to produce a total of 18.5 megawatts of electricity when it is fully operational by the end of 2021. It is estimated that the hybrid power plant could reduce carbon emissions by over 40,000 metric tons per year once it is fully operational (Boulogiorgou & Ktenidis, 2020). This is a remarkable achievement and demonstrates Algeria's commitment to becoming a global leader in renewable energy production. In addition to reducing its carbon footprint, the hybrid power plant will also help reduce the country's dependence on fossil fuels and bring a much-needed source of electricity to the region. The hybrid power plant is not only beneficial to the environment, but it will also provide a much-needed boost to Algeria's economy, as it creates jobs in the region and brings in new sources of income through the sale of electricity to other countries.

The hybrid power plant is a remarkable achievement, demonstrating Algeria's commitment to renewable energy production and commitment to reducing the country's dependence on fossil fuels while also providing a much-needed source of electricity to the region. In addition to the environmental and economic benefits, this project demonstrates Algeria's commitment to becoming a leader in renewable energy production and its desire to join the global movement toward sustainable energy sources by leading the charge in innovative energy production solutions. The hybrid power plant will be a significant milestone for Algeria, proving its dedication to the fight against climate change and paving the way for other countries to follow suit in their efforts to reduce emissions and create a more sustainable and healthier environment for future generations.

The hybrid power plant is a bold step forward for Algeria and a shining example of the nation's commitment to renewable energy production and its dedication to preserving the natural environment for future generations. It is a remarkable example of how Algerian ingenuity and forward thinking can lead to lasting positive changes for the country and the world at large. Not only will the project bring ecological, economic, and environmental benefits to Algeria, but it will also serve as a beacon of hope for other countries looking for innovative solutions to the global climate change crisis. Through this project, Algeria will set a standard for sustainable development and provide an inspiring example of what can be accomplished with innovation, dedication, and commitment to reducing emissions and protecting the environment. This project is a true testament to the nation's willingness to take decisive action in combating climate change, will serve as an example for other nations around the world to follow, and will demonstrate to the world how a small nation can have a significant impact on a global scale.

The Algerian Government's commitment to the Algiers Carbon Capture and Storage Project is a testament to the country's willingness to step up and make a positive difference in the world by investing in a project that will not only benefit the people of Algeria but will also have positive implications for the global environment and the world as a whole. This project demonstrates the importance of investing in sustainable development and should serve as a reminder to other countries of what can be accomplished when nations come together to combat climate change and work toward creating a healthier, more sustainable future for generations to come. The Algiers Carbon Capture and Storage Project is a clear example of the Algerian government's dedication to making a positive impact on the environment, both within their own nation and around the world. By investing in this project, Algeria will set an example for other countries to follow and prove that a small nation can have a significant impact on a global scale in regard to environmental protection and sustainability initiatives.

By investing in this project, Algeria is sending a clear message to the world that it is committed to doing its part in regard to creating a better future for the planet and its inhabitants. Algeria is sending a clear message that it is dedicated to leading the way in addressing climate change and ensuring a more sustainable future for generations to come. Algeria's investment in the Algiers Carbon Capture and Storage Project not only has global implications but also provides direct benefits for their own people by allowing them to develop and use renewable energy resources, reducing the environmental impact of burning fossil fuels, and creating new jobs that will help diversify the country's economy and provide more economic stability for its citizens. While these are all important considerations, the most powerful impact of this project is its symbolic significance as a demonstration of Algeria's commitment to environmental progress and a sustainable future for the world.

The Algiers Carbon Capture and Storage Project is a major step forward for the country of Algeria and serves as a beacon of hope to other nations across the globe that are looking for ways to reduce their emissions and be more responsible stewards of the environment. By investing in this project, Algeria has made it clear that it is dedicated to combating climate change and creating a more sustainable future for both its own citizens and the world as a whole. This investment will have long-term ramifications for not only Algeria but also the world as a whole, as the country's example can be followed by others looking to reduce their emissions and make responsible choices toward a sustainable future. As such, the Algiers Carbon Capture and Storage Project is a powerful symbol of Algeria's commitment to protecting its environment and safeguarding the future of the world for generations to come.

5.4.3.4 South African Projects Utilizing Solar and Thermal

Construction of the 50 MW Touwsrivier CPV Solar Project concentrated photovoltaic (CPV) power plant began in December 2014 in Touwsrivier, South Africa's Western Cape Province, and the project was officially completed in October 2015 (Fatoki, 2020). With an estimated cost of 5.5 billion Rand, the Touwsrivier CPV Solar Project is the first commercial-scale CPV power plant in Africa and is expected to generate enough energy to power 40,000 households for a period of twenty years (Fatoki, 2020). The Touwsrivier CPV Solar Project is an impressive feat of engineering that has changed the way South Africans view renewable energy and serves as an example to other countries of what can be achieved when renewable energy is harnessed efficiently and effectively. The Touwsrivier CPV Solar Project is a milestone for Africa, as it is not only the first large-scale CPV power plant on the continent but also a step toward renewable energy sources becoming the primary source of electricity for African countries.

In addition, the Touwsrivier CPV Solar Project is a testament to the power of collaboration and cooperation, as it was a joint venture between the Industrial Development Corporation of South Africa, Tata Power Solar, and other partners who worked together to make this ambitious project a reality. On September 13, 2013, Kalkbult, in the Northern Cape, inaugurated a 75 MW solar power plant (implemented by Scatec), which has the capacity to provide enough electricity to meet the needs of 70,000 households. The same company finished construction on two additional PV plants in 2014, making the total amount of electricity provided by the three plants a total of 197.6 MW, enough to power up to 160,000 households in South Africa. This project marks a significant moment for South Africa, as the country's energy sector is transitioning from largely fossil fuel-based to more renewable energy sources, and this project will provide a significant contribution toward achieving the goal of 13.8 GW from renewable energy sources by 2030. These are situated in the sun-drenched Linde region in the Northern Cape and Dreunberg in the Eastern Cape, which both have some of the best conditions for solar power production in the world. This project is particularly meaningful for South Africa, as it will not only provide enough electricity to power nearly a quarter of the households in the country but also contribute significantly toward meeting their 2030 renewable energy goals.

Additionally, the three PV plants are part of a larger trend of transitioning away from fossil fuels and toward renewable sources of energy in South Africa, which is an important step toward helping the country achieve energy independence, create green jobs, and reduce its greenhouse gas emissions. Approximately 90,000 homes in South Africa receive their electricity from these three plants on a daily basis, and the plants are estimated to save up to 140,000 tons of CO_2 emissions per year. This project is a milestone for South Africa in its efforts to reduce its dependence on fossil fuels and move toward renewable energy sources. The completion of the three PV plants is a major success for South Africa, not just in terms of providing enough electricity to power thousands of households but also as a step toward achieving their 2030 renewable energy goals and providing a viable, sustainable alternative to fossil fuels.

The South African government has set a target to produce 42% of its electricity from renewable sources by 2030, and the successful completion of these three solar PV plants is an important part of reaching this goal. In addition to providing electricity for up to 90,000 households, these plants are also responsible for significantly reducing CO_2 emissions in the atmosphere and contributing to a healthier environment for the people of South Africa (Fatoki, 2020). The successful completion of these three solar PV plants is a major milestone for South Africa in its efforts to reduce its dependence on fossil fuels, move toward renewable energy sources, and create a more sustainable future for its citizens. The success of these three plants marks a major step forward in the journey toward achieving South Africa's 2030 renewable energy goals and is a testament to the country's commitment to taking action against climate change and creating a greener, more sustainable environment for its citizens.

In Northern Cape province, Sonnedix and Juwi plan to build an 86-MW solar photovoltaic (PV) plant that is expected to produce enough electricity to power over 90,000 households and contribute to reducing CO_2 emissions in the atmosphere by approximately 94,000 tons per year (Fatoki, 2020). The Mulilo Sonnedix plant in Prieska, which was chosen through the third round of South Africa's Renewable Energy Independent Power Producer Program, will be larger than any PV plant currently operational on the African continent (REIPPP). The project's financial close is anticipated for July 2014, and commissioning is set for the second half of 2016. Additionally, the two companies have also signed an agreement with the South African government to build a 17-MW photovoltaic plant in De Aar, located in the Northern Cape Province.

The Mulilo Sonnedix plant and the De Aar photovoltaic plant will have a considerable impact on South Africa's commitment to renewable energy, as they are expected to produce a combined 103 MW of electricity over the lifetime of the projects. Together, these two projects will produce enough electricity to power the equivalent of 87,000 South African households, or a total of 217,000 people. The expected completion date for both the Mulilo Sonnedix plant and the De Aar photovoltaic plant is July 2016. Meanwhile, the Ivanpah Solar Electric Generating System, located in California's Mojave Desert and managed by BrightSource Energy, consists of three 450-foot solar towers that can each generate up to 77 MW of electricity (Fatoki, 2020).

In total, the Ivanpah Solar Electric Generating System has a production capacity of more than 400 MW, enough to power 140,000 homes. That is, an incredible amount of electricity, enough to power a total of 357,000 South African households (Fatoki, 2020). With such a large capacity, the Ivanpah Solar Electric Generating System is projected to significantly reduce South Africa's carbon emissions and contribute to its growing renewable energy industry. As the Ivanpah Solar Electric Generating System continues to grow and its production capacity increases, it is estimated that it will produce an estimated 2.3 million megawatt hours of renewable energy annually, significantly contributing to South Africa's efforts toward becoming energy independent and reducing its carbon emissions.

According to a press release from Iberdrola, the 96-MW Jasper Solar Energy Project solar photovoltaic (PV) power plant is now complete, the largest plant in Africa, and is expected to produce enough clean energy to power approximately 130,000 South African households.

Over 325,000 PV modules were installed during the operation, which took place in an area of desert with high temperatures, making it the perfect location for the largest PV power plant in Africa. The Jasper Solar Energy Project is a great addition to South Africa's renewable energy portfolio, and its output has the potential to have a significant impact on the country's carbon emissions. This PV plant is situated in a remote, semidesert area of the Northern Cape province. The area it covers is equal to 205 football fields. The project started in 2013, and since then, it has provided more than 310 MW of solar energy to the region while avoiding an estimated 560,000 tons of CO_2 annually. In addition to the environmental benefits, this project has created jobs for the local community, with approximately 730 people employed during its construction phase (Fatoki, 2020). In total, the PV plant is estimated to have created 2500 direct and indirect jobs in the region during its construction phase, and it is expected to have an even greater impact on the local economy in the future. Since its completion, the solar energy project has had a large positive impact on the region and the environment. It is estimated that the project will offset 1.6 million tons of CO_2 annually, and the total amount of electricity produced is equivalent to the annual consumption of approximately 460,000 households (Fatoki, 2020).

In December 2014, Acciona inaugurated the Sishen solar photovoltaic (PV) plant in the Northern Cape Province of South Africa. The plant is the largest of its kind on the African continent and, when fully operational, will have a total installed capacity of 100 MW and a maximum output of 94.3 megawatts (MWp)—74 nominal MW, providing enough electricity to power over 460,000 households. By tracking the sun's path across the sky, its horizontal tracking structures enable the PV panels to capture more radiation than a fixed-tilt system. The plant, which is part of South Africa's Renewable Energy Independent Power Producer Program (REIPPP), is expected to offset 1.6 million tons of CO_2 emissions annually and contribute significantly to reducing the country's dependence on fossil fuels. The plant will produce enough electricity to supply 420,000 people, or approximately 100,000 South African households, with electricity (the 2011 South African household size is 3.6 people) (Fatoki, 2020). The plant is the largest of its kind on the African continent and, when fully operational, will have a total installed capacity of 100 MW and a maximum output of 94.3 megawatts (MWp)—74 nominal MW, providing enough electricity to power a small town for nearly 20 years (Fatoki, 2020).

The plant will be the largest of its kind on the African continent, providing clean energy that will help reduce South Africa's dependence on fossil fuels and contribute significantly to reducing CO2 emissions. A total of 593 MW generated by PV plants was connected to the grid as of January 2015. The plant is expected to have a lifespan of approximately 20 years, with minimal maintenance costs and no emissions. Additionally, the plant is expected to create employment for over 200 people, both directly and indirectly. The plant is estimated to generate 1.2 gigawatts of electricity, enough to power over 800,000 homes in South Africa and beyond. As a result, the PV plant is expected to provide clean, sustainable energy to over 800,000 households while simultaneously reducing South Africa's dependence on nonrenewable energy sources.

A 50-MW concentrating solar power (CSP) project in South Africa's Northern Cape was successfully commissioned by a group led by ACWA Power International (Saudi Arabia) in December 2015. The project, which costs approximately $740 million, is South Africa's first solar thermal energy power plant and is estimated to reduce carbon dioxide emissions by over 690,000 metric tons each year. The 1.2 gigawatts of electricity generated by the PV plant will therefore join the 50 MW CSP plant to significantly reduce South Africa's reliance on nonrenewable energy sources and provide much-needed clean, sustainable energy to millions of households. As such, CSP and PV plants, which together cost almost $1.5 billion, are expected to have a large positive impact on South Africa's energy future while significantly reducing its reliance on nonrenewable energy sources.

The 50-MW CSP project is projected to provide clean energy to over 100,000 households while also reducing South Africa's reliance on nonrenewable energy sources by over 690,000 metric tons annually. The joint venture of the 50 MW CSP and 1.2 gigawatt PV projects is South Africa's largest renewable energy project and will no doubt bring significant benefits to the country. By providing clean, sustainable energy to millions of households, the 50 MW CSP and 1.2 gigawatt PV projects are expected to have a huge positive impact on South Africa's energy future. Furthermore, a 100-MW photovoltaic (PV) plant is currently being constructed in De Aar in the Northern Cape that is expected to reduce emissions of carbon dioxide (CO_2) by more than 100,000 metric tons a year. The 50 MW CSP project is a key part of South Africa's ambitious renewable energy program, and its success will depend on the joint efforts of all stakeholders involved in the project (Fatoki, 2020). The joint venture of the 50-MW CSP and 1.2-gigawatt PV projects is expected to not only provide clean and sustainable energy to millions of households but also create job opportunities and foster economic development in the area by enabling economic development, creating job opportunities, and reducing carbon emissions.

The 50-MW CSP and 1.2-gigawatt PV projects are expected to significantly improve energy access and security in South Africa, providing clean energy to millions of households and creating job opportunities while also reducing carbon emissions by more than 100,000 metric tons a year. The De Aar PV plant is expected to be completed in 2018, which will make South Africa a leader in renewable energy production. Bokpoort CSP will be able to meet demand during peak hours between 5 and 9 p.m. due to its 9.3 hours of thermal energy storage capacity, allowing electricity to be produced when the sun is not shining. The joint venture of the 50-MW CSP and 1.2-gigawatt PV projects is expected to bring many benefits to the people of South Africa, not just in terms of energy but also for economic development and job creation (Fatoki, 2020).

The joint venture of the 50-MW CSP and 1.2-gigawatt PV projects is expected to provide much-needed benefits to South Africa, such as energy access, job opportunities, and economic development. The two projects combined are estimated to bring an additional 5.5 billion South African rand into the economy annually, generating a total of 5800 jobs directly and indirectly during the construction phase. Additionally, the projects are expected to bring social benefits to the people of South Africa, such as access to energy for communities that have not had access before and the potential for businesses to thrive in areas previously unable to benefit from renewable energy production.

According to the ACWA, the largest thermal storage system ever constructed for a CSP plant of this size and capacity is this one in South Africa. Furthermore, these two renewable energy projects are estimated to offset 3.2 million tons of CO_2 annually, further contributing to South Africa's goal of becoming a more sustainable and responsible producer of electricity (Fatoki, 2020). A new African record for uninterrupted, nonstop electricity supply has been set by the South African solar power plant. The Khi Solar One and Xina Solar One projects in South Africa have already begun to make a large difference in the country's commitment to renewable energy. Both projects have the ability to store energy for up to seven and a half hours without interruption, making them South Africa's largest solar thermal storage systems.

The Khi Solar One and Xina Solar One projects are two large-scale renewable energy projects that have revolutionized the way South Africa produces electricity by relying on the sun's energy to power the country instead of relying on nonrenewable resources. The two projects are projected to provide a total of 9.3 gigawatts of power when completed and will power more than 1.8 million households in the country once the projects are completed. This renewable energy system is capable of providing electricity to more than 150,000 households and businesses in the South African provinces of the Northern Cape and Free State through its innovative solar thermal storage technology, which uses molten salt to store thermal energy.

Abengoa is constructing two CSP plants in the Northern Cape province of South Africa, and they are expected to be completed in 2019. Khi Solar One is a 50-MW solar power plant located close to Upington, and Kaxu Solar One is a 100-MW parabolic trough plant located close to Pofadder with a 3-hour storage capacity (Fatoki, 2020). Upon completion, the two projects will provide a total of 9.3 gigawatts of power and be capable of providing electricity to more than 1.8 million households and businesses in the country. Pofadder is also close to Xina Solar One, which has a 100 MW overall installed capacity and will be the largest solar power plant in South Africa. This renewable energy system has a major impact on the lives of people in South Africa, providing sustainable electricity for up to 150,000 households and businesses (Fatoki, 2020).

Abengoa solar energy plants are expected to reduce the country's greenhouse gas emissions by approximately 300,000 tons annually by utilizing the sun's energy. In addition, these plants are expected to create up to 1500 jobs during the construction phase and hundreds of other permanent jobs for the maintenance and operation of the plants. Abengoa solar energy plants have the potential to revolutionize renewable energy in South Africa (Fatoki, 2020), providing a clean, reliable, and affordable source of energy to the population while at the same time providing long-term economic benefits through job creation and investment in the local economy. The Abengoa solar energy plants represent an important step toward the development of a sustainable energy infrastructure in South Africa, one that will provide a clean and reliable source of energy for years to come.

5.5 Smart Grid Technology

Increasing electric grid access in Africa is a key component of the continent's transition to net-zero emissions by 2050 and is essential for the development of African economies. Improving access to electricity is a crucial step in reducing emissions and meeting climate goals in Africa, as it would allow households to switch from traditional sources of energy, such as burning wood and kerosene, to cleaner and more efficient electricity alternatives, such as solar or wind, which both produce energy with zero emissions and are also more cost-effective than traditional energy sources. Additionally, access to electricity would allow households to benefit from technological advances such as access to the Internet, which can lead to increased education and economic opportunities and help countries become more competitive in the global economy. By increasing access to electricity, African households and communities can become more connected, efficient, and able to reduce emissions while also improving educational and economic opportunities and achieving the ultimate goal of net-zero emissions by 2050. As such, investments in infrastructure and technology to improve access to electricity across the African continent are essential for countries to reduce emissions, become more connected and prosperous, and reach net-zero emissions by 2050. As the world works to reduce emissions and meet its climate goals, improving access to electricity in Africa is critical for African countries to benefit from technological advances, reduce emissions, and make progress toward net-zero emissions by 2050. This will require a significant investment in infrastructure and technology, but the benefits far outweigh the costs in terms of economic and social progress, climate stability, and improved quality of life for the African population.

In general, investment in the African electricity grid is essential if African countries are to reduce emissions, become more connected and prosperous, and meet their climate targets in a sustainable and cost-effective manner. Investment in infrastructure and technology to improve access to electricity in Africa is thus crucial for its development and should be prioritized by countries as they look to reduce emissions, make progress toward net-zero emissions by 2050, and increase the quality of life for the African population. A major challenge for African countries is the lack of access to reliable, affordable electricity, which can hinder economic growth, limit job opportunities, and reduce access to services such as education and healthcare. Therefore, there is an urgent need for investment in energy infrastructure and renewable energy technologies to improve access to electricity, reduce emissions, and ultimately increase the economic and social welfare of African countries.

To meet these goals, African countries must work together with the international community to ensure that investments are made in energy infrastructure and renewable energy technologies that are financially and technologically feasible, as well as environmentally sound and socially equitable. African countries must also invest in research and development to develop and adopt renewable energy technologies that are suited to local conditions and can be deployed quickly, such as solar PV or mini-grids. To achieve these goals, African countries need to cooperate with the international community to ensure that any investments are financially and technologically feasible while also ensuring that they are environmentally sound and socially equitable. This means that African countries must focus on developing an integrated strategy for energy infrastructure and renewable energy technologies that are not only cost-effective but also have the potential to reduce emissions, promote economic growth, and improve access to essential services. African countries must collaborate with the international community to identify and secure funding for these investments and develop strategies for transitioning to a low-carbon economy that can support their economic development while reducing the emissions of climate-altering gases.

African countries must also focus on developing an integrated energy strategy and plan for renewable energy technologies that include considerations for cost-effectiveness, environmental sustainability, economic growth, social equity, and access to essential services while also ensuring that the investments made are technically feasible and financially affordable for African countries. This collaboration and integration of strategies must be accompanied by the commitment of African countries to implement their energy plans to improve energy access, reduce emissions, and promote economic growth to ensure a successful transition to a low-carbon economy that can sustainably develop their countries for years to come. African countries must strive to build an effective and well-rounded energy strategy that takes into account the various considerations of cost-effectiveness, environmental sustainability, economic growth, social equity, and access to essential services and ensures that the investments made and strategies developed are both technically feasible and financially affordable for their citizens.

To achieve this, African countries must work in partnership with the international community to obtain the financial resources needed to invest in renewable energy technologies and develop an integrated energy strategy that is tailored to their individual countries' needs. In addition, African countries must prioritize the human aspect of their energy strategy, including access to clean and affordable energy services for all citizens, the adoption of renewable energy technologies, the promotion of jobs and entrepreneurship in the energy sector, and the development of a strong regulatory framework that encourages investment and innovation in renewable energy technologies and energy efficiency measures. These strategies should be implemented in a manner that prioritizes the inclusion of women, who are often at the forefront of providing energy access to their families and communities, while also taking into account the impact on vulnerable populations, such as those in rural areas who may be unable to access energy due to poor infrastructure or limited resources and those in developing countries who are most affected by climate change and its associated risks.

Ultimately, African countries must build a holistic energy strategy that is tailored to their individual needs and considers all factors of cost-effectiveness, environmental sustainability, economic growth, social equity, access to essential services, and the impact on vulnerable populations to ensure that they can continue to make progress on their development goals and provide an equitable and secure future for all. This can only be achieved by partnering with the international community and investing in renewable energy technologies, as well as prioritizing access to clean and affordable energy services for all citizens, promoting jobs and entrepreneurship in the sector, implementing a strong regulatory framework that encourages investment and innovation in renewable energy technologies, and ensuring that the strategy is tailored to the individual needs of each African country and considers all factors of cost-effectiveness, environmental sustainability, economic growth, social equity, access to essential services, and the impact on vulnerable populations to ensure that no one is left behind.

5.6 Modern Bioenergy

In 2050, 25% of the world's primary energy supply will come from bioenergy, which includes liquid biofuels, solid biomass, biogas, and biomethane, according to IRENA's 1.5 °C scenario (Aklin, 2022). This percentage is more than double the current use of bioenergy sources, which only account for 11% of energy usage. With more people looking for clean, renewable energy sources to reduce emissions, the use of bioenergy will continue to rise. This rise in bioenergy usage can be attributed to both government initiatives and private companies working together to increase research into more efficient and sustainable bioenergy sources. In the past, bioenergy has been used primarily in small-scale applications such as cooking and heating, but advances in technology have allowed for larger-scale applications such as electricity generation and transportation fuels.

In 2018, the International Energy Agency estimated that bioenergy accounts for 17% of global energy consumption, which is a large increase from the 10% estimated in 2006 (Aklin, 2022). These technological advances and increased investment have enabled bioenergy to become an important contributor to the global energy mix, and with more renewable energy sources becoming available, it is expected that the usage of bioenergy will continue to increase in the coming years.

As a result of these advances, bioenergy is becoming increasingly attractive to both governments and private companies due to its environmental benefits, cost-effectiveness, and ability to help meet renewable energy targets. In the past decade, bioenergy has seen a rapid expansion in its application and usage, driven by technological advancements and increased investment from both the public and private sectors, which has enabled the generation of bioenergy to become cost-competitive with other energy sources such as fossil fuels.

This expansion has enabled bioenergy to become a viable alternative to other forms of energy, and its increasing usage has resulted in improved air quality and reduced carbon emissions, leading to a decrease in global warming. In addition, bioenergy is becoming increasingly attractive to governments and private companies because it can help meet renewable energy targets and offer environmental benefits such as improved air quality, reduced carbon emissions, and a decrease in global warming, making it a viable, cost-effective, and attractive option for energy production. The numerous benefits of bioenergy have led to its increasing popularity, with governments and private companies increasingly turning to it as a viable alternative energy source. In general, the rapid growth of bioenergy over the last decade has been driven by technological advances, increased investment, and its numerous environmental benefits, such as reduced reliance on fossil fuels, improved air quality, and reduced global warming, making it an appealing and viable option for energy production. All of these factors have contributed to the increasing popularity and widespread use of bioenergy in recent years, making it an attractive and viable option for energy production in many countries around the world.

Despite these many advantages, there are still some drawbacks to the use of bioenergy that must be addressed if bioenergy continues to remain a viable energy source in the future. For example, bioenergy can often be more expensive than fossil fuels, and the production of bioenergy can have negative impacts on land, water, and air quality, as well as lead to deforestation due to the clearing of land for energy production facilities. There are also concerns about the long-term sustainability of bioenergy, given that it is dependent on the availability of biomass resources, which are finite and vulnerable to climate change. Furthermore, the production of bioenergy involves the burning of biomass, which produces emissions that can contribute to climate change as well as air pollution If these issues are not addressed, it is possible that bioenergy could become a less desirable option for renewable energy generation in the future. Consequently, it is important to consider the environmental impacts of bioenergy production to ensure that it can be done sustainably and safely.

To ensure the sustainability of bioenergy, it is important that proper management strategies are employed to maximize its efficiency and minimize the negative environmental impacts associated with it. Such strategies could include the implementation of technologies that increase the efficiency of bioenergy production, as well as measures to reduce the emissions generated from burning biomass and the development of bioenergy sources that utilize more environmentally friendly feedstocks such as organic waste or agricultural residues, as well as other techniques to ensure the sustainability of bioenergy resources such as crop rotation, fertilization, and conservation practices. These management strategies are essential to ensuring that bioenergy can remain a viable source of renewable energy in the future while minimizing the environmental impact of its production and consumption. Additionally, careful land use and resource management practices should be employed to ensure that land is used efficiently, reducing the amount of land required for bioenergy production and thereby reducing the amount of land devoted to bioenergy production and consequently reducing any potential adverse environmental impacts associated with its production.

According to IRENA's 1.5 °C scenario, by 2050, bioenergy, which includes solid biomass, biogas and biomethane, and liquid biofuels, would account for 25% of the world's primary energy supply (Aklin, 2022). These green energy resources are required to support the African continent's transition to net zero while still ensuring that it has access to the energy and other resources it needs to develop economically and improve the lives of its citizens. To achieve this ambitious goal, it is essential that governments and policymakers implement effective management strategies for the production and use of bioenergy in Africa. This includes strategies such as developing land-use policies that protect ecosystems and ensure sustainability, establishing renewable energy standards and incentives for bioenergy production, and providing adequate financing for bioenergy production projects on the continent.

These strategies should also prioritize the adoption of technologies that are proven to be environmentally and socially beneficial, such as those that make use of sustainably sourced feedstock and energy efficiency improvements, while encouraging the development of new and innovative solutions that will benefit both African nations and the environment. In addition, it is important to ensure that these management strategies are transparent and accountable to both local and international stakeholders so that there is confidence that they are being implemented with the best interests of both the environment and people in mind. To achieve this goal, governments and policymakers must develop and implement a comprehensive suite of management strategies that prioritize sustainability, renewable energy standards, incentives for bioenergy production, adequate financing for bioenergy projects on the continent, and the adoption of proven technologies that are environmentally and socially beneficial while also encouraging the development of new innovative solutions that will benefit both African nations and the environment. While these strategies must be developed and implemented to ensure the environmental and social well-being of African nations, it is equally important that they are tailored to address the needs of individual countries. Ultimately, these strategies must be implemented in a way that respects the needs of local communities, especially those living in and around bioenergy production areas.

All of these strategies should be accompanied by clear communication and education campaigns that provide resources to African citizens to help them understand the importance of implementing these management strategies to ensure a more sustainable and equitable future for African nations and the world at large. To achieve successful implementation of these strategies, governments and policymakers must engage in collaborative efforts with local communities, industry leaders, researchers, and civil society organizations to ensure that these strategies are not only implemented effectively but also take into account the perspectives and needs of African citizens and local communities. This will help ensure that the implementation of these strategies is informed by the needs and values of local communities and that African citizens are able to have a meaningful stake in their own development while also respecting the rights and needs of those living in and around bioenergy production areas.

Additionally, it is essential that governments and policymakers prioritize local capacity building and technology transfer to ensure that African citizens have access to the necessary resources to effectively implement these management strategies and achieve a more sustainable and equitable future for African nations and the global community as a whole. This requires a significant investment of time, resources, and commitment from all stakeholders involved in the process, but if done correctly, it can help to ensure that the bioenergy management strategies proposed are successfully implemented and result in improved livelihoods and socioeconomic outcomes for African citizens while also promoting long-term ecological sustainability and global climate resilience. This will also ensure that African nations are able to develop and maintain their own energy production capabilities rather than relying on outside sources of assistance or investment, which could lead to a lack of self-reliance, economic independence, and security for African nations in the long term and, as a result, create lasting and positive change for African nations, the global community, and their citizens.

This collaborative effort between governments and policymakers, local communities, industry leaders, researchers, and civil society organizations is critical to ensuring the successful implementation of these strategies while also taking into account African citizens' perspectives and needs and promoting a more equitable and sustainable energy future for African nations, perspectives that differ significantly from the ancient author's and the global community as a whole. This will ensure that the implementation of these strategies is carried out in a way that takes into account local communities' needs and values and allows African citizens to meaningfully participate in their own development while also respecting the rights and needs of those living in and around the bioenergy projects, thus creating a truly inclusive and equitable process for everyone involved.

In general, to ensure successful implementation of bioenergy management strategies, governments and policymakers must prioritize local capacity building and technology transfer while engaging in meaningful collaboration with local communities, industry leaders, researchers, and civil society organizations to create a strong foundation of knowledge, expertise, and infrastructure necessary to carry out the proposed strategies in a way that is sustainable, equitable, and beneficial for African nations and their citizens in unprecedented ways to promote economic stability, autonomy, and security in African nations. This will in turn provide African nations with the opportunity to develop and maintain their own energy production capabilities while also promoting ecological sustainability and global climate resilience as well as socioeconomic outcomes that are beneficial for everyone involved while still respecting the perspectives, values, and rights of those living in and around bioenergy projects and taking into account the perspectives and needs of African citizens, an effort that must be undertaken in an equitable and sustainable way that enables African nations to build upon existing local energy resources to achieve energy security, autonomy, and economic stability in a way that respects their unique perspectives and needs.

This collaborative effort will be essential to ensuring that the strategies are implemented in a way that is sustainable, equitable, and beneficial for African nations while still achieving the overarching goal of promoting sustainable and equitable bioenergy management in Africa. By approaching the implementation of bioenergy management strategies in this way, African nations can truly create a framework that enables them to promote sustainable and equitable bioenergy management in an inclusive, equitable, and respectful manner that is beneficial for the environment and all stakeholders involved while also promoting social and economic justice and development for African nations in unprecedented ways. This will be an invaluable and challenging endeavor that, if successful, has the potential to revolutionize African energy production in a way that is beneficial for African nations while promoting ecological sustainability, global climate resilience, and socioeconomic outcomes that are equitable and beneficial for all involved. This will provide African nations with the opportunity to gain true autonomy over their own energy production capabilities and build a strong foundation of knowledge, expertise, and infrastructure necessary to pursue these strategies in a way that is sustainable, equitable, and beneficial for all those involved while addressing the global climate crisis.

5.7 Carbon Capture and Removal

Due to the difficulty of decarbonizing hard-to-abate sectors such as heavy industry, shipping, and aviation, some emissions from the use of fossil fuels and industrial processes will still be present in 2050. Thus, the development of carbon capture and storage (CCS) technologies, CO_2 removal strategies, and technologies for atmospheric CO_2 removal will be required to compensate for emissions that cannot be avoided if the global temperature rise is to be kept below 2 degrees Celsius. CCS technologies such as direct air capture, which traps CO_2 from the atmosphere, and bioenergy with CCS, which involves growing biomass to generate energy while trapping and storing its resulting CO_2 emissions, are two options for removing CO_2 from the atmosphere. To reduce emissions and reach climate goals, these measures need to be combined with other strategies, such as energy efficiency, renewable energy deployment, and low-carbon fuel substitution.

These measures will be costly, but the long-term economic and environmental benefits of reducing emissions and avoiding the worst impacts of climate change will far outweigh the costs of implementing these technologies. Therefore, if the world is to succeed in achieving its climate goals, investments in CCS and CDR technologies must be made to minimize the amount of CO_2 in the atmosphere and maximize the chances of a more sustainable future. Governments and industries need to cooperate and develop a comprehensive set of climate policies to facilitate the transition to a low-carbon economy and provide adequate resources for the implementation of CCS and CDR technologies to ensure their successful adoption.

To do this, governments and industries must collaborate on a well-defined policy framework that outlines the respective roles and responsibilities of each party in driving the transition to low-carbon technologies and provides clear guidance on how to finance the implementation of CCS and CDR technologies as well as incentivizing their adoption and use to reduce the impacts of climate change and preserve a sustainable future for generations to come. In addition to providing adequate resources, governments should also strive to create an enabling environment that encourages the private sector and civil society organizations to invest in CCS and CDR technologies through financial incentives and other supportive measures, such as providing access to low-cost financing and tax benefits, as well as ensuring that adequate capacity building and public awareness activities are undertaken to ensure that these technologies are deployed effectively and efficiently on a global scale to achieve the desired outcomes. This will help to ensure that these investments have the maximum impact in reducing emissions and achieving global climate goals while also making a real contribution toward achieving economic and social objectives, such as poverty alleviation, job creation, and improved public health and well-being.

In general, governments and industries must work together to create a comprehensive set of climate policies to facilitate the transition to a low-carbon economy and provide adequate resources for the successful adoption of CCS and CDR technologies to reduce emissions and mitigate the effects of climate change while at the same time ensuring that these investments lead to tangible economic and social benefits for the people and communities in which they are employed and to ensure that the entire process is fair and equitable for all involved parties to ensure a sustainable future for all.

These CCS and CDR technologies have the potential to be transformative in both reducing emissions and providing the necessary resources to build a more sustainable future, but only if adequate resources are made available and the necessary policies are adopted to support the African continent's transition to net-zero emissions by 2050 and beyond. In general, the transition to a low-carbon economy in Africa is possible if governments are willing to make the necessary investments in CCS and CDR technologies and create an enabling environment that encourages the private sector and civil society organizations to do their part in achieving the goals set forth in the Paris Agreement and other international climate change agreements while still ensuring that these investments benefit the African people and their communities by creating economic and social opportunities such as employment, poverty reduction, improved public health and well-being, and a fair and equitable process for all those involved to ensure the success of the transition and a sustainable future for all.

In general, the transition to a low-carbon economy in Africa is possible, but only if governments are willing to make the necessary investments and create the right enabling environment that encourages the private sector and civil society organizations to work together toward achieving global climate goals while also ensuring that these investments benefit the African people and their communities and lead to a fair and equitable process for all those involved to ensure a sustainable future for all. This will not only help to reduce emissions and mitigate the effects of climate change but will also have numerous social and economic benefits, including job creation, poverty alleviation, improved public health, and a more equitable process for all involved parties.

In fact, it is clear that governments and industries must work together to create a comprehensive set of climate policies to facilitate the transition to a low-carbon economy through the deployment of CCS and CDR technologies in Africa and to ensure that the investments benefit African communities, create economic and social opportunities, and ultimately lead to a fair and equitable process for all those involved to make this transition successful and help to secure a sustainable future for all. By creating an enabling environment that encourages the private sector, civil society organizations, and governments to work together toward global climate goals, Africa can make the necessary investments in CCS and CDR technologies and reap both economic and social benefits for its people while also helping to reduce emissions and mitigate the effects of climate change to make this transition successful and ensure that the investments benefit African communities and lead to a fair and equitable process for all to secure a sustainable future.

African governments and industries must work together to create a comprehensive set of climate policies to facilitate the transition to a low-carbon economy through the deployment of CCS and CDR technologies to ensure that the necessary investments are made, that African communities benefit and that the process is equitable and just for all those involved so that the transition to a low-carbon economy is successful and sustainable. This will ultimately help to ensure that African communities benefit and enjoy improved public health, poverty alleviation, job creation, and a more equitable process for all while reducing emissions and mitigating the effects of climate change to make the transition to a low-carbon economy successful and sustainable despite all of its difficulties. African countries must implement comprehensive climate policies to effectively deploy CCS and CDR technologies, which will require substantial investments from both the public and private sectors to ensure that the investments are put to good use and that the transition is fair, equitable, and just for all stakeholders while also providing improved public health, poverty alleviation, job creation, and an equitable transition process to secure a successful, sustainable transition to a low-carbon economy while simultaneously reducing emissions and mitigating the effects of climate change for the first time. African countries must work together to ensure that the necessary investments are made and that those investments benefit their communities and lead to a fair and equitable process for all to make this transition successful and secure a sustainable future while minimizing the impacts of climate change on the African continent.

This will require substantial and comprehensive planning, investments, and collaborations in order to ensure that African governments, industries, and communities are able to benefit from the transition to a low-carbon economy while mitigating the effects of climate change, while at the same time ensuring that the transition process is equitable and just for all stakeholders in order to make sure the transition is just and fair and that all the necessary investments are put to good use and CDR technologies and create an enabling environment that encourages the private sector and civil society organizations to do their part in achieving the goals set forth in the Paris Agreement and other international climate change agreements while still ensuring that these investments benefit the African people and their communities by creating economic and social opportunities such as employment, poverty reduction, improved public health and well-being, and a fair and equitable process for all those involved to ensure the success of the transition and a sustainable future for all. In general, the transition to a low-carbon economy in Africa is possible, but only if governments are willing to make the necessary investments and create the right enabling environment that encourages the private sector and civil society organizations to work together toward achieving global climate goals while also ensuring that these investments benefit the African people and their communities and lead to a fair and equitable process for all those involved to ensure a sustainable future for all.

This will not only help to reduce emissions and mitigate the effects of climate change but will also have numerous social and economic benefits, including job creation, poverty alleviation, improved public health, and a more equitable process for all involved parties. In fact, it is clear that governments and industries must work together to create a comprehensive set of climate policies to facilitate the transition to a low-carbon economy through the deployment of CCS and CDR technologies in Africa and to ensure that the investments benefit African communities, create economic and social opportunities, and ultimately lead to a fair and equitable process for all those involved to make this transition successful and help to secure a sustainable future for all.

By creating an enabling environment that encourages the private sector, civil society organizations, and governments to work together toward global climate goals, Africa can make the necessary investments in CCS and CDR technologies and reap both economic and social benefits for its people while also helping to reduce emissions and mitigate the effects of climate change to make this transition successful and ensure that the investments benefit African communities and lead to a fair and equitable process for all to secure a sustainable future. African governments and industries must work together to create a comprehensive set of climate policies to facilitate the transition to a low-carbon economy through the deployment of CCS and CDR technologies to ensure that the necessary investments are made, that African communities benefit and that the process is equitable and just for all those involved so that the transition to a low-carbon economy is successful and sustainable.

This will ultimately help to ensure that African communities benefit and enjoy improved public health, poverty alleviation, job creation, and a more equitable process for all while reducing emissions and mitigating the effects of climate change to make the transition to a low-carbon economy successful and sustainable despite all of its difficulties. African countries must implement comprehensive climate policies to effectively deploy CCS and CDR technologies, which will require substantial investments from both the public and private sectors to ensure that the investments are put to good use and that the transition is fair, equitable, and just for all stakeholders while also providing improved public health, poverty alleviation, job creation, and an equitable transition process to secure a successful, sustainable transition to a low-carbon economy while simultaneously reducing emissions and mitigating the effects of climate change for the first time.

African countries must work together to ensure that the necessary investments are made and that those investments benefit their communities and lead to a fair and equitable process for all to make this transition successful and secure a sustainable future while minimizing the impacts of climate change on the African continent. This will require substantial and comprehensive planning, investments, and collaborations to ensure that African governments, industries, and communities are able to benefit from the transition to a low-carbon economy while mitigating the effects of climate change while at the same time ensuring that the transition process is equitable and just for all stakeholders to ensure that the transition is just and fair and that all the necessary investments are put to good use. This can be achieved through the implementation of comprehensive climate policies, such as carbon capture and storage (CCS) and carbon dioxide removal (CDR) technologies.

Such policies must also be tailored to the unique needs of each African nation, taking into account local and regional contexts, so that they are able to drive progress in the right direction. African governments, industries, and communities must work together to develop comprehensive climate policies that enable effective deployment of CCS and CDR technologies while prioritizing the needs of local communities and protecting their rights and interests. This includes the creation of long-term and innovative financing mechanisms, such as public–private partnerships and international grant funding, to ensure that investments are made in a sustainable and equitable manner that are necessary for the successful implementation of CCS and CDR technologies across the continent. The implementation of comprehensive climate policies across Africa is a difficult but achievable goal that has the potential to create widespread and lasting benefits for the continent's people and environment.

The establishment of a coordinated regional strategy is essential to this effort, as it can help facilitate the development of coherent policies and regulations that are tailored to the particular needs and characteristics of each African nation and enable the effective deployment of CCS and CDR technologies while also incentivizing collaboration and sharing of expertise between governments, industries, and communities throughout the continent. To ensure the success of these policies, it is also important for African nations to develop plans for the scaling-up of CCS and CDR technologies in an effective manner while engaging with a broad spectrum of stakeholders, including scientists, engineers, academics, and industry representatives, to create policies that will effectively and efficiently address the challenges posed by climate change in Africa. To this end, African nations must prioritize investments in the research, development, and deployment of CCS and CDR technologies and ensure that the necessary resources are available for their implementation, such as funding from international organizations, technical assistance from external partners, and financial incentives for industry stakeholders.

While the goal of comprehensive climate policies in Africa is ambitious, it is attainable with the potential to create tremendous benefits for African nations and their people and help them create a future of prosperity that is more resilient to the impacts of climate change. By focusing on the implementation of a coordinated regional strategy that takes into account the individual needs and characteristics of each African nation, Africa can effectively deploy CCS and CDR technologies while engaging stakeholders from multiple sectors in policy development and decision-making to ensure the most successful outcomes and ensure that all of the steps needed for their successful implementation, from research to funding, are taken. African nations must also ensure that adequate monitoring, evaluation, and feedback mechanisms are in place to track the success of their climate policies and adjust them if needed to ensure their ongoing effectiveness and sustainability to achieve the set goals to address the global climate crisis and prevent the potential negative effects of climate change on Africa's people, environment, and economy to maximize their impact on mitigating climate change.

African nations must commit to this bold vision and prioritize investments in the research, development, and deployment of CCS and CDR technologies if they are to succeed in the task of achieving a prosperous and sustainable future that is resilient to the impacts of climate change while safeguarding the health, safety, and well-being of African citizens. African nations must embrace this ambitious vision of comprehensive climate policies to address the global climate crisis, ensure their own long-term sustainability and prosperity, and secure the future of their citizens in a meaningful and effective way. African nations must come together to create a unified strategy that is tailored to the needs and goals of each nation while taking into account the realities of their current economic and environmental conditions and aligning them with the global goals of mitigating climate change and transitioning to a more sustainable and resilient society if they wish to achieve the highest level of success in meeting the goals of reducing emissions and avoiding the most severe impacts of climate change in the short term and long term. African nations must commit to taking decisive action to ensure their success in meeting the global challenge of climate change.

In general, African nations must commit to comprehensive climate policies and coordinated investments in research, development, and deployment of carbon capture and removal technologies to secure their future and the future of their citizens in a meaningful way, reduce emissions, avoid the worst effects of climate change, and create a more sustainable and resilient society. To achieve these ambitious goals, African nations must develop and implement a unified strategy to coordinate their efforts to achieve maximum success while leveraging the support of international partners. This strategy must include an agreement on which technologies and practices are best suited for each nation, investment in the necessary infrastructure to support the deployment of these technologies, adequate financial and technical support from other countries or international organizations, and capacity building to ensure that individuals and communities are informed about and engaged in the process to effectively implement the strategies required to address climate change and its associated impacts.

This strategy must also include the development of long-term plans that reflect a commitment to sustainable practices and include policies that support the use of renewable energy sources, the conservation and protection of ecosystems, and incentives for individuals and businesses to reduce their carbon footprint. In addition, African nations must collaborate and cooperate with one another to effectively manage their natural resources, address cross-border issues, and ensure that they are using the most effective strategies to address the climate crisis as a continent overall. Furthermore, these plans should also include investments in developing technologies and practices that can be used to reduce emissions and remove carbon from the atmosphere, such as carbon capture and storage or sequestration technologies, and the implementation of programs to reduce poverty and promote social equity to create an environment in which African nations can more effectively and efficiently address the impacts of climate change. They must also collaborate with other nations, the private sector, and civil society to create economic incentives for the adoption of such policies, strengthen the capacity of vulnerable countries to take action on climate change, and help build a clean energy future for Africa and the world.

All of these strategies should ultimately be rooted in an overarching strategy to transition African nations to a low-carbon, climate-resilient economy to ensure that they are not only capable of managing the challenges posed by climate change but also to position them to benefit from the economic opportunities afforded by transitioning away from a high-carbon economy. In sum, African nations must come together to collaborate and cooperate on an array of solutions for effectively addressing the climate crisis to mitigate the worst effects of climate change and develop a prosperous, sustainable future for the continent and its people, as well as the world at large. They must work together to create comprehensive plans to implement renewable energy, energy efficiency, and conservation measures to reduce their carbon emissions and transition away from fossil fuels. They must also provide economic, social, and legal frameworks for countries to transition away from high-carbon energy sources, incentivize the use of clean energy solutions, and fund research and development initiatives for green technologies while also investing in adaptation measures to prepare for the climate shocks that are already unavoidable.

These measures must be coupled with investments in resilience, adaptation, and disaster risk reduction strategies so that African countries are better prepared to face the climate-related challenges they are already facing and that they will likely face in the future. Finally, African countries must create cross-border strategies and policies that encourage regional collaboration and coordination to promote the flow of capital, technology, resources, knowledge, and expertise needed for a successful transition and ensure that their economic growth and development are not jeopardized by the effects of climate change. These measures require political will, commitment, and leadership from the African governments to implement effective strategies and build the necessary infrastructure, laws, and regulations that will support these changes. This calls for a holistic approach to tackling climate change in Africa, where governments must take the lead in formulating and implementing policies that incentivize the move toward low-carbon development pathways while also creating mechanisms that drive investment, both public and private, in low-carbon development projects on the continent.

This will ensure that African countries can benefit from the low-carbon economy that is needed to mitigate and adapt to climate change while simultaneously encouraging economic growth and development. African countries must therefore embrace a two-pronged approach to tackling climate change: one that simultaneously focuses on reducing emissions while also creating conditions that enable African countries to economically benefit from the low-carbon transition that is taking place in the world. This includes developing regulatory frameworks that provide incentives for both private and public sector investments in low-carbon development, reforming fiscal policies to support the shift to low-carbon economies, investing in renewable energy infrastructure and technology, and promoting research and capacity-building initiatives. This will require careful planning, a well-thought-out strategy, and a strong commitment from African governments to ensure that the continent does not fall behind in the global transition toward a low-carbon economy that promotes greater understanding and innovation in the field of low-carbon development.

Furthermore, African countries need to ensure that their populations benefit from the shift to a low-carbon economy by providing them with the education and resources needed to actively participate in it and leveraging the resources of the international community to promote greater access to climate financing, technology transfer, and capacity building initiatives. This will enable African countries to create the necessary infrastructure and resources to take full advantage of the opportunities associated with a low-carbon economy while also mitigating their own risk of being exposed to adverse climate impacts. By working together in a coordinated fashion, African countries can leverage their collective resources to develop the necessary strategies and policies that promote low-carbon economic development and investment while also protecting their citizens and the environment from the risks associated with climate change. To be successful in this endeavor, African governments need to take a proactive approach to facilitate the transition toward a low-carbon economy while simultaneously taking the necessary steps to protect their vulnerable populations from the risks associated with climate change through climate financing, technology transfer, and capacity-building initiatives.

This will enable African countries to create the necessary infrastructure and resources to take full advantage of the opportunities associated with a low-carbon economy while also mitigating their own risk of being exposed to adverse climate impacts. By working together in a coordinated fashion, African countries can leverage their collective resources to develop the necessary strategies and policies that promote low-carbon economic development and investment while also protecting their citizens and the environment from the risks associated with climate change. To be successful in this endeavor, African governments need to take a proactive approach to facilitate the transition toward a low-carbon economy while simultaneously taking the necessary steps to protect their vulnerable populations from the risks associated with climate change. African governments have an important role to play in ensuring that their populations can benefit from a low-carbon economy without sacrificing their access to basic human needs or their ability to protect themselves from the adverse effects of climate change.

To do this, African governments must build an effective coalition of stakeholders who can work together to develop effective policies and strategies that promote low-carbon economic development while also protecting vulnerable populations from the risks associated with climate change. Such a coalition should include representatives from all levels of government, civil society organizations, and the private sector to ensure that there is a balanced approach to achieving low-carbon economic development while protecting vulnerable populations from the risks of climate change. This coalition of stakeholders should be responsible for identifying and addressing the most pressing issues related to transitioning to a low-carbon economy and helping vulnerable populations cope with climate change impacts. The coalition should also work to create strong partnerships between government and nongovernmental organizations, private sector actors, and local communities to ensure that the solutions developed by the coalition are comprehensive, inclusive, and socially just. In addition to these efforts, the coalition should focus on capacity building among government, civil society organizations, and the private sector to ensure that all stakeholders are well equipped to lead and participate in this transition.

To facilitate this process, the coalition should also work on developing a set of clear objectives and indicators that outline the goals and outcomes of transitioning to a low-carbon economy. This would enable the coalition to measure progress, ensure that all stakeholders are held accountable for their actions, and ensure the success of the transition in the long run. The coalition should also focus on building strong relationships with citizens to ensure that the transition to a low-carbon economy is widely accepted by the population and provide citizens with a clear understanding of the benefits of the transition, such as improved air quality and economic savings. This should be achieved through the creation of citizen-centric campaigns and initiatives, as well as by fostering dialogue between all stakeholders to ensure that citizens are provided with a platform to have their voices heard and to raise any concerns that they may have.

5.8 Conclusion

Renewable energy sources have the potential to significantly reduce energy poverty and enhance people's standards of living in Africa. They can reduce dependency on costly and unreliable grid electricity, improve living standards in a cost-effective way, and reduce emissions by substituting emissions-intensive energy sources. In recent years, technological paths toward net-zero goals in Africa have accelerated, offering promising pathways to decarbonize Africa's energy sector, create green jobs, and reduce poverty while also benefiting public health and the environment.

African countries are taking advantage of new technologies and business models to create economic incentives for renewable energy investments, which can bring numerous benefits to African nations, such as job creation, economic diversification, improved access to energy, and reduced greenhouse gas emissions.

Africa is investing in renewable energy sources, energy efficiency, and forest conservation to meet its net-zero goals and promote a sustainable future. It is also pursuing economic and social reforms to protect the environment and reduce carbon emissions. African countries are leveraging technologies such as renewable energy, big data, and blockchain to drive digital transformation and create pathways to more efficient use of natural resources, diversified sources of energy, and improved access to services. These technologies have the potential to not only address climate change but also increase economic opportunity, create jobs, and deliver a greater quality of life for African citizens. To realize these opportunities, African nations must create an enabling environment for digital infrastructure to take hold and improve access to financial services, healthcare, education, and other essential services.

To mitigate these risks, they must ensure citizens have access to the right education and resources, develop a regulatory framework that allows digital technology companies to develop, grow, and innovate while maintaining trust and accountability, invest in digital infrastructure, and develop strong local digital ecosystems. Additionally, they should develop regional partnerships and collaborations to share best practices and resources. African nations must take a responsible and meaningful approach to digital transformation, recognizing the power of technology and harnessing it in a way that is respectful of local culture and traditions. To achieve net-zero emissions in Africa by 2050, governments must invest in green technologies and infrastructure, such as solar, wind, and hydropower, and invest in training and education to ensure citizens are equipped with the knowledge and skills necessary to participate in the green economy. This shift to a green economy will require collaboration among African governments, businesses, and citizens to realize the potential of technological innovation and renewable energy sources to reduce emissions, build a more resilient and sustainable future, and ensure that economic development is coupled with social justice.

This chapter examines the strategies that African governments can pursue to achieve their goals and outlines the specific actions that need to be taken by governments, businesses, and citizens for this transition to occur successfully. To achieve this, African governments must invest in green technologies and infrastructure, as well as training and education opportunities. Additionally, businesses must be encouraged to embrace green technologies and energy sources, and citizens must take ownership of their roles in the transition by being aware of their environmental impact and advocating for green policies and sustainable practices. This chapter illustrates the importance of working together to create a more sustainable future for the continent and to develop renewable energy sources that will drive economic growth, reduce emissions, and create a better future for generations to come. African governments, businesses, and citizens must work together to create a brighter future for Africa by developing clean technologies, investing in renewable energy sources, encouraging sustainable practices, and advocating for green policies and sustainable practices.

To achieve this, African governments must invest in green technologies and infrastructure, businesses must embrace green technologies and energy sources, citizens must take ownership of their roles in the transition, and incentives and policies must be created to support renewable energy sources and clean technologies. In general, a successful transition to a more sustainable future for Africa can only be achieved if all three entities take part in the process.

References

Ahjum, F., Merven, B., Stone, A., & Caetano, T. (2018, September 21). Road transport vehicles in South Africa towards 2050: Factors influencing technology choice and implications for fuel supply. *Journal of Energy in Southern Africa, 29*(3), 33–50. https://doi.org/10.17159/2413-3051/2018/v29i3a5596

Akinyemi, O., Osabuohien, E., Alege, P. O., & Ogundipe, A. A. (2015). Energy security, trade and transition to green economy in Africa. *SSRN Electronic Journal.* https://doi.org/10.2139/ssrn.2700026

Aklin, M. (2022). Political economies of energy transition: Wind and solar power in Brazil and South Africa. *Political Science Quarterly, 137*(2), 424–426. https://doi.org/10.1002/polq.13323

Akrofi, M. M. (2021, February 2). Correction to: An analysis of energy diversification and transition trends in Africa. *International Journal of Energy and Water Resources, 5*(2), 235–236. https://doi.org/10.1007/s42108-020-00107-z

Aquila Jesuleye, O., Foluso, A. O., & Oluwaseun, A. A. (2020, April 7). Factors influencing solar photovoltaic utilization for energy services among Residentials in Akure, Nigeria. *Journal of Energy Research and Reviews*, 14–28. https://doi.org/10.9734/jenrr/2020/v4i330128

Asante, D., He, Z., Mintah Ampaw, E., Gyamerah, S., Ankrah Twumasi, M., Opoku-Mensah, E., Kyere, F., Asante, B., & Afia Akyia, E. (2021, November). Renewable energy technology transition among small-and-medium scale firms in Ghana. *Renewable Energy, 178*, 549–559. https://doi.org/10.1016/j.renene.2021.06.111

Baiyegunhi, L., & Hassan, M. (2014, June). Rural household fuel energy transition: Evidence from Giwa LGA Kaduna state, Nigeria. *Energy for Sustainable Development, 20*, 30–35. https://doi.org/10.1016/j.esd.2014.02.003

Baker, L. (2017, December 1). Commercial-scale renewable energy in South Africa and its Progress to date. *IDS Bulletin, 48*(5–6). https://doi.org/10.19088/1968-2017.165

Boulogiorgou, D., & Ktenidis, P. (2020, February). TILOS local scale technology innovation enabling low carbon energy transition. *Renewable Energy, 146*, 397–403. https://doi.org/10.1016/j.renene.2019.06.130

Clark, S., Van Niekerk, J., Petrie, J., & McGregor, C. (2022, September 26). The role of natural gas in facilitating the transition to renewable electricity generation in South Africa. *Journal of Energy in Southern Africa, 33*(3). https://doi.org/10.17159/2413-3051/2022/v33i3a12742

Ekeh, C. (2020, August). How COVID-19 is accelerating energy transition in developing oil-rich countries: Nigeria as a case study. *Global Energy Law and Sustainability, 1*(2), 234–240. https://doi.org/10.3366/gels.2020.0034

Emodi, N. V., Haruna, E. U., Abdu, N., Aldana Morataya, S. D., Dioha, M. O., & Abraham-Dukuma, M. C. (2022, September). Urban and rural household energy transition in sub-Saharan Africa: Does spatial heterogeneity reveal the direction of the transition? *Energy Policy, 168*, 113118. https://doi.org/10.1016/j.enpol.2022.113118

Fatoki, O. (2020, January 1). Factors influencing the purchase of energy-efficient appliances by young consumers in South Africa. *Foundations of Management, 12*(1), 151–166. https://doi.org/10.2478/fman-2020-0012

Field, T. L. (2021, February 26). A just energy transition and functional federalism: The case of South Africa. *Transnational. Environmental Law, 10*, 1–25. https://doi.org/10.1017/s2047102520000436

Ghazi, F. E., Sedra, M. B., & Akdi, M. (2021, April 10). Energy transition from fossil to renewable sources in North Africa: focus on the renewable electricity generation in Morocco. *International Journal of Energy Economics and Policy, 11*(3), 236–242. https://doi.org/10.32479/ijeep.11036

Giglmayr, S., Brent, A. C., Gauché, P., & Fechner, H. (2015, November). Utility-scale PV power and energy supply outlook for South Africa in 2015. *Renewable Energy, 83*, 779–785. https://doi.org/10.1016/j.renene.2015.04.058

Hendam, M., Schittekatte, T., Abdel-Rahman, M., & Kamh, M. Z. (2022, October). Rethinking electricity rate design: Fostering the energy transition in North Africa. *Energy Policy, 169*, 113172. https://doi.org/10.1016/j.enpol.2022.113172

Isah Samaila Nitte. (2023, March 13). Global energy transition and its implications on energy security in Nigeria: A critical review. *EPRA International Journal of Climate and Resource Economic Review*, 1–9. https://doi.org/10.36713/epra12607

Isihak, S., Akpan, U., & Bhattacharyya, S. (2022, August). Evolution of GIS-based rural electrification planning models and an application of OnSSET in Nigeria. *Renewable and Sustainable Energy Transition, 2*, 100019. https://doi.org/10.1016/j.rset.2022.100019

Israel-Akinbo, S., & Fraser, J. S. G. (2018). The energy transition patterns of low-income households in South Africa: An evaluation of energy programme and policy. *Journal of Energy in Southern Africa, 29*(3). https://doi.org/10.17159/2413-3051/2017/v29i3a3310

Joubert, E., Hess, S., & Van Niekerk, J. (2016, November). Large-scale solar water heating in South Africa: Status, barriers and recommendations. *Renewable Energy, 97*, 809–822. https://doi.org/10.1016/j.renene.2016.06.029

Kanu, M. I. (2016). Energy transition & geopolitics in sub-Saharan Africa: Outlining a legal framework to scale the energy infrastructure financing conundrum. *SSRN Electronic Journal*. https://doi.org/10.2139/ssrn.2831175

Masekela, M. E., & Semenya, K. (2021, September 19). Factors influencing the use of firewood postelectrification in rural South Africa: The case of Ga-Malahlela village. *Journal of Energy in Southern Africa, 32*(3). https://doi.org/10.17159/2413-3051/2021/v32i3a7781

McEwan, C. (2017, January). Spatial processes and politics of renewable energy transition: Land, zones and frictions in South Africa. *Political Geography, 56*, 1–12. https://doi.org/10.1016/j.polgeo.2016.10.001

Mosaka, T. B. M., Mararakanye, N., & Bekker, B. (2021, December 13). International procurement policies influencing renewable energy siting–Implications for South Africa. *Journal of Energy in Southern Africa, 32*(4), 58–68. https://doi.org/10.17159/2413-3051/2021/v32i4a8397

Netshipise, L. F., & Semenya, K. (2022, June 17). Evaluating factors influencing firewood consumption in households at the Thulamela local municipality, South Africa. *Journal of Energy in Southern Africa, 33*(2), 48–62. https://doi.org/10.17159/2413-3051/2022/v33i2a9741

Nwokolo, S. C., Amadi, S. O., Obiwulu, A. U., Ogbulezie, J. C., & Eyibio, E. E. (2022a). Prediction of global solar radiation potential for sustainable and cleaner energy generation using improved angstrom-Prescott and Gumbel probabilistic models. *Cleaner Engineering and Technology, 6*(January), 100416. https://doi.org/10.1016/j.clet.2022.100416

Nwokolo, S. C., Obiwulu, A. U., & Ogbulezie, J. C. (2022b). Machine learning and analytical model hybridization to assess the impact of climate. *Physics and Chemistry of the Earth, 130*(February), 103389. https://doi.org/10.1016/j.pce.2023.103389

Nwokolo, S. C., & Ogbulezie, J. C. (2018a). A qualitative review of empirical models for estimating diffuse solar radiation from experimental data in Africa. In Renewable and Sustainable Energy Reviews. https://doi.org/10.1016/j.rser.2018.04.118

Nwokolo, S. C., & Ogbulezie, J. C. (2018b). A qualitative review of empirical models for estimating diffuse solar radiation from experimental data in Africa. *Renewable and Sustainable Energy Reviews, 92*. https://doi.org/10.1016/j.rser.2018.04.118

Nwokolo, S. C., & Ogbulezie, J. C. (2018c). A quantitative review and classification of empirical models for predicting global solar radiation in West Africa. *Beni-Suef University Journal of Basic and Applied Sciences, 7*(4), 367–396. https://doi.org/10.1016/j.bjbas.2017.05.001

Nwokolo, S. C., Ogbulezie, J. C., & Obiwulu, A. U. (2022c). Impacts of climate change and meteo-solar parameters on photosynthetically active radiation prediction using hybrid machine learning with physics-based models. *August., 70*, 3614. https://doi.org/10.1016/j.asr.2022.08.010

Nwokolo, S., & Ogbulezie, J. (2017a). A critical review of theoretical models for estimating global solar radiation between 2012-2016 in Nigeria. *International Journal of Physical Research, 5*(2), 60. https://doi.org/10.14419/ijpr.v5i2.8160

Nwokolo, S., & Ogbulezie, J. (2017b). Modelling the influence of cloudiness on diffuse horizontal irradiation under various sky conditions in Nigeria. *International Journal of Physical Research, 5*(2), 91. https://doi.org/10.14419/ijpr.v5i2.8312

Nwokolo, S., & Otse, C. (2019). Impact of sunshine duration and clearness index on diffuse solar radiation estimation in mountainous climate. *Trends in Renewable Energy, 5*(3), 307–332. https://doi.org/10.17737/tre.2019.5.3.00107

Nwokolo, S. C., Obiwulu, A. U., Amadi, O., & Ogbulezie, J. C. (2023a). Assessing the impact of soiling, tilt angle, and solar radiation on the performance of solar PV systems. *Trends in Renewable Energy, 9*(2), 121–137. https://doi.org/10.17737/tre.2023.9.2.00156

Nwokolo, S. C., Ogbulezie, J. C., & Ushie, O. J. (2023b). A multimodel ensemble-based CMIP6 assessment of future solar radiation and PV potential under various climate warming scenarios. *Optik, 285*(May), 170956. https://doi.org/10.1016/j.ijleo.2023.170956

Nwokolo, S. C., Proutsos, N., & Meyer, E. L. (2023c). Machine learning and physics-based hybridization models for evaluation of the effects of climate change and urban expansion on photosynthetically active radiation. *Atmosphere, 14*(687), 1–36. https://doi.org/10.3390/atmos14040687

Obiwulu, A. U., Chendo, M. A. C., Erusiafe, N., & Nwokolo, S. C. (2020a). Implicit meteorological parameter-based empirical models for estimating back temperature solar modules under varying tilt-angles in Lagos, Nigeria. *Renewable Energy, 145*, 442. https://doi.org/10.1016/j.renene.2019.05.136

Obiwulu, A. U., Erusiafe, N., Olopade, M. A., & Nwokolo, S. C. (2020b). Modelling and optimization of back temperature models of mono-crystalline silicon modules with special focus on the effect of meteorological and geographical parameters on PV performance. *Renewable Energy, 154*, 404. https://doi.org/10.1016/j.renene.2020.02.103

Obiwulu, A. U., Erusiafe, N., Olopade, M. A., & Nwokolo, S. C. (2022). Modelling and estimation of the optimal tilt angle, maximum incident solar radiation, and global radiation index of the photovoltaic system. *Heliyon, 8*(6), e09598. https://doi.org/10.1016/j.heliyon.2022.e09598

Ojukwu, K. (2020). What strategies and policies can save Nigeria in the energy transition? *SSRN Electronic Journal.* https://doi.org/10.2139/ssrn.3701350

Okoh, A. S., & Okpanachi, E. (2023, December). Transcending energy transition complexities in building a carbon-neutral economy: The case of Nigeria. *Cleaner Energy Systems, 6*, 100069. https://doi.org/10.1016/j.cles.2023.100069

Omaye, S. O., & Sa'ad, S., Hamma Adama, A., & Dotti, R. (2022, December). Energy consumption, economic growth and energy transition in Africa: A cross-sectional dependence analysis. *OPEC Energy Review, 46*(4), 502–514. https://doi.org/10.1111/opec.12271

Soile, I. (2016). What are the factors influencing demand for energy in Nigeria: A sectoral analysis. *SSRN Electronic Journal.* https://doi.org/10.2139/ssrn.2817621

Sokona, Y., Mulugetta, Y., & Gujba, H. (2012, June). Widening energy access in Africa: Towards energy transition. *Energy Policy, 47*, 3–10. https://doi.org/10.1016/j.enpol.2012.03.040

Todd, I., & McCauley, D. (2021). Assessing policy barriers to the energy transition in South Africa. *Energy Policy, 158*, 112529. https://doi.org/10.1016/j.enpol.2021.112529

Unuigbe, M., Zulu, S. L., & Johnston, D. (2022, August 7). Exploring factors influencing renewable energy diffusion in commercial buildings in Nigeria: A grounded theory approach. *Sustainability, 14*(15), 9726. https://doi.org/10.3390/su14159726

Chapter 6
Decarbonizing Hard-to-Abate Sectors in Africa

6.1 Introduction

It is imperative that Africa decarbonize its hard-to-abate sectors to ensure a cleaner and healthier environment for its citizens and those around the world. The African continent must take urgent action to decarbonize its hard-to-abate sectors, such as transportation, heating and cooling, and industry, if it is to meet its development objectives and achieve a low-carbon future that will enable it to adapt to and mitigate the effects of climate change to maximize the potential of its renewable energy sources while simultaneously managing its overall carbon footprint and safeguarding the health of its citizens and environment. In addition to embracing a range of new and innovative technologies, Africa needs to develop appropriate policy frameworks and regulatory incentives that will ensure that the decarbonization of its hard-to-abate sectors is carried out in an economically, socially, and environmentally responsible manner. This will require major investments in energy efficiency and clean energy technologies, as well as supporting infrastructure such as access to low-carbon transport and buildings that are designed to reduce energy consumption and emissions. This can be done by encouraging the development of renewable energy sources, such as solar and wind, as well as exploring other alternatives, such as energy storage and decentralized energy solutions.

Furthermore, education and awareness programs need to be implemented that promote the understanding of climate change and its effects while also encouraging citizens to become more energy conscious and active participants in their own energy management. This in turn will create jobs (Nwokolo et al., 2022a), stimulate economic growth (Nwokolo et al., 2022a), and help deliver the African Union's (AU) Agenda 2063 commitment to achieve "an environmentally sustainable and climate resilient continent." By taking these steps, the AU will be able to reduce emissions while increasing access to clean energy (Nwokolo et al., 2022c) and transitioning away from the current use of fossil fuels (Nwokolo et al., 2023a), which are largely responsible for the current global warming and climate crisis

S. C. Nwokolo et al., *Africa's Path to Net-Zero*, CSR, Sustainability, Ethics & Governance, https://doi.org/10.1007/978-3-031-44514-9_6

facing the world today (Nwokolo et al., 2023b). To effectively address the climate crisis (Nwokolo et al., 2023c), African countries must focus on transitioning away from their reliance on fossil fuels and toward more renewable sources of energy such as solar, wind, and hydro while also investing in energy efficiency programs to reduce overall energy consumption.

These investments in renewable energy and efficiency measures will help reduce African countries' reliance on fossil fuels, create more job opportunities for the local population, stimulate economic growth, and achieve the AU's Agenda 2063 goals of providing an environmentally sustainable (Nwokolo & Ogbulezie, 2018a) and climate-resilient continent (Nwokolo & Ogbulezie, 2018a). In addition to the measures outlined above, African countries can also consider transitioning to cleaner sources of fuel for electricity production and transportation (Nwokolo & Ogbulezie, 2018b), as well as investing in carbon offsetting measures such as reforestation and mangrove restoration to encourage the development of new technologies that can facilitate the transition to renewable energy (Nwokolo & Ogbulezie, 2017a) and help African countries reduce their emissions (Nwokolo & Ogbulezie, 2017b) and become more resilient to climate change. This will not only benefit African countries in the short term but also help protect them from the worst effects of climate change in the long term (Nwokolo & Otse, 2019).

By transitioning to renewable energy sources, investing in energy efficiency programs (Obiwulu et al., 2020a), switching to cleaner sources of fuel for electricity production (Obiwulu et al., 2020b) and transportation, and investing in carbon offsetting measures (Obiwulu et al., 2022), African countries can significantly reduce their reliance on fossil fuels while also providing economic and environmental benefits that will ensure a sustainable future for generations to come. African countries have an opportunity to make a real difference in the fight against climate change by investing in renewable energy and efficiency measures and supporting innovative technologies that can help reduce their emissions and increase their resilience to the impacts of climate change. This is not an impossible task, but it does require concerted effort, commitment, and resources from African governments to make the necessary changes that will be required to make a meaningful contribution to mitigating the effects of climate change both domestically and globally. African countries should take the necessary steps to move toward a low-carbon and climate-resilient economy for the benefit of their citizens and the planet as a whole.

By investing in renewable energy sources and energy efficiency measures, African countries have the potential to reduce their reliance on fossil fuels, improve access to reliable electricity, create jobs, and stimulate economic growth while simultaneously making meaningful contributions to the global effort to mitigate climate change (Samuel, 2017). African governments should pursue renewable energy sources and energy efficiency measures while also investing in innovative technologies that can reduce emissions and increase resilience to climate change in their countries. Moreover, these efforts can be further bolstered by enacting supportive policies such as incentives for renewable energy and energy efficiency investments (Ogbulezie et al., 2017), strengthened building codes and standards (Agbor et al., 2023a), more efficient transportation networks, and increased investment in research and development of climate-resilient solutions (Agbor et al., 2023b).

Ultimately, the successful transition to a low-carbon and climate-resilient economy in Africa is essential to protect the planet from the effects of climate change and ensure a prosperous future for its citizens and the global community. This can be achieved through the coordinated action of African governments, businesses, and civil society stakeholders to create and implement effective strategies to increase the adoption of renewable energy, reduce emissions, and increase climate resilience across the continent (Chukwujindu et al., 2022). To achieve this, African countries should focus on creating policies and regulations that promote energy efficiency, encouraging private sector investments in renewable energy projects, and driving public–private partnerships for climate change initiatives and programs.

Furthermore, African countries should implement climate risk management and adaptation strategies to help communities become more resilient to the effects of climate change and should also strive to create regional and international partnerships to foster collaboration and knowledge-sharing between African countries to increase the capacity for meaningful climate action across the continent (Azadnia et al., 2023). As the world's most vulnerable region to the impacts of climate change, Africa has an urgent need to take meaningful action toward a low-carbon and climate-resilient economy for its own sustainability (Bokka & Lau, 2023) and for the future of its people and its environment. In general, the African continent needs to take bold and immediate steps to mitigate the effects of climate change and create an environmentally friendly and resilient future for its citizens.

To accomplish this, African countries must prioritize investing in renewable energy, diversifying their energy resources, implementing climate risk management strategies and public–private partnerships for climate change initiatives, and creating regional and international collaboration networks that can share resources and knowledge across borders to fully prepare for the impacts of climate change and ensure that African citizens are not disproportionately affected by its consequences. African countries must be proactive in creating a sustainable, resilient future by investing in renewable energy, diversifying their energy resources, implementing climate risk management strategies, and partnering with the private sector and other countries to share resources and knowledge to adequately respond to the effects of climate change on the continent (Callister & McLachlan, 2023).

These steps must be taken as soon as possible for the continent to reduce its vulnerability to climate change and ensure that it does not become disproportionately burdened with the costs of adapting to a changing environment (Fragkos, 2022). The African continent is in a unique and precarious position, as it is particularly vulnerable to the effects of climate change due to its high population growth, lack of access to resources, and geographic location in a rapidly warming world. However, it is also equipped with the potential to become a leader in developing solutions and adapting to the effects of climate change through a proactive approach that incorporates investing in renewable energy sources (Hassan et al., 2021), diversifying energy resources (Hassan et al., 2022), risk management strategies, and collaborative networks with other African countries as well as the international community to build a more sustainable and resilient future for African citizens.

Therefore, it is vital that African countries take immediate action to prepare for the impacts of climate change by investing in renewable energy sources, diversifying their energy resources, implementing risk management strategies, creating public–private partnerships for climate initiatives, and creating regional and international partnerships. To protect themselves, African countries must become leaders in tackling climate change by making substantial investments in renewable energy sources, diversifying their energy resources (Lah, 2017), implementing climate risk management strategies, and creating regional and international collaborative networks to develop solutions to climate challenges to protect their citizens and environment from the detrimental effects of climate change and ensure a better future for the people and environment of the continent.

This leaves African countries with the daunting task of having to prepare for and mitigate the effects of climate change in an effective and timely manner to ensure the long-term sustainability and resilience of their citizens, environments, and economies while also ensuring economic and social development while also leading the way in developing innovative solutions to address climate change and adapting to its impacts on the continent (Rumayor et al., 2020). As African countries continue to face the challenges of climate change, it is imperative that they become proactive in finding innovative solutions to address these challenges and take immediate action to create and implement climate adaptation strategies, build climate resilience (Prussi & Chiaramonti, 2022), and foster partnerships between African countries and the international community to strengthen their capacity to protect their citizens from the impacts of climate change.

This chapter discusses various hard-to-abate sectors in Africa, such as the heavy, shipping, and aviation industries (Rumayor et al., 2020), which contribute significantly to air pollution and carbon emissions across the continent. Despite their importance to the African economy, these sectors have typically seen less investment than other sectors, leading to inefficient and highly polluting practices that are difficult to curb without proper governmental support and regulatory enforcement (Prussi & Chiaramonti, 2022). This can be a major hurdle in the continent's ability to transition to cleaner sources of energy, reduce its dependence on fossil fuels, and bring down air pollution and carbon emissions to safe levels.

This chapter also discusses solutions to the problems associated with reducing hard-to-abate sectors in Africa, such as incentivizing industries to transition to renewable energy sources and investing in low-carbon technologies, among other measures. To achieve the goal of reducing air pollution and carbon emissions in Africa, it is essential that governments provide better investment opportunities in hard-to-abate sectors, offer incentives for businesses to transition to renewable energy sources to create a cleaner, more sustainable future, and invest in low-carbon technologies that can help Africa reduce its dependence on fossil fuels and create a more sustainable future that has grown over the years.

6.2 Hard-to-Abate Sectors in Africa

Regarding climate change externalities in Africa, the heavy, shipping, and aviation industries are some of the most difficult sectors to mitigate due to their large carbon footprint, a corresponding lack of economic resources (Santos et al., 2021), and an insufficient regulatory environment to implement meaningful change. Furthermore, Africa is expected to be one of the continents most adversely impacted by climate change in the coming years due to its limited resources (Superchi et al., 2023) and infrastructure, making it more difficult to respond to increased natural disasters (Superchi et al., 2022), rising sea levels, and other climate-related effects (Zhang et al., 2022). The sheer scale of the issue, combined with a lack of resources, makes addressing climate change in Africa a difficult task requiring substantial international collaboration and investment. To combat the negative externalities of climate change in Africa, substantial efforts must be undertaken both domestically and globally to encourage investment in the green energy infrastructure, support the development of new regulatory frameworks, and promote climate change mitigation strategies that can mitigate the effects of climate change on African economies and populations and help them become more resilient and adaptive to the new environmental challenges they face.

African governments and international partners must work together to facilitate an effective response to climate change and the mitigation of its externalities. For example, the establishment of regional partnerships and initiatives, such as the African Renewable Energy Initiative and the Global Environment Facility (Nwokolo et al., 2022a), is a step in the right direction, as these programs provide African countries with access to the funds and expertise needed to develop and implement climate change mitigation strategies that are adapted to their particular needs and circumstances (Nwokolo et al., 2022b). Additionally, the promotion of public–private partnerships, increased investments in green energy infrastructure, and the sharing of best practices can all help African countries become more resilient and adaptive to climate change and its associated externalities.

These investments in green energy infrastructure, regulatory frameworks, and climate change mitigation strategies are essential for African countries to combat the negative externalities of climate change and ensure a more sustainable future. The sheer scale of the issue, combined with a lack of resources (Nwokolo et al., 2023a), makes addressing climate change in Africa a difficult task requiring substantial international collaboration and investment. To combat the negative externalities of climate change in Africa, substantial efforts must be undertaken both domestically and globally to encourage investment in the green energy infrastructure, support the development of new regulatory frameworks (Nwokolo et al., 2022c), and promote climate change mitigation strategies that can mitigate the effects of climate change on African economies and populations and help them become more resilient and adaptive to the new environmental challenges they face. African governments and international partners must work together to facilitate an effective response to climate change and the mitigation of its externalities.

For example, the establishment of regional partnerships and initiatives, such as the African Renewable Energy Initiative and the Global Environment Facility, is a step in the right direction, as these programs provide African countries with access to the funds and expertise needed to develop and implement climate change mitigation strategies that are adapted to their particular needs and circumstances. Additionally, the promotion of public–private partnerships, increased investments in green energy infrastructure, and the sharing of best practices can all help African countries become more resilient and adaptive to climate change and its associated externalities. These investments in green energy infrastructure, regulatory frameworks, and climate change mitigation strategies are essential for African countries to combat the negative externalities of climate change and ensure a more sustainable future.

Furthermore, the international community must work in tandem with African governments to ensure that they receive the necessary support and resources to implement such initiatives and programs and to develop capacity-building programs that can help local actors understand the complexities of climate change and its associated externalities, as well as the importance of taking proactive measures to mitigate them. These efforts should be complemented by initiatives that provide education and resources to the local populations to raise awareness and encourage them to participate in climate change mitigation efforts at the local and national levels. As such, it is imperative for the international community to prioritize the advancement of programs and initiatives that are designed to support African countries in mitigating the externalities associated with climate change and promoting green energy solutions that will result in a more sustainable future for African countries and their people.

In general, while the effects of climate change can be profound and far-reaching, African countries have the potential to make significant progress in mitigating the externalities associated with this global phenomenon if the international community and African governments work together to implement initiatives, programs, and capacity-building initiatives (Nwokolo et al., 2023c) that are designed to help African countries protect their people and environment from the potentially devastating effects of climate change and offer them the tools and resources necessary to build a more resilient future for themselves and for the generations to come.

Generally, it is clear that the international community and African governments must join forces to address the externalities of climate change in African countries by focusing on advancing initiatives that are designed to support African countries in developing green energy solutions, increasing access to climate-resilient resources, and providing education and resources to local populations so that they are empowered to participate in local and national climate change mitigation efforts to create a more sustainable future. This should be done to help African countries build a more sustainable and resilient future while also giving the people of these countries the necessary tools to protect themselves and their environment from the potentially devastating effects of climate change and to ensure that these efforts are undertaken in a way that is equitable and respects the rights of all individuals and communities impacted by climate change. It is also important for African countries and the international community to collaborate to establish policies and regulations that are designed to protect vulnerable populations from being exposed to environmental pollution, poor working conditions, and other dangers associated with climate change (Obiwulu et al., 2022).

This will ensure that African countries are provided with the resources and technical expertise necessary to mitigate climate change and its associated externalities while at the same time mitigating any negative environmental, economic, and social effects that these initiatives may have on the people of these countries. In conclusion, it is imperative that the international community and African governments come together to create initiatives, programs, and capacity-building initiatives that are designed to help African countries protect their people and environment from the potentially devastating effects of climate change while at the same time promoting a more equitable, sustainable, and resilient future for all. These initiatives should be well thought out and developed with the input of local communities, civil society organizations, and other stakeholders to ensure that they are both effective and equitable and that the voices of the people most impacted by climate change are taken into account to create meaningful and lasting change for African countries.

6.2.1 Heavy Industry

Heavy industry is one of the hardest sectors to mitigate in regard to the externalities of climate change in Africa. With large-scale and sometimes inefficient production, the heavy industry sector emits large amounts of greenhouse gases, significantly contributing to climate change in the region. Moreover, the cost of mitigating the emissions from this sector is too high for African countries to bear, making the future of the heavy industry sector in Africa uncertain. This uncertainty is compounded by the fact that the technology used in most heavy industries in Africa is outdated and inefficient, further adding to its emissions and making it even harder to reduce the sector's emissions due to a lack of capital and expertise needed to upgrade the infrastructure. This situation is an urgent problem for African countries, as it increases the risk of further climate change, thereby affecting not only the environment but also food security and economic growth in the region.

With the looming climate crisis, it is imperative that African countries focus on modernizing and decarbonizing their heavy industry sectors to reduce their emissions while also taking into consideration the cost that they can afford and the expertise they have access to. Therefore, African countries must adopt a multipronged approach to mitigating emissions from the heavy industry sector that involves both modernizing and decarbonizing the sector, investing in sustainable technologies, and training local personnel to build their capacity for industry growth while also reducing emissions and supporting economic growth in the region. This multipronged approach will provide African countries with the tools to address the urgent problem of heavy industry sector emissions, as it enables them to reduce their emissions while also reducing costs and supporting economic growth in the region through the use of sustainable technologies and personnel training programs while still taking into consideration the cost that they can afford and the expertise that they have access to sustainably grow the sector and create a future in which they can reap the rewards of their hard work without suffering from the effects of climate change in unprecedented ways. African countries must invest in more efficient technologies and provide incentives to encourage heavy industry firms to adopt more environmentally friendly and sustainable solutions to reduce emissions and create a future in which they can sustainably grow the sector without being impacted by climate change.

Despite the efforts of countries to reduce emissions through regulatory measures, the sector still accounts for a substantial proportion of global greenhouse gas emissions, and it is essential for African countries to take decisive action to mitigate the emissions from this sector and create a brighter future for themselves, their citizens, and the world as a whole. Therefore, African countries must adopt a multipronged approach to mitigating emissions from the heavy industry sector that involves both modernizing and decarbonizing the sector, investing in sustainable technologies, and training local personnel to build their capacity for the use of such technologies to reduce emissions from the sector and create a future in which they can reap the rewards of their hard work without being adversely impacted by the effects of climate change in an unprecedented way. African countries can reduce emissions and promote economic growth through the use of sustainable technologies and personnel training programs. This multipronged approach will enable them to develop more sustainable and efficient industries that can provide them with long-term economic growth and sustainable development while also allowing them to reap the rewards of their hard work without being adversely impacted by the effects of climate change in the long term.

This approach of modernizing and decarbonizing the heavy industry sector, investing in sustainable technologies, and training local personnel has the potential to result in substantial benefits for African countries in the form of economic growth, job creation, and improved air quality while also helping to mitigate the effects of climate change on the continent. Additionally, by engaging in an ambitious renewable energy strategy and creating policies that encourage the use of renewable energy sources, African countries can reduce their reliance on fossil fuels, increase access to clean energy, create jobs, and promote economic development while simultaneously reducing emissions and contributing to the global fight against climate change.

All of these strategies for reducing emissions and promoting economic growth are essential for African countries to successfully navigate the twenty-first century, as they will benefit both their economies and the environment in the long term. While these solutions may be ambitious, they are achievable with the right support from the international community and commitment from African countries to pursue them. While the challenges African countries face in regard to decarbonizing and modernizing their heavy industry sector, investing in sustainable technologies, and training local personnel may seem daunting, it is essential that they take on these tasks to mitigate the effects of climate change, which will disproportionately affect their populations if left unchecked. To meet these goals, African countries must prioritize investment in clean energy and green technology, create policies that support the transition away from fossil fuels and the growth of renewable energy industries, increase access to energy efficiency programs to reduce the energy intensity of their economies, and simultaneously expand access to basic services, such as electricity and clean cooking fuels, to support their populations in their transition to a low-carbon economy.

Furthermore, African countries must develop robust monitoring and evaluation systems to ensure the effectiveness of their strategies, as well as cultivate domestic financial resources and attract foreign investment to finance the transition. These solutions are ambitious but achievable and will be essential to achieving the necessary emissions reductions to avert catastrophic climate change. African countries are capable of meeting these goals if they remain committed to the task and continue to build on the advances made by their current initiatives. To achieve success in these areas, African countries will need strong institutional capacity and collaboration between government, industry, and civil society to develop, implement, and manage programs to reduce emissions and promote sustainable development. However, this will require sustained international support for capacity building and technology transfer to ensure that African countries are able to build the infrastructure and access the resources necessary for success.

These initiatives will be crucial to ensuring that African countries are able to make the necessary changes to transition to a low-carbon and climate-resilient future and create a sustainable future for their people. The path to success for African countries is clear: prioritize and invest in low-carbon development strategies, build institutional capacity, and foster collaborative efforts between governments, industry, and civil society to create successful and lasting change to be successful. As a result, it is essential that the international community provide support for capacity building, technology transfer, and access to resources to ensure that African countries can transition to a low-carbon and climate-resilient future to truly make the most of the opportunities presented by this transition.

6.2.2 Shipping

The shipping industry is one of the hardest sectors to mitigate in regard to the externalities of climate change in Africa due to a lack of resources and advanced technology, as well as the constraints of international regulations and policies. Despite this, the shipping industry has taken strides toward environmental sustainability, such as investing in cleaner fuels, creating efficient ships and green port networks to reduce their carbon footprint, and undertaking initiatives such as cutting GHG emissions and waste management programs that have enabled the industry to become more sustainable in recent years. The efforts have been mostly successful and are continuing to become even more effective, but much work still needs to be done to reduce the environmental footprint of the shipping industry in Africa and ensure that it is as green and sustainable as possible.

The shipping industry in Africa should strive to further reduce its environmental impact by continuing to invest in cleaner technologies, ships, and green port networks; cutting emissions and waste more effectively; and implementing stricter international regulations. The shipping industry has taken strides toward environmental sustainability, such as investing in cleaner fuels, creating efficient ships and green port networks to reduce their carbon footprint, and undertaking initiatives such as cutting GHG emissions and waste management programs that have enabled the industry to become more sustainable in recent years. The efforts have been mostly successful and are continuing to become even more effective, but much work still needs to be done to reduce the environmental footprint of the shipping industry in Africa and ensure that it is as green and sustainable as possible. The shipping industry in Africa should strive to further reduce its environmental impact by continuing to invest in cleaner technologies, ships, and green port networks; cutting emissions and waste more effectively; and implementing stricter international regulations. To further these efforts, it is essential that the shipping industry in Africa has access to reliable and advanced technology, as well as sufficient resources to ensure that their initiatives are effective and able to sustain a greener and more sustainable future for the industry.

The shipping industry in Africa has a great opportunity to make a positive impact on the environment by investing in cleaner and more efficient technologies, ships, and green port networks, increasing the implementation of stricter international regulations, and reducing emissions and waste even more effectively. In addition to investing in more sustainable practices and initiatives, the shipping industry in Africa must continue to prioritize and invest in research and development so that it has access to more reliable, advanced, and innovative technology that can facilitate greener operations and better enable the industry to meet its environmental goals. Additionally, adequate funding must be provided to the shipping industry in Africa so that the necessary resources are available to ensure that their initiatives are effective and able to produce tangible results in creating a more sustainable future for the industry and the African continent as a whole. These resources should include but not be limited to capital, personnel, and research and development, as these are all crucial components in allowing the shipping industry in Africa to implement the necessary initiatives and practices that will contribute toward a more sustainable future for both the industry and the African continent.

It is also important for the African shipping industry to collaborate with organizations, companies, and government agencies to ensure better access to the best available technologies and resources and create a stronger, more united front in advocating for the greener practices that are essential to their industry and to the future of our planet as a whole. In addition to investing in technology, ships, and green port networks, African shipping companies should also focus on education and collaboration with other industry stakeholders. In addition, investments in renewable energy sources and the development of new, greener methods of transportation should be explored to further reduce the industry's environmental footprint. By investing in resources, collaborating with other organizations and government agencies, focusing on education, and exploring renewable energy sources, African shipping companies can work together to create a more sustainable future for the industry and for the continent as a whole. African shipping companies need to take responsibility for their impact on the environment by taking action to reduce their carbon footprint and conserve resources while also creating new opportunities for growth and development and encouraging innovation in the sector.

This is the only way to ensure that Africa's shipping industry remains competitive and sustainable in the long term to ensure better access to the best available technologies and resources and create a stronger, more united front in advocating for the greener practices that are essential to their industry and to the future of our planet as a whole. In addition to investing in technology, ships, and green port networks, African shipping companies should also focus on education and collaboration with other industry stakeholders. In addition, investments in renewable energy sources and the development of new, greener methods of transportation should be explored to further reduce the industry's environmental footprint. By investing in resources, collaborating with other organizations and government agencies, focusing on education, and exploring renewable energy sources, African shipping companies can work together to create a more sustainable future for the industry and for the continent as a whole.

African shipping companies need to take responsibility for their impact on the environment by taking action to reduce their carbon footprint and conserve resources while also creating new opportunities for growth and development and encouraging innovation in the sector. This is the only way to ensure that Africa's shipping industry remains competitive and sustainable in the long term. Furthermore, shipping companies should work to create a culture of environmental awareness and responsibility among their employees and promote best practices that encourage the efficient use of resources, as well as the adoption of environmentally friendly technologies and the development of energy-efficient systems and operations, to promote sustainability and growth in African shipping.

Additionally, African shipping companies should collaborate with government agencies and other organizations, such as the International Maritime Organization, to help ensure that regulations and standards are in place to protect the environment and create a more sustainable industry. African shipping companies have a responsibility to create and maintain a sustainable industry that prioritizes both environmental conservation and economic growth to achieve this over the long term. This is essential to ensure that the industry continues to remain a viable option for African countries and regions, which is why it is important for shipping companies to take proactive measures in regard to environmental protection and sustainability measures. In conclusion, African shipping companies should strive to create a culture of environmental responsibility and awareness among their employees to promote sustainability and growth while also collaborating with other organizations to ensure that the necessary regulations and standards are in place to challenge the status quo and push for recognition of the industry as a viable option for Africa's economic future.

This can be achieved by instituting policies and procedures that promote the preservation of natural resources, reduce pollution, and encourage responsible practices among employees as well as customers, partners, and suppliers. This will ensure that African shipping companies remain competitive and resilient in a global economy that is increasingly reliant on environmental protection and sustainability measures and ensure that they are able to capitalize on the potential of African shipping in a sustainable way. This, in turn, will help to create a more positive perception of African shipping companies and their ability to contribute to the global economy and provide meaningful employment opportunities for people living in the region.

6.2.3 Aviation

The aviation industry is one of the hardest sectors to mitigate in regard to the externalities of climate change in Africa due to the high energy consumption and carbon emissions associated with air travel, which can adversely affect the climate of many African countries and increase their vulnerability to climate-related risks such as increased drought and water scarcity as well as extreme weather events. The magnitude of climate change in African countries is compounded by their lack of technological capacity, infrastructure, and resources to mitigate its effects. Furthermore, the cost of air travel in Africa is often unaffordable for most citizens, and there is a lack of awareness about how to make sustainable choices in regard to travelling by plane. Additionally, air transport can contribute to local air pollution and noise in African communities, leading to negative environmental and health impacts. Consequently, there is an increasing need to invest in alternative technologies, such as electric- and hydrogen-powered aircraft, to reduce carbon emissions and energy consumption.

As such, it is imperative that African countries devise and implement comprehensive strategies to address the impacts of aviation on the climate by encouraging the development and adoption of cleaner, more sustainable air travel solutions. To mitigate the effects of air travel on the climate, African countries should invest in technological capacity and infrastructure that can facilitate greener and more sustainable modes of transport. African governments need to devise strategies that focus on both energy efficiency and the shift toward cleaner fuels, such as sustainable aviation fuels (SAFs). These strategies should also include initiatives to increase the access and affordability of air travel so that the general public can benefit from the economic and social opportunities it provides while at the same time taking action to reduce its environmental impact.

Furthermore, African countries should also promote public–private partnerships to spur innovation and the development of low-carbon air travel solutions such as the adoption of electric-powered aircraft and other energy-efficient technologies. This will enable African countries to increase their resilience to climate change and facilitate their transition to a low-carbon economy while also providing increased access to air travel and its associated benefits for their citizens. The strategies African governments should implement should also promote the development of environmentally friendly transportation infrastructure, such as electric vehicle charging stations, high-speed rail networks, and improved public transportation.

In addition, African governments need to ensure that energy policies are financially viable for airlines and other stakeholders while also having a positive impact on the environment. As part of these efforts, African governments should provide incentives to airlines and other stakeholders to promote the adoption of new technologies such as electric-powered aircraft and other energy-efficient technologies. African governments should also establish and enforce strict regulations to ensure that airlines and other stakeholders comply with the necessary environmental standards and take the necessary measures to reduce emissions and air pollution. In conclusion, African governments should adopt and implement policies and strategies to promote the development of environmentally friendly transportation infrastructure and establish energy policies that are financially viable for airlines and other stakeholders while also having a positive impact on the environment to create a cleaner, healthier future for the African continent.

Furthermore, African governments should invest in the research and development of new and efficient energy sources that are clean, renewable, and cost-effective to power the future of transportation, reduce emissions, and ensure a sustainable future for the continent. This could be achieved through the establishment of financial incentives, such as tax credits and subsidies, to incentivize the adoption of these new technologies, as well as the promotion of public–private partnerships that encourage collaboration between the private sector and governments to stimulate investment and innovation in the sector and ensure the success of these initiatives. African governments should also ensure that international and local airlines, airports, aircraft, and other stakeholders comply with all necessary international regulations to protect the environment.

In addition, African governments should promote the development of green airports and invest in new technologies such as electric aircraft and fuel-efficient engines that are not only more efficient but also more cost-effective for airlines than traditional methods of air travel to reduce the environmental impact of air travel, make it more affordable for the general public, and create job opportunities for skilled labor in the aviation and energy sectors. The steps outlined above would allow African governments to ensure a cleaner and more sustainable future for the continent and reduce the environmental impacts of air travel while also increasing economic development, creating employment opportunities, and improving the standard of living for their citizens by making the air travel sector more effective, cost-effective, and sustainable. Africa is a continent that is increasingly recognizing the importance of sustainability and environmental protection, and this should be reflected in its policies to promote the development of sustainable aviation technologies and the implementation of stricter regulations for airlines, airports, and other stakeholders in the aviation sector to reach these goals and protect the environment for future generations.

African governments must also ensure that international and local airlines, airports, aircraft, and other stakeholders comply with all necessary international regulations to protect the environment and ensure that the air travel sector is operating in a safe and sustainable manner while also improving their economies and creating jobs in the aviation sector to make the air travel sector more sustainable and environmentally friendly for their people in the limelight, going beyond simply acting on the stage and educating the general public. In addition, African governments should invest in green airports and electric aircraft, as well as fuel-efficient engines, to reduce the environmental impact of air travel and create employment opportunities in the aviation sector while also reducing the cost of air travel and improving passenger comfort for the benefit of the African people. African governments should take steps to support the development of sustainable aviation technologies, promote stronger environmental regulations for airlines and airports, ensure compliance with international environmental regulations, and invest in green airports, electric aircraft, and fuel-efficient engines to reduce the environmental impact of air travel, improve economies, create employment opportunities in the aviation sector, and make air travel more affordable and comfortable for their people.

All of these efforts should go hand in hand with raising public awareness about the importance of sustainable aviation and environmental protection as well as the need for people to be mindful of their own travel habits and behaviors to reduce their own personal impact on the environment to make a meaningful contribution to improving the environment and providing a better quality of life for their citizens while making air travel more accessible and comfortable for their people. All in all, the African governments need to ensure that the air travel sector is operating in a safe and sustainable manner while also improving their economies and creating jobs in the aviation sector so that they can continue to provide the public with affordable, safe, and comfortable air travel services while reducing their environmental footprint and protecting the environment for future generations. By taking these steps, African governments can contribute to creating a more sustainable aviation industry, reduce their environmental impact, improve the quality of life for their citizens, and develop a more efficient and cost-effective air travel sector that is attractive to businesses and individuals alike while also making air travel more accessible and comfortable for their people, all while simultaneously providing economic and environmental benefits to their countries and society.

6.3 Remedies to the Challenges of Mitigating Hard-to-Abate Sectors in Africa

6.3.1 Mitigation Cost

One of the biggest issues in Africa is the cost of mitigating hard-to-abate sectors such as transportation, buildings, and industrial processes, which are responsible for the majority of their greenhouse gas emissions due to the continent's lack of resources and access to capital. This presents a huge challenge for the continent in their fight against climate change and will require creative solutions to tackle, such as building a sustainable infrastructure and deploying advanced renewable energy technologies that utilize the continent's abundance of natural resources such as the sun, wind, and hydropower to support a more sustainable and resilient economy. This is a major challenge for African countries, as the cost of transitioning to low-carbon technologies is often too expensive for their limited resources to afford.

Therefore, it is important to find ways to make the transition to low-carbon technologies more affordable and accessible through international investments and partnerships that can provide access to finance, technology, and knowledge needed to accelerate the transition to a more sustainable and resilient economy for African countries. Many African countries, as well as international organizations such as the UN and the World Bank, have been working hard to address this issue and promote the development of clean energy technologies to reduce greenhouse gas emissions, improve the health of people living in African countries (Nwokolo et al., 2022a), and mitigate the impacts of climate change on the continent by creating incentives, subsidies, and other forms of support for businesses, governments, and communities to invest in green technologies that are capable of reducing emissions and increasing access to renewable energy sources.

These initiatives include providing financial support to developing countries, such as providing access to concessional loans and grants to help them purchase clean energy technology and reduce their dependence on fossil fuels, and developing innovative financing schemes such as green bonds and carbon taxes that can help reduce the cost of transitioning to low-carbon energy sources while also generating revenues to help fund these projects, which will pay for themselves over time. Additionally, many African countries have begun implementing renewable energy policies, such as mandating that a certain percentage of their electricity comes from renewable sources. Additionally, international organizations are working with African countries to develop innovative models for clean energy deployment, such as carbon trading, market-based incentives, and renewable energy auctions, which can provide additional sources of financing for clean energy projects and help make the transition to low-carbon energy sources more affordable and attractive for countries and businesses.

Additionally, the UN and other international organizations are providing technical assistance to African countries to build capacity in clean energy technology, such as improving grid systems and investing in research and development for new renewable energy technologies (Nwokolo et al., 2022a), while also helping to create new economic opportunities in the renewable energy sector. These efforts are essential to helping African countries transition to low-carbon energy sources, reduce their dependence on fossil fuels, and provide access to more affordable and reliable electricity while also creating jobs and other economic opportunities in the region. However, these efforts remain insufficient, and there is a need for more financial support and resources to be devoted to achieving the goal of transitioning to low-carbon energy sources and increasing access to affordable and reliable electricity in Africa.

To achieve this goal, more international organizations need to dedicate additional resources and financial support to this effort to help African countries achieve the goal of transitioning to low-carbon energy sources, increasing access to affordable and reliable electricity, and ensuring the sustainability of their energy systems in the future. In addition to these technical assistance initiatives, the international community must be willing to provide greater financial support and resources for African countries to transition from fossil fuels to renewable energy sources and for African countries to create and sustain an environment of economic opportunity for their citizens by investing in infrastructure and technology, providing capacity-building training, and providing financial support for local businesses that can capitalize on the potential of renewable energy sources and help drive economic development in the region. These efforts are important not only for the transition to low-carbon energy sources and for access to affordable and reliable electricity in Africa but also to create jobs and promote economic development in the region that could ultimately improve the quality of life for all citizens in Africa. Ultimately, the goal of transitioning to low-carbon energy sources and increasing access to affordable and reliable electricity in Africa is attainable, but it will require an immense effort on the part of international organizations, governments, and private entities to provide the financial resources and technical assistance needed for African countries to achieve this goal.

These efforts must be both broad and targeted to ensure success. They must focus on providing technical assistance to African countries to help them develop their own renewable energy infrastructure, providing capacity-building training, and supporting local businesses that can capitalize on the potential of renewable energy sources to create economic growth and employment opportunities for African citizens. Additionally, these efforts should also be supplemented with economic incentives for companies to invest in Africa's renewable energy infrastructure and policy reforms that incentivize the development of renewable energy sources while also encouraging African countries to embrace low-carbon energy sources to move away from carbon-intensive energy sources and to increase access to affordable and reliable electricity.

This combination of financial resources, technical assistance, economic incentives, and policy reforms will be necessary to effectively support the transition to low-carbon energy sources and increase access to affordable and reliable electricity in Africa. The goal of these efforts is to enable African countries to develop their own renewable energy infrastructure and access affordable, reliable electricity while also enabling them to reduce their carbon footprint, increase economic growth, and create employment opportunities for African citizens. These initiatives are critical to support the African continent's transition to renewable energy sources and its goal of providing sustainable and reliable access to electricity for its citizens while also protecting the environment and mitigating the effects of climate change. By investing in renewable energy infrastructure and supporting policy reforms, African countries can move away from carbon-intensive energy sources, reduce their carbon footprint, increase access to affordable and reliable electricity, and create economic opportunities while reducing poverty and improving the quality of life for their citizens. By focusing on the transition to renewable energy sources and investing in the infrastructure, economic incentives, policy reforms, and technical assistance necessary to facilitate this transition, African countries can effectively meet their goals of providing sustainable and reliable access to electricity while also protecting the environment and mitigating the effects of climate change.

African countries have made considerable efforts in recent years to invest in renewable energy sources and infrastructure, as well as the technical assistance and policy reforms required to support this transition and ensure its success. This focus on renewable energy sources is critical to ensure that the African continent achieves its ambitious energy and climate goals while also providing its citizens with access to affordable and reliable electricity. In addition, investing in renewable energy infrastructure has the potential to create new economic opportunities and jobs, stimulate economic growth, and reduce poverty levels across the continent.

By investing in renewable energy sources and infrastructure, African countries can not only achieve their ambitious energy and climate goals but also ensure that citizens have access to reliable and affordable electricity and reduce their dependence on fossil fuels, which can have adverse effects on the environment and public health while simultaneously driving economic growth and reducing poverty levels. Additionally, renewable energy sources, such as solar and wind power, can help to create a more sustainable and secure future for the African continent by reducing its dependence on finite fossil fuels and increasing its reliance on renewable energy sources, creating new jobs and economic opportunities in the process. These investments in renewable energy sources and infrastructure can have a positive impact on the African continent in a variety of ways, from achieving ambitious energy and climate goals to providing access to reliable and affordable electricity to reducing poverty levels, creating jobs, stimulating economic growth, and driving long-term economic development and sustainability throughout the continent.

6.3.2 Technological Limitations

One of the difficulties in Africa's hard-to-abate sectors is technological limitations, which prevent many African countries from being able to access or deploy the most up-to-date technologies for energy and water efficiency that would help reduce their emissions and promote sustainable development. This difficulty is compounded by a lack of financial resources and knowledge, making it even more difficult for African countries to transition to clean energy sources and remain competitive in the global economy. This makes it particularly challenging for African countries to transition to clean energy sources since these technological advancements require significant investments in terms of both money and resources that African countries may not have the capacity to make at this time.

Currently, African countries are in a difficult position because of the gap between what they can access and what is necessary to meet the demands of transitioning to clean energy sources while remaining competitive in the global economy. While African countries face significant difficulties in accessing the technologies and resources necessary for transitioning to clean energy sources, it is important to recognize that this is only one aspect of the challenges they face in regard to transitioning to clean energy sources and remaining competitive in the global economy. In addition to the aforementioned difficulties, African countries also face structural and cultural issues that impede their progress in transitioning to clean energy sources and in their ability to stay competitive in the global economy.

These issues include, but are not limited to, corruption, poor infrastructure, and an inadequate education system, all of which have contributed to a lack of capital and access to resources that would enable African countries to transition to clean energy sources and remain competitive in the global economy. Despite these significant challenges, African countries have been making progress in transitioning to clean energy sources and remaining competitive in the global economy through collaborative efforts between African nations and the international community, including foreign investment and technical assistance as well as increased regional cooperation and the implementation of effective policies that support the development of clean energy sources and encourage innovation to meet the challenges of transitioning to clean energy sources and remain competitive in the global economy.

This demonstrates that there are many obstacles for African countries in their efforts to transition to clean energy sources and remain competitive in the global economy, but with increased collaboration between the international community and African nations, as well as effective policies that support the development of clean energy sources and encourage innovation, it is possible for African countries to overcome these challenges and succeed in transitioning to clean energy sources while remaining competitive in the global economy. African countries have thus demonstrated that, with proper support from the international community and effective policies in place, it is possible for them to overcome the many obstacles in their paths and succeed in transitioning to clean energy sources while remaining competitive in the global economy. Despite the significant challenges that African countries face, such as corruption, poor infrastructure, and an inadequate education system, it is possible for African countries to transition to clean energy sources and remain competitive in the global economy with the help of foreign investment and technical assistance, as well as increased regional cooperation and the implementation of effective policies that support the development of clean energy sources and encourage innovation.

In conclusion, the transition to clean energy sources and remaining competitive in the global economy is an achievable goal for African countries with increased collaboration between African nations and the international community, including foreign investment and technical assistance as well as increased regional cooperation and the implementation of effective policies that support the development of clean energy sources and encourage innovation. African countries have the potential to overcome these obstacles and remain competitive in the global economy, but they will need support from foreign investors and technical assistance from the international community, as well as increased regional cooperation and effective policies that promote clean energy sources and support innovation, to achieve their goals in unprecedented ways, effectively navigate the obstacles they face, and complete the transit successfully.

6.3.3 *Political and Regulatory Challenges*

Political and regulatory issues are one of the challenges in Africa's hard-to-abate sectors, such as energy and transport, which require creative solutions and collaboration between governments, businesses, civil society, and international partners to transition toward a low-carbon economy. These issues are not unique to Africa; however, due to its lack of access to technology and resources, the continent is especially vulnerable to the effects of climate change and could be greatly impacted by failure to address these challenges quickly and effectively. To reduce carbon emissions and mitigate climate change, it is essential for African countries to implement targeted solutions that address both the challenges and the underlying causes of these political and regulatory issues to achieve their ambitious climate goals.

One of the most important initiatives in this regard is the establishment of an African Climate Change and Sustainable Development Alliance, which would bring together experts from all relevant sectors to develop a comprehensive strategy to address climate change-related issues and create a collective platform for decision makers, businesses, civil society organizations, and international partners to share best practices and coordinate action on climate adaptation and mitigation measures. This alliance should also serve to promote collaboration between stakeholders and support African countries in the transition to a low-carbon economy by providing access to the latest technological advancements, capacity building, and financial resources needed for the implementation of sustainable energy solutions at the national, regional, and international levels.

In general, the African continent is particularly vulnerable to the effects of climate change due to its lack of access to resources and technology and its disproportionate reliance on natural resources for economic and social development. However, for African countries to meet their ambitious climate goals and reduce emissions, it is essential to develop tailored solutions that address the underlying causes of these political and regulatory issues that limit the ability of African countries to address climate change while also creating job opportunities, increasing access to energy, and unlocking the potential of Africa's vast renewable energy resources. To achieve these goals, the African Climate Change and Sustainable Development Alliance should actively collaborate with local governments and civil society organizations to identify areas of need, develop policies and regulations that support climate action, and create financing mechanisms that facilitate investments in low-income communities and renewable energy projects at the regional and national levels.

In this way, the Alliance can help to close the gaps between African countries and the global community in terms of access to energy, technological advances, and financial resources needed for transitioning to a low-carbon economy. Additionally, the Alliance should support capacity-building efforts within African countries to build the knowledge and skills needed for sustainable climate action that is critical for transitioning to a low-carbon economy while ensuring that economic growth, job opportunities, and sustainable development are at the core of Africa's climate ambitions. By working with African countries, the Alliance can support the development of innovative and tailored solutions to address climate change while also creating opportunities for economic growth, jobs, and sustainable development that benefit all of Africa's citizens and beyond. By working together to create policies, regulations, and financing mechanisms that support African countries in their climate goals, the African Climate Change and Sustainable Development Alliance can play an important role in ensuring that all African countries are able to benefit from a low-carbon future.

The African Climate Change and Sustainable Development Alliance (ACC-SDA) can provide a platform for collaboration between African countries and global actors to ensure the successful transition of African countries to a low-carbon economy while enabling access to the resources and technologies needed to build resilient communities and economies (Nwokolo et al., 2022b). The ACC-SDA can facilitate dialogue and promote best practices, knowledge exchange, and capacity building to ensure that African countries can transition to a low-carbon future in an equitable and sustainable way. The ACC-SDA should also focus on developing policies and regulations that will incentivize investment in clean energy and low-carbon solutions, as well as create a framework for African countries to access finance for their climate action plans and projects. In this way, the African Climate Change and Sustainable Development Alliance can act as a catalyst for global action on climate change while simultaneously driving economic growth and job creation in Africa.

In addition, the ACC-SDA can provide a platform for African countries to come together and share their experiences, identify common challenges, and develop joint strategies to address climate change and build a more sustainable future for the continent. By utilizing the ACC-SDA's broad range of activities and initiatives, African countries can make a successful transition to a low-carbon economy, access the resources and technologies needed to build resilient communities and economies, create policies and regulations that incentivize the uptake of green investments and technologies, and help ensure that Africa is at the forefront of global climate action and collaboration. As a result, the ACC-SDA can have an immense impact on Africa's present and future by helping African countries build a more sustainable and resilient future, one that is equipped to tackle the challenges of climate change while also leveraging green investments and technologies to drive economic growth, create jobs, and improve the quality of life for Africans across the continent.

The ACC-SDA's ability to foster collaboration, share best practices, and provide access to resources and technology makes it an invaluable resource for African countries looking to build a more sustainable and resilient future in the face of climate change. To achieve this, the ACC-SDA provides a wide range of activities and initiatives that allow African countries to access the resources and technologies they need to make a successful transition to a low-carbon economy. These activities and initiatives include supporting research and development, providing access to finance and investment opportunities, building technical capacity, promoting education and awareness campaigns, and strengthening collaboration between African countries to create an enabling environment for successful green investments and technologies. The ACC-SDA's support for research and development is particularly crucial, as it helps African countries stay ahead of the curve in regard to developing innovative solutions to address the impacts of climate change and access the cutting-edge green technologies needed to mitigate and adapt to its effects. The ACC-SDA also works to build technical capacity in African countries by providing access to training, information, and resources that empower African countries with the knowledge and expertise needed to make effective use of green technologies and help them transition to a low-carbon economy.

Additionally, the ACC-SDA offers education and awareness campaigns that focus on educating citizens about climate change and its effects, as well as empowering them with the information they need to make informed decisions about their energy consumption and the ways in which they can contribute to mitigating the effects of climate change. By providing these initiatives and activities, the ACC-SDA seeks to create an enabling environment for green investments and technologies to be successful and help African countries move toward a low-carbon, climate-resilient, and sustainable development path that is beneficial to both current and future generations. The ACC-SDA's support is invaluable, as it helps African countries stay abreast of the latest developments and take advantage of opportunities to benefit from green technology while also creating a more informed and engaged public that is better able to understand the importance of their decisions in contributing to global climate action.

The ACC-SDA's initiatives have the potential to transform African countries and make them more competitive in a global economy that is increasingly focused on sustainability and environmental sustainability. By empowering citizens to understand the effects of climate change and the importance of making sustainable decisions, the ACC-SDA's education and awareness campaigns can help African countries become better prepared for a future with a changing climate. In addition to providing educational and awareness activities, the ACC-SDA also supports green investments and technologies in African countries by providing direct support to African entrepreneurs, researchers, and innovators who are working to develop new solutions to tackle climate change and promote sustainable development on the continent. The ACC-SDA is an invaluable asset to African countries in the fight against climate change and the promotion of sustainable development. The ACC-SDA's efforts have the potential to have a lasting impact on African countries and make them more competitive in an ever-changing global economy. By providing support for African entrepreneurs, researchers, and innovators to develop solutions for climate change, the ACC-SDA is giving African countries the opportunity to take ownership of their future and become leaders in sustainability.

Furthermore, the ACC-SDA is helping to ensure that African countries can benefit from the latest green technologies and investments, enabling them to participate in a global market that increasingly demands eco-friendly solutions to tackle the most pressing environmental issues facing the world today. The ACC-SDA is also working to raise awareness of the devastating effects of climate change on African countries and to empower communities with the tools they need to adapt and survive in the face of these changes. Ultimately, the ACC-SDA is playing a crucial role in transforming Africa into an environmentally friendly region and giving its citizens the capacity to confront environmental challenges and build a more secure, prosperous future. Through these efforts, the ACC-SDA is paving the way for a more prosperous and sustainable future for Africa, with the potential to transform African countries into leading players in global economic growth and development. The ACC-SDA is focused on strengthening partnerships between African governments and the private sector to secure the necessary investments and resources needed to achieve green development goals and create economic opportunities that contribute to the development of resilient, resource-efficient, and low-carbon societies across the continent.

In addition, the ACC-SDA is actively advocating for international support and resources to address climate change in Africa, ensuring that African countries receive their fair share of the global efforts toward a low-carbon economy. The ACC-SDA is also committed to raising awareness about environmental issues and solutions among African citizens, encouraging participation in sustainable development initiatives, and promoting behavioral changes that lead to improved environmental outcomes. This focus on partnership, advocacy, and awareness is critical to the success of the ACC-SDA's mission to improve the environmental and economic resilience of African societies and enable sustainable development on the continent.

As such, the ACC-SDA provides a crucial platform for African countries to develop their green economies and to achieve their targets in terms of sustainable development and climate change mitigation and adaptation. This in turn helps African countries take full advantage of global climate action and benefit from the new economic opportunities that arise from a low-carbon transition. By engaging in collaborations and partnerships with government, civil society organizations, and the private sector, the ACC-SDA can help ensure that African nations are not left behind in terms of environmental protection and sustainable development and contribute to the global effort to build a more sustainable future.

In doing so, the ACC-SDA plays a critical role in bringing together stakeholders to work together toward achieving African nations' sustainable development goals. The ACC-SDA is therefore an important platform for Africa to ensure that it is able to leverage climate action and new economic opportunities while also protecting the environment and contributing to global efforts toward a more sustainable future. With its multilateral approach, the ACC-SDA is uniquely positioned to make a tangible contribution to the fight against climate change and ensure that African nations are able to build resilience, drive economic growth, and provide improved standards of living for their citizens by collaborating in the areas of climate change adaptation, mitigation, and disaster risk management. This collaboration allows African nations to benefit from global knowledge and experience in the area of sustainable development and provides a platform for stakeholders to develop innovative solutions and share best practices across the continent.

This makes the ACC-SDA an essential institution for Africa, as it serves as a vital bridge between African nations and the global community in regard to taking meaningful action on climate change and achieving sustainable development goals. The ACC-SDA is an invaluable asset for Africa, helping it to address the multiple challenges of climate change, economic growth, and sustainable development while providing opportunities for African nations to engage in meaningful dialogue with global partners, build relationships, and coordinate activities to create lasting solutions that will benefit the continent and its citizens in the long term. By leveraging the combined power of African nations, the ACC-SDA is able to make an unparalleled contribution to the fight against climate change and sustainable development throughout the continent.

6.4 Conclusion

African countries must take urgent action to decarbonize their hard-to-abate sectors, such as transportation, heating and cooling, and industry, to meet their development objectives. To reduce emissions, African countries must adopt a multipronged approach that involves modernizing and decarbonizing the sector, investing in sustainable technologies, and training local personnel. By investing in renewable energy, diversifying their energy resources, implementing climate risk management strategies, and creating regional and international collaboration networks, African countries can reduce their reliance on fossil fuels, increase access to clean energy, create jobs, and promote economic development while simultaneously reducing emissions and contributing to the global fight against climate change. The international community must provide support for capacity building, technology transfer, and access to resources to ensure that African countries can transition to a low-carbon and climate-resilient future and create a sustainable future for their people. The shipping industry in Africa must prioritize and invest in research and development to facilitate greener operations and better enable the industry to meet its environmental goals.

African governments should promote the development of environmentally friendly transportation infrastructure and establish energy policies that are financially viable while also having a positive impact on the environment. To mitigate the effects of air travel on the climate, African countries should invest in technological capacity and infrastructure and promote public–private partnerships to spur innovation and the development of low-carbon air travel solutions.

African countries are facing a major challenge in their fight against climate change due to their lack of resources and access to capital. To address this, international organizations such as the UN and World Bank have been working to promote the development of clean energy technologies to reduce greenhouse gas emissions, improve the health of people living in African countries, and mitigate the impacts of climate change. These initiatives include providing financial support to developing countries, developing innovative financing schemes, and mandating that a certain percentage of their electricity comes from renewable sources. Additionally, the UN and other international organizations are providing technical assistance to build capacity in clean energy technology, such as improving grid systems and investing in research and development for new renewable energy technologies. These efforts are essential to helping African countries transition to low-carbon energy sources, reduce their dependence on fossil fuels, and provide access to more affordable and reliable electricity while also creating jobs and other economic opportunities.

However, there is a need for more financial support and resources to be devoted to this cause to achieve the goal. These efforts must focus on technical assistance, capacity-building training, and supporting local businesses, as well as economic incentives for companies to invest in renewable energy infrastructure and policy reforms to incentivize the development of renewable energy sources. African countries are investing in renewable energy infrastructure and policy reforms to meet their goals of providing sustainable and reliable access to electricity while also protecting the environment and mitigating the effects of climate change. However, they face technological limitations and a lack of financial resources and knowledge, as well as structural and cultural issues such as corruption, poor infrastructure, and an inadequate education system. Despite these challenges, African countries have been making progress through collaborative efforts between African nations and the international community.

With proper support from the international community and effective policies, it is possible to transition to clean energy sources and remain competitive in the global economy. The African continent is particularly vulnerable to the effects of climate change due to its lack of access to resources and technology and its disproportionate reliance on natural resources. To reduce carbon emissions and mitigate climate change, African countries need to implement targeted solutions that address both the challenges and the underlying causes of these issues. To achieve this, the African Climate Change and Sustainable Development Alliance (ACC-SDA) should collaborate with local governments and civil society organizations to identify areas of need, develop policies and regulations that support climate action, and create financing mechanisms that facilitate investments in low-income communities and renewable energy projects. The ACC-SDA can act as a catalyst for global action on climate change while simultaneously driving economic growth and job creation in Africa.

References

Agbor, M., Udo, S., Ewona, I., Nwokolo, S., & Ogbulezie, J. A. (2023a). Potential impacts of climate change on global solar radiation and PV output using the CMIP6 model in West Africa. *Cleaner Engineering and Technology, 13*, 100630. https://doi.org/10.1016/j.clet.2023.100630

Agbor, M., Udo, S., Ewona, I., Nwokolo, S., Ogbulezie, J., Amadi, S., & Billy, U. (2023b). Effects of angstrom-Prescott and Hargreaves-Samani coefficients on climate forcing and solar PV technology selection in West Africa. *Trends in Renewable Energy, 9*(1), 78–106. https://doi.org/10.17737/tre.2023.9.1.00150

Azadnia, A. H., McDaid, C., Andwari, A. M., & Hosseini, S. E. (2023, August). Green hydrogen supply chain risk analysis: A European hard-to-abate sectors perspective. *Renewable and Sustainable Energy Reviews, 182*, 113371. https://doi.org/10.1016/j.rser.2023.113371

Bokka, H. K., & Lau, H. C. (2023, July). Decarbonizing Borneo's power and industry sectors by carbon capture and storage. *Geoenergy Science and Engineering, 226*, 211796. https://doi.org/10.1016/j.geoen.2023.211796

Callister, P., & McLachlan, R. (2023, May 31). Decarbonizing Aotearoa New Zealand's aviation sector: Hard to abate, but even harder to govern. *Policy Quarterly, 19*(2), 9–18. https://doi.org/10.26686/pq.v19i2.8232

Chukwujindu, S., Umunnakwe, A., & Ogbulezie, J. C. (2022). Hybridization of statistical machine learning and numerical models for improving beam, diffuse and global solar radiation prediction. *Cleaner Engineering and Technology, 9*(May), 100529. https://doi.org/10.1016/j.clet.2022.100529

Fragkos, P. (2022, December 19). Decarbonizing the international shipping and aviation sectors. *Energies, 15*(24), 9650. https://doi.org/10.3390/en15249650

Hassan, M. A., Bailek, N., Bouchouicha, K., Ibrahim, A., Jamil, B., Kuriqi, A., & Nwokolo, S. C. (2022). Evaluation of energy extraction of PV systems affected by environmental factors under real outdoor conditions. *Theoretical and Applied Climatology, 150*, 715. https://doi.org/10.1007/s00704-022-04166-6

Hassan, M. A., Bailek, N., Bouchouicha, K., & Nwokolo, S. C. (2021). Ultrashort-term exogenous forecasting of photovoltaic power production using genetically optimized nonlinear autoregressive recurrent neural networks. *Renewable Energy, 171*(February), 191–209. https://doi.org/10.1016/j.renene.2021.02.103

Lah, O. (2017, September 12). Decarbonizing the transportation sector: Policy options, synergies, and institutions to deliver on a low-carbon stabilization pathway. *Wiley Interdisciplinary Reviews: Energy and Environment, 6*(6), e257. https://doi.org/10.1002/wene.257

Nwokolo, S. C., Amadi, S. O., Obiwulu, A. U., Ogbulezie, J. C., & Eyibio, E. E. (2022a). Prediction of global solar radiation potential for sustainable and cleaner energy generation using improved Angstrom-Prescott and Gumbel probabilistic models. *Cleaner Engineering and Technology, 6*(January), 100416. https://doi.org/10.1016/j.clet.2022.100416

Nwokolo, S. C., Obiwulu, A. U., & Ogbulezie, J. C. (2022b). Machine learning and analytical model hybridization to assess the impact of climate. *Physics and Chemistry of the Earth, 130*(February), 103389. https://doi.org/10.1016/j.pce.2023.103389

Nwokolo, S. C., & Ogbulezie, J. C. (2018a). A qualitative review of empirical models for estimating diffuse solar radiation from experimental data in Africa. *Renewable and Sustainable Energy Reviews., 92*, 353. https://doi.org/10.1016/j.rser.2018.04.118

Nwokolo, S. C., & Ogbulezie, J. C. (2018b). A quantitative review and classification of empirical models for predicting global solar radiation in West Africa. *Beni-Suef University Journal of Basic and Applied Sciences, 7*(4), 367–396. https://doi.org/10.1016/j.bjbas.2017.05.001

Nwokolo, S. C., Ogbulezie, J. C., & Obiwulu, A. U. (2022c). Impacts of climate change and meteo-solar parameters on photosynthetically active radiation prediction using hybrid machine learning with physics-based models. *Advances in Space Research., 70*, 3614. https://doi.org/10.1016/j.asr.2022.08.010

Nwokolo, S., & Ogbulezie, J. (2017a). A critical review of theoretical models for estimating global solar radiation between 2012-2016 in Nigeria. *International Journal of Physical Research, 5*(2), 60. https://doi.org/10.14419/ijpr.v5i2.8160

Nwokolo, S., & Ogbulezie, J. (2017b). Modelling the influence of cloudiness on diffuse horizontal irradiation under various sky conditions in Nigeria. *International Journal of Physical Research, 5*(2), 91. https://doi.org/10.14419/ijpr.v5i2.8312

Nwokolo, S., & Otse, C. (2019). Impact of sunshine duration and clearness index on diffuse solar radiation estimation in mountainous climate. *Trends in Renewable Energy, 5*(3), 307–332. https://doi.org/10.17737/tre.2019.5.3.00107

Nwokolo, S. C., Obiwulu, A. U., Amadi, O., & Ogbulezie, J. C. (2023a). Assessing the impact of soiling, tilt angle, and solar radiation on the performance of solar PV systems. *Trends in Renewable Energy, 9*(2), 121–137. https://doi.org/10.17737/tre.2023.9.2.00156

Nwokolo, S. C., Ogbulezie, J. C., & Ushie, O. J. (2023b). A multimodel ensemble-based CMIP6 assessment of future solar radiation and PV potential under various climate warming scenarios. *Optik, 285*(May), 170956. https://doi.org/10.1016/j.ijleo.2023.170956

Nwokolo, S. C., Proutsos, N., & Meyer, E. L. (2023c). Machine learning and physics-based hybridization models for evaluation of the effects of climate change and urban expansion on photosynthetically active radiation. *Atmosphere, 14*(687), 1–36. https://doi.org/10.3390/atmos14040687

Obiwulu, A. U., Chendo, M. A. C., Erusiafe, N., & Nwokolo, S. C. (2020a). Implicit meteorological parameter-based empirical models for estimating back temperature solar modules under varying tilt-angles in Lagos, Nigeria. *Renewable Energy, 145*, 442. https://doi.org/10.1016/j.renene.2019.05.136

Obiwulu, A. U., Erusiafe, N., Olopade, M. A., & Nwokolo, S. C. (2020b). Modelling and optimization of back temperature models of mono-crystalline silicon modules with special focus on the effect of meteorological and geographical parameters on PV performance. *Renewable Energy, 154*, 404. https://doi.org/10.1016/j.renene.2020.02.103

Obiwulu, A. U., Erusiafe, N., Olopade, M. A., & Nwokolo, S. C. (2022). Modelling and estimation of the optimal tilt angle, maximum incident solar radiation, and global radiation index of the photovoltaic system. *Heliyon, 8*(6), e09598. https://doi.org/10.1016/j.heliyon.2022.e09598

Ogbulezie, J. C., James, U. O., & Chukwujindu, N. S. (2017). A review of regression models employed for predicting diffuse solar radiation in North-Western Africa. *Trends in Renewable Energy*. https://doi.org/10.17737/tre.2017.3.2.0042

Paltsev, S., Gurgel, A., Morris, J., Chen, H., Dey, S., & Marwah, S. (2022, August). Economic analysis of the hard-to-abate sectors in India. *Energy Economics, 112*, 106149. https://doi.org/10.1016/j.eneco.2022.106149

Paltsev, S., Morris, J., Kheshgi, H., & Herzog, H. (2021, October). Hard-to-abate sectors: The role of industrial carbon capture and storage (CCS) in emission mitigation. *Applied Energy, 300*, 117322. https://doi.org/10.1016/j.apenergy.2021.117322

Prussi, M., & Chiaramonti, D. (2022, December 1). Alternative fuels for hard-to-abate sectors: A carbon intensity assessment. *Journal of Physics: Conference Series, 2385*(1), 012044. https://doi.org/10.1088/1742-6596/2385/1/012044

Rumayor, M., Dominguez-Ramos, A., & Irabien, A. (2020, July 22). Toward the Decarbonization of hard-to-abate sectors: A case study of the soda ash production. *ACS Sustainable Chemistry & Engineering, 8*(32), 11956–11966. https://doi.org/10.1021/acssuschemeng.0c01598

Samuel Chukwujindu, N. (2017). A comprehensive review of empirical models for estimating global solar radiation in Africa. *Renewable and Sustainable Energy Reviews, 78*(October), 955–995. https://doi.org/10.1016/j.rser.2017.04.101

Santos, D. A., Dixit, M. K., Pradeep Kumar, P., & Banerjee, S. (2021, November). Assessing the role of vanadium technologies in decarbonizing hard-to-abate sectors and enabling the energy transition. *IScience, 24*(11), 103277. https://doi.org/10.1016/j.isci.2021.103277

Superchi, F., Mati, A., Carcasci, C., & Bianchini, A. (2023, July). Techno-economic analysis of wind-powered green hydrogen production to facilitate the decarbonization of hard-to-abate sectors: A case study on steelmaking. *Applied Energy, 342*, 121198. https://doi.org/10.1016/j.apenergy.2023.121198

Superchi, F., Mati, A., Pasqui, M., Carcasci, C., & Bianchini, A. (2022, December 1). Techno-economic study on green hydrogen production and use in hard-to-abate industrial sectors. *Journal of Physics: Conference Series, 2385*(1), 012054. https://doi.org/10.1088/1742-6596/2385/1/012054

Zhang, K., Lau, H. C., Bokka, H. K., & Hadia, N. J. (2022, June). Decarbonizing the power and industry sectors in India by carbon capture and storage. *Energy, 249*, 123751. https://doi.org/10.1016/j.energy.2022.123751

Chapter 7
Impacts of Climate Change in Africa

7.1 Introduction

Due to poverty, poor adaptation capabilities, and reliance on natural resources such as water, forests, and agriculture, Africa is vulnerable to climate change. African countries face serious difficulties due to rising temperatures, droughts, floods, crop failures, displacement, and conflict due to limited resources. This situation is further exacerbated by the fact that African nations have limited financial and technological resources available to address this urgent issue.

According to some experts, Africa is the world's most susceptible region to climate change due to issues such as poverty, sluggish economies (Combating Climate Change in Africa: Hearing Before the Subcommittee on Africa and Global Health of the Committee on Foreign Affairs, 2010), and limited access to resources (Berck et al., 2018), all of which make it difficult to respond to (Addaney & Dumenu, 2023) and adapt to changing climate conditions (Adepoju, 2021). Africa faces severe impacts from climate change, including food insecurity (Ajala & Chagwiza, 2022), water scarcity (Anekwe et al., 2023), displacement (Dietz & Engels, 2018), and increased infectious diseases (Asafu-Adjaye, 2010). These consequences will have severe consequences for African populations and the global community (DiMento & Doughman, 2007).

Rapid urbanization, overuse of natural resources, and unsustainable economic practices can reduce resilience to extreme weather events (Gbenga, 2014). These factors exacerbate the deterioration of communities' ability to recover from such events. In addition, the lack of access to essential services such as health care, education, and financial assistance can lead to increased poverty and greater insecurity. Factors such as biodiversity loss (Guzmán, 2019), pollution (Igbokwe & Mkpado, 2011), soil erosion (Zewdie, 2014), deforestation (Adefisan, 2022), and desertification exacerbate existing environmental challenges (Joseph et al., 2020), affecting ecosystems' ability to provide essential goods and services (Kang, 2022). This increases vulnerability and worsens the impacts of climate change.

S. C. Nwokolo et al., *Africa's Path to Net-Zero*, CSR, Sustainability, Ethics & Governance, https://doi.org/10.1007/978-3-031-44514-9_7

The degradation of ecosystems exacerbates climate change and poverty, weakening local communities' ability to adapt and recover from extreme weather events and creating a vicious cycle of interrelated impacts (Letcher, 2009). This has created a complex and devastating situation as local communities become trapped in a cycle of poverty and environmental degradation while also struggling to cope with the increasing frequency and intensity of climate-related disasters that can have a direct and devastating impact on their lives and livelihoods.

Climate change threatens African communities and economies by affecting agricultural production (Li et al., 2022), food security (Moreland & Smith, 2013), water resources (Ndhlovu et al., 2020), and ecosystem services (Njenga et al., 2013). Rising temperatures, changing rainfall patterns, and sea level rise all contribute to more frequent and intense extreme weather events, which affect crop yields, livestock health, and soil fertility, resulting in lower food quality and quantity.

The IPCC predicted in 2007 that agricultural productivity and food security in African nations and regions would be significantly impacted by climate change and climate variability (Oguntunde et al., 2017). Africa's high rate of poverty, reliance on climate-sensitive industries (Ojewola et al., 2021), and lack of adaptive capacity all contribute to the continent's vulnerability to climate change (Pauleit et al., 2015). Strategies for risk reduction and adaptation must be incorporated into ecosystem management and agricultural production systems to manage this risk. Populations may be forced to relocate on a large scale as a result of environmental degradation and declining food security brought on by climate change and variability.

Sustainable agricultural methods such as agroforestry (Perkins, 2013), conservation tillage (Pittock, 2005), crop diversification, and water harvesting are promoted to lower the risk of climate change. Agroforestry enhances soil fertility, water retention, erosion control, and pest management while producing both wood and nonwood goods (Romm, 2018). Additionally, it lowers the likelihood that crops will fail due to extreme weather, boosting small-scale farmers' productivity and adaptability. In turn, this lessens hunger and poverty in the area and enhances livelihoods.

Nearly the entire surface of the Earth is anticipated to warm due to climate change in the near future, and average global rainfall will rise as a result. This increase in temperature and rainfall will lead to longer and more frequent droughts (Sokona, 2001), higher sea levels, more intense storms, stronger storm surges, and larger wildfire areas across many regions, resulting in more frequent and severe floods, an increase in the intensity of droughts, changes to ocean temperature and acidity, and a decline in arctic sea ice and glacier mass (Sokona & Denton, 2001). These changes will have numerous implications for ecosystems, human health, water resources, food and energy production, coastal communities, and global economic systems (Strydom & Savage, 2017).

Due to the faster rate of global warming, Africa is more susceptible to the effects of climate change. This is brought on by high rates of poverty, a lack of resources for adaptation, and a reliance on natural resources for economic growth and subsistence. Africa is an important region for adaptation and mitigation due to its extreme poverty levels and reliance on natural resources (United States Congress, 2018). Furthermore, Africa is highly dependent on its agricultural production for food security and nutrition, making it very vulnerable to the increasingly unpredictable weather patterns and changing rainfall patterns caused by climate change. Many African countries struggle with climate change response due to limited resources, weak institutions, technology access, and inadequate skills for planning and implementing adaptation measures (Wright & Kapwata, 2023).

The swift impacts of global warming may render vast areas of the continent uninhabitable, which would have catastrophic consequences for human health, nutrition security, and deprivation. Moreover, this has the potential to displace millions of people, potentially leading to new waves of migration and further economic instability that would further complicate the lives of those affected and their surrounding communities and escalating existing geopolitical tensions. As the effects of climate change become increasingly evident, they will have an enormous impact on the global population, both directly through natural disasters and changes in air quality as well as indirectly by destabilizing economies and societies. These changes, if left unmitigated, will cause serious disruptions to both the human and natural worlds. Thus, it is clear that climate change will have far-reaching effects and is already beginning to have an impact on the world (Ziervogel et al., 2014).

Change signs at specific sites are slightly inevitable, but regional effects on rainfall in the tropics are expected to be more spatially variable (Hassan et al., 2021). Variability in precipitation across the tropics may be attributed to the existence of numerous wet and dry regions, and varying rainfall amounts are anticipated across different regions (Hassan et al., 2022). To further complicate the picture, different regions may respond to changing climate conditions in different ways due to factors such as ocean circulation (Nwokolo et al., 2022a), topography (Nwokolo et al., 2022b), and land cover changes (Nwokolo et al., 2022c). Precipitation patterns in the tropics are likely to become more variable as climate change progresses (Nwokolo et al., 2023a), leading to uncertain and disparate outcomes across regions (Nwokolo et al., 2023b) and further complicating efforts to predict and plan for climate-related changes (Nwokolo et al., 2023c). To accurately assess regional precipitation patterns, comprehend the effects of climate change in particular areas, and create effective adaptation plans to address the effects of the change as it happens, localized assessments are essential.

Surface temperatures in Africa have increased by approximately 1 °C between the late 19th and early 21st centuries, with local temperatures reaching up to 3 °C (United States Congress, 2018). This rise is attributed to factors such as urbanization, deforestation, and land-use changes, which have led to increased greenhouse gas emissions (Obiwulu et al., 2020a). This has impacted African ecosystems, increasing the frequency and intensity of extreme weather events (Obiwulu et al., 2020b), decreasing biodiversity, and altering the frequency and location of seasonal monsoonal rains (Obiwulu et al., 2022). This has led to increased drought (Samuel, 2017), desertification (Ogbulezie et al., 2017), and water scarcity (Nwokolo & Ogbulezie, 2017a). Climate change poses a significant challenge to the sustainability of African ecosystems and the well-being of its people. It is crucial for African nations to reduce greenhouse gas emissions, implement solutions to mitigate climate change effects, and ensure the future sustainability of their ecosystems.

Due to its reliance on subsistence agriculture, which is impacted by weather patterns (Nwokolo & Ogbulezie, 2017b), and its tropical location (Nwokolo & Ogbulezie, 2018a), Kenya faces high vulnerability to the effects of climate change (Nwokolo & Ogbulezie, 2018b). Due to limited access to resources such as technology and finance, as well as a limited capacity for adaptation and resilience (Nwokolo & Ogbulezie, 2018c), the rural population, which is heavily reliant on natural resources (Nwokolo & Otse, 2019), is also vulnerable.

Kenya is already facing a decrease in agricultural production, which is affecting food security for the country's population. These impacts are likely to worsen in the future if climate change continues to accelerate, as it is predicted to do, unless significant steps are taken to mitigate and adapt to these changes. Kenya needs to act right away to prepare for and lessen the effects of climate change. It should concentrate on a low-carbon economy, make the switch away from fossil fuels, spend money on renewable energy sources such as solar and wind power, and improve its infrastructure to accommodate the changing climate. In the upcoming years, this will shield its populace from the worsening effects of climate change.

To combat the effects of climate change, regional actors are putting into action a variety of adaptation plans, such as the SADC Policy Paper on Climate Change and the adaptation plan for the water sector (United States Congress, 2018). However, there is a need for these strategies to be accompanied by effective implementation plans with clear and measurable goals as well as appropriate budget allocations to ensure that they are effectively put into action and to ensure their success. Such implementation plans must include robust monitoring and evaluation systems to track progress and ensure that the strategies are effective in addressing regional climate change adaptation challenges. Additionally, there is a need for strong collaboration between government departments and stakeholders.

By bolstering national strategies, encouraging regional cooperation, and exchanging knowledge, information, experiences, and best practices, the Programme on Climate Change Adaptation and Mitigation in Eastern and Southern Africa (COMESA-EAC-SADC) aims to improve African countries' capacity to manage climate change (Strydom & Savage, 2017). The program focuses on increasing regional capacity in legal and institutional frameworks, strengthening regional networks for early warning systems, and sharing expertise to effectively manage climate change.

With the success of such programs, African countries can better manage, build resilience, and achieve sustainable development goals. The program is a positive step in the right direction, providing African countries with an important tool for responding to the increasingly urgent threat posed by climate change and its associated impacts. This program will ensure that African countries are better able to anticipate and share, as well as build capacity for legal and policy frameworks, scientific research, and early warning systems to help them.

The African Union proposed 47 climate change objectives and actions in a 2014 draft report for its 55 member states. These objectives and actions were designed to address the pressing issue of climate change, which has had a devastating impact on the African continent. The report emphasized the need for member states to take urgent action to reduce greenhouse gas emissions and increase their resilience to the impacts of climate change. Among the proposed actions were measures to promote renewable energy, improve energy efficiency, and increase access to clean water and sanitation. The report also called for increased investment in sustainable agriculture and forestry practices, as well as efforts to protect biodiversity and ecosystems. In addition, it highlighted the importance of promoting education and awareness about climate change among African citizens. Overall, the African Union's proposed objectives and actions represent a comprehensive approach to tackling one of the greatest challenges facing our planet today. By working together to implement these measures, member states can help ensure a sustainable future for generations to come.

Through the implementation of this plan, African countries have already seen significant progress in improving the resilience of their infrastructure and curbing the effects of climate change. This plan has become an example to the rest of the world on how to effectively address climate-related issues and achieve a sustainable, low-carbon future. The African Union's "2030 Energy Plan for Sustainable Development" was an ambitious blueprint, and its success has been a testament to its foresight and the hard work of the African countries that have adopted it. The results of the plan's implementation speak for themselves: renewable energy sources are becoming increasingly commonplace, infrastructure is more resilient than ever, and economic stability has been greatly improved as a result of this plan.

In addition to the progress that African countries have made, the successful implementation of this plan has also had an impact beyond their borders. Around the world, other countries have looked to the African Union's 2030 Energy Plan as a template for achieving their own sustainable and low-carbon goals. By highlighting the successes of the African Union's 2030 Energy Plan, other countries have been inspired to take action and join in the effort to reduce emissions and adopt renewable energy sources.

The African Union's 2030 Energy Plan was truly ground-breaking in its ambition, and its successes have been a model for countries around the world looking to implement their own sustainable energy plans. This is a remarkable accomplishment and one that stands as a testament to the hard work and commitment of the African Union and its member states in helping to fight climate change on a global scale and leading the way in transitioning to a net zero economy.

As part of his efforts to combat climate change, the UN Secretary General stresses cooperation with the African Union. This close collaboration with the African Union is necessary to address the environmental, economic, and social issues facing African countries today, given their unique geographic, climatic, and economic challenges resulting from their dependence on natural resources and their relative lack of technological advancements (Nwokolo et al., 2022a). To assist African countries in dealing with these unique challenges, the union has made a commitment to work closely with the African Union and other regional organizations to ensure that African countries can transition toward a greener economy while maintaining strong economic growth.

This commitment involves providing financial and technical assistance to African countries, as well as promoting sustainable development practices that can help reduce the negative impact of natural resource extraction on the environment. Additionally, the union is working to promote greater regional integration among African countries, which can help facilitate the sharing of resources and expertise and lead to more effective solutions to common challenges. Ultimately, the goal is to create a more sustainable and prosperous future for all African nations, one that is built on a foundation of environmental stewardship and economic growth. By working together with African leaders and communities, the union believes that it can help create a brighter future for this important region of the world.

Achieving the Sustainable Development Goals in Africa would require an estimated $1.3 trillion in funding each year due to the continent's rising population (United States Congress, 2018). This funding would need to be allocated toward a variety of initiatives, including improving access to education and healthcare, promoting economic growth and job creation, and investing in sustainable infrastructure. Additionally, efforts must be made to address the root causes of poverty and inequality, such as corruption and political instability. To ensure that progress is sustained over the long term, it will also be necessary to prioritize environmental sustainability and combat climate change. Achieving these goals will require collaboration between governments, international organizations, civil society groups, and the private sector. By working together toward a common vision for a more equitable and sustainable future, we can help ensure that Africa's growing population has access to the resources it needs to thrive.

The International Monetary Fund projects that $50 billion may be required to cover the costs of climate adaptation, highlighting the necessity for African governments across the continent to invest in adaptation and mitigation measures. Future generations will be guaranteed a secure future as a result. To better comprehend the science of climate change and create creative solutions to mitigate its effects, long-term research and development are essential. To secure a stable future for generations to come, African governments must prioritize the development of climate mitigation and adaptation strategies that are effective, equitable, and financially viable and that promote sustainable development and environmental stewardship. While the cost of investing in these strategies may seem high in comparison to other priorities, the cost of not investing could be far more severe. It is critical that African governments take urgent action to protect the future of their citizens and safeguard their economies and natural resources from the devastating consequences of a warming planet. For these reasons, African governments must urgently invest in climate mitigation and adaptation strategies that are socially responsible, financially sound, and environmentally sustainable.

In this chapter, the causes of climate change in Africa are discussed, including natural factors such as ocean cycles and El Niño events as well as human-driven causes such as emissions from burning fossil fuels. Additionally, it looks at the elements that make Africa susceptible to climate change, discusses how it affects Africa, and then looks at how to lessen its effects. Climate change is a global issue that affects every corner of the world, and Africa is no exception. The continent is particularly vulnerable to its effects due to a combination of factors such as poverty, weak infrastructure, and limited access to resources. Natural factors such as ocean cycles and El Niño events exacerbate the situation by causing extreme weather conditions such as droughts and floods. Human-driven causes such as emissions from burning fossil fuels also contribute significantly to climate change in Africa. These emissions are mainly from industries, transportation, and deforestation. The effects of climate change on Africa are widespread and devastating, including food insecurity, water scarcity, loss of biodiversity, and increased health risks. To lessen its effects, there is a need for immediate action by both African governments and the international community. This includes implementing policies that promote sustainable development practices such as renewable energy sources, reducing greenhouse gas emissions, improving infrastructure resilience, and investing in adaptation strategies. It is essential to recognize that climate change is not just an environmental issue but also a social one.

The chapter emphasizes collective action and a multifaceted approach to tackle climate change's negative impacts in Africa, emphasizing the need for collective action. With this multifaceted approach, African governments must have access to sufficient financial and technical resources, as well as an understanding of the science and technology behind climate change solutions and the ability to craft effective policies. In addition, African governments must have effective policies and regulations in place to ensure that investments are aligned with the goals of reducing emissions while also incentivizing the implementation of low-carbon and resilient development pathways. Additionally, effective stakeholder engagement and local partnerships are essential for the success of any climate change mitigation or adaptation strategies implemented in Africa.

7.2 Africa's Effects of Changing Rainfall Patterns and Rising Temperatures

As a result of human activity, Africa's climate is changing, as evidenced by the continent's rising temperatures and changing rainfall patterns. In many regions of the continent, these changes have an effect on food production, water availability, and human health. Due to a decrease in sources of safe drinking water, decreased crop yields, and increased vulnerability to extreme weather events such as floods and droughts, people in African countries are at an increased risk of malnutrition, waterborne illnesses, and diseases such as malaria as the effects of climate change worsen. As the world's temperatures rise and the climate changes, these effects are predicted to worsen, causing more people to live in poverty, lose their livelihoods, and be uprooted from their homes in African nations.

As the impacts of climate change on the continent and its people continue to worsen, African nations are likely to experience previously unheard-of pressure on their ecosystems and resources. Unless precautionary measures are taken to reduce greenhouse gas emissions, mitigate the effects of climate change, and ensure access to basic services such as safe drinking water and healthcare, this climate crisis could completely upend the lives of millions, if not billions, of people living in Africa. African countries must act now to avert the looming crisis of climate change by developing and putting into practice regulations that lower greenhouse gas emissions, support clean energy sources and technologies, and make investments in adaptation strategies that will make people and ecosystems more tolerant to the effects of climate change, such as floods, droughts, extreme temperatures, and sea level rise.

Governments in Africa must place a high priority on equitable and sustainable development, ensuring that everyone, regardless of race or socioeconomic status, has access to resources and can profit from economic growth without endangering the environment. This is a difficult and complex task that African countries must undertake to ensure that their people have access to the resources they need to survive and thrive in the face of climate change and to ensure the health and safety of their citizens. African countries must develop and put into action comprehensive plans that center on lowering emissions, promoting clean energy sources and technologies, and investing in adaptation measures to protect their environment and guarantee equal access to resources for all citizens if they are to secure their future and ensure the prosperity of their citizens for future generations. To guarantee that their communities are not excluded from the future effects of climate change, they must also turn to international partners for funding and expertise. To ensure that the most vulnerable communities are not left behind as they travel the challenging road of preparing for and adapting to the impacts of climate change, cooperation, and dialogue between African nations and the international community are crucial.

In general, the future of African nations depends on the successful implementation of an ambitious and comprehensive climate action plan, which will not only help them protect their communities from the devastating impacts of climate change but also ensure that they are able to take advantage of any opportunities that may arise in the future as the world moves toward a more sustainable and climate-resilient future. This must include investment in infrastructure, technology, and education so that African nations can not only adapt to climate change but also capitalize on any new economic opportunities that may arise as a result. However, given the immense financial, technological, and educational resources required to implement ambitious climate action plans and ensure that African nations are resilient in the face of climate change, it is essential that they look to international partners for assistance.

7.3 Clear Indicators of Africa's Changing Climate

Clear indicators of Africa's changing climate can be seen throughout the continent, with desertification in the north, drought, and changing weather patterns in the south, and severe flooding along the West African coast being some of the most prominent examples. These changes have had devastating effects on agriculture and livelihoods, forcing many to leave their homes in search of new opportunities and putting immense pressure on countries' natural resources, infrastructure, and services. The changing climate conditions have further exacerbated the effects of poverty and environmental degradation, leading to a vicious cycle in which those most vulnerable are hurt the most in agriculture and livelihoods, forcing many to leave their homes in search of new opportunities and putting immense pressure on countries' natural resources, infrastructure, and services.

Further compounding these issues, Africa's population is expected to double by 2050, with experts forecasting that over 90% of the population growth will occur in cities and towns as an increasing number of rural people migrate to urban areas in search of economic stability, job opportunities, and access to services such as education and healthcare (Nwokolo et al., 2023a). This mass migration is likely to put an even greater strain on already limited resources, with growing cities and towns struggling to keep up with rising demand for food, water, and energy, leading to a significant rise in food insecurity, water scarcity, and the risk of droughts. Consequently, it is evident that poverty and other vulnerable communities can ultimately have devastating economic, social, and environmental implications for a wide range of stakeholders if the current trends of environmental degradation and poverty are not addressed urgently and drastically reduced in the coming decades.

To mitigate these problems, there is an urgent need for governments and other relevant stakeholders to invest in adaptation measures to better protect those living in poverty and vulnerable communities, as well as to implement long-term solutions such as sustainable development initiatives that focus on improving agricultural practices, preserving water resources, and investing in renewable energy sources to reduce the risk of climate-related disasters in Africa and other areas of the world affected by these changes. While efforts must be made to promote sustainable development and adaptive measures, such efforts should also be accompanied by social protection initiatives that target those living in poverty and other vulnerable communities to provide them with greater access to more resilient and prosperous lives, going against the societal norms of their time.

Despite the many challenges posed by climate change, it is evident that there is still hope for the future of those living in poverty and other vulnerable communities if the appropriate strategies are implemented in a timely and efficient manner and they benefit from the sustainable development initiatives that are currently in place in a way that enables them to actively participate in the process of reducing the risk of climate-related disasters and ultimately ensuring that those living in poverty and other vulnerable communities can be provided with a better quality of life.

This section discusses the clear indicators of Africa's changing climate, including agricultural communities, desertification, crop failures, water shortages, increased poverty and poor governance, and devastating effects on the continent's population. All of these consequences are stark reminders of the very real danger that climate change poses to Africa and, indeed, to the world, which is why we must take action now to address this critical global challenge. By acting now, we can help the most vulnerable populations in Africa adapt to the changing climate, make the necessary investments to strengthen their agricultural and water systems, and improve their ability to cope with the new realities (Nwokolo et al., 2022a). With tens of millions of people in Africa at risk of displacement and extreme poverty due to loss of livelihoods, increased droughts, and other climate-related disasters, in general, climate change poses a grave threat to African nations, their people, and the world as a whole, and we must not delay in taking decisive action to address this pressing issue if we are to protect the future of Africa and our planet.

7.3.1 Agricultural Communities

The impacts of rising temperatures and changing patterns of rainfall in Africa are especially problematic for agricultural communities in Africa, as they depend heavily on natural climate conditions to produce food for their own sustenance and for the wider economy due to their limited access to modern agricultural technologies such as irrigation and fertilizers. Modern technologies are necessary to ensure sustenance and economic growth in agricultural communities in the face of the threat posed by climate change. Limited access to technologies hinders their ability to fully compensate for climate change shifts, leaving communities vulnerable and potentially facing uncertain futures if measures are not taken soon.

Innovative approaches must be put into practice to lessen the effects of climate change on agricultural communities while enabling them to protect themselves, maintain food security, and maintain economic stability. Governments in Africa must aid these communities in creating adaptation plans for the region's changing climes while also preserving their cultural and traditional knowledge. In doing so, sustainable resource management and fair resource allocation will be ensured.

This includes providing them with access to modern agricultural technologies, ensuring they have the capacity to produce enough food for their own sustenance and the wider economy, supporting local climate adaptation initiatives, and helping them develop strategies that enable them to use traditional knowledge of their land to ensure their food security, economic stability, and cultural preservation in a way that is sustainable and equitable for the first time in history. Going forward, governments must take responsibility for helping communities in Africa by providing them with the necessary resources, knowledge, and technological solutions to ensure their food security, economic stability, and cultural preservation in a sustainable and equitable way while also preserving the environment and their traditional knowledge of the land so that they can continue to enjoy their livelihoods and the land in a way that honors the traditions of their ancestors. Ultimately, this is the only way to ensure the future success and sustainability of agricultural communities in Africa, both for them and for future generations.

7.3.2 Desertification, Crop Failures, and Water Shortages

These changes have already caused problems such as desertification, crop failures, and water shortages, which have further increased poverty levels in many African countries and will continue to have far-reaching implications if left unchecked. Millions of people are displaced and put under stress as a result of climate change in African countries, which also results in extreme poverty, destruction of livelihoods, and escalating conflict. This has been particularly true in countries such as Somalia and South Sudan, where recent droughts have caused mass displacement of rural communities searching for food and shelter, causing a significant refugee crisis in the region and leading to a growing humanitarian crisis and an increasing demand for international aid and support.

Millions of people in Africa suffer from poverty and forced migration as a result of climate change (Nwokolo et al., 2023b). This trend will continue if immediate action is not taken, having a negative impact on millions of lives. To lessen the effects of climate change and guarantee vulnerable communities access to resources and a better standard of living, the international community must cooperate. Africa must address both surface effects and root causes if it is to move toward a sustainable and prosperous future.

Africa needs aid and assistance to cut carbon emissions, invest in renewable energy, and improve agricultural practices to combat climate change. Future generations will be able to live in a sustainable and climate-resilient future, protecting their way of life and opening up economic opportunities. The international community must work together to aid African countries in reducing the effects of climate change and putting policies in place to ensure survival and prosperity in the face of it, ultimately raising their standard of living.

The international community ought to make investments in enhancing the capacity of African nations to adapt to the effects of climate change, upgrade infrastructure, and offer essential services to vulnerable populations. Long-term quality of life will be enhanced, and resilience will increase. Together, the international community can assist African nations in combating climate change, ensuring that they are well prepared to address its effects and forge a sustainable future. To build infrastructure that is climate resilient, transition to renewable energy sources, and improve the quality of life for citizens, it is necessary to provide the necessary resources, technical know-how, and financial support.

This, in turn, will reduce their vulnerability to climate change and improve their capacity to respond effectively and efficiently to the threats that come with its increasingly extreme weather patterns and other associated risks. Additionally, by working together to promote and implement sustainable agricultural practices in African countries, the international community can increase food security and production levels while also helping to protect the environment and reduce the reliance of farmers on traditional methods that have a negative environmental impact, such as deforestation and overcultivation of land. This would also help to prevent soil erosion, reduce water and air pollution, conserve water resources, and protect biodiversity, which are essential components of an adaptive and resilient society.

Citizens in African nations who have access to climate-related information and education are better able to use natural resources wisely and lessen the impact of climate change on their lives and way of life. This could lead to increased agricultural productivity and resilience, reduced poverty levels, improved public health outcomes, and greater overall socioeconomic development in African countries. In sum, by investing in sustainable agriculture and promoting access to climate-related information and education for citizens in African countries, the international community can contribute to reducing poverty levels and improving public health outcomes while also conserving natural resources and protecting the environment.

7.3.3 Increased Poverty and Poor Governance

These changes are likely to have serious implications for many African countries, particularly those that are already struggling with the impacts of poverty and poor governance. As temperatures continue to rise, the lack of access to resources and social services such as water, electricity, and health care will become more acute and further limit the ability of African governments to cope with the challenges posed by climate change externalities. An increased risk of floods, droughts, and other natural disasters, combined with poverty and poor governance, could exacerbate existing social tensions, resulting in an increase in civil unrest and conflicts across the continent. These have significant economic and social consequences for Africa in the years ahead, particularly for the most vulnerable and disadvantaged communities already suffering from poverty and marginalization.

This is why it is so important for African countries to prioritize the development of resilient and adaptive policies that enable them to better manage their resources. Such policies must include measures to improve access to social services and infrastructure, strengthen disaster risk management capabilities, build the capacity of local governments, enhance food security, ensure the equitable distribution of resources, support economic diversification, and create employment opportunities for vulnerable groups.

These measures are critical for Africa's sustainable development, and the international community must provide assistance to African countries to ensure that their efforts are successful. African countries need to develop resilient and adaptive policies to better manage their resources, improve access to social services and infrastructure, strengthen disaster risk management capabilities, build the capacity of local governments, ensure the equitable distribution of resources, and support economic diversification.

Furthermore, investments must be made in renewable energy sources and clean technologies to mitigate and adapt to the effects of climate change, reduce greenhouse gas emissions, and create employment opportunities for vulnerable groups. Moreover, to ensure the successful implementation of these policies and their effectiveness, assistance from the international community is essential by providing capacity-building support, technical assistance, and financial resources to African countries to enable them to implement effective and resilient policies and programs.

7.3.4 Devastating Effect on the African Population

These changes have a devastating effect on the African population, which is already struggling with poverty, malnutrition, and limited access to resources. A lack of investment in infrastructure and government initiatives to improve living standards results in a further decrease in the quality of life for many people across the continent, making it difficult for them to respond and adapt to the changes quickly and effectively. By investing in disaster preparedness and risk mitigation, as well as providing access to clean energy and water, African governments can help create a more resilient population that is better equipped to face future generations and is given the best chance to lead successful and prosperous lives.

Despite these difficult circumstances, there is still hope for a brighter future for African nations if African governments make the necessary investments in infrastructure and government initiatives. This could help create a more resilient population that is better prepared for climate change and more capable of responding and adapting to changes quickly and effectively. This would not only help reduce the risk of natural disasters and climate-related impacts on people's lives but also spur economic development and create opportunities for the African continent to become a more prosperous and equal region in the world. Despite these daunting challenges, there are opportunities to use technology, innovation, and collaboration to combat climate change externalities and create a brighter future for African nations and their people. African governments must make the necessary investments to support innovation and development in the continent's infrastructure, government initiatives, and clean energy sources to create a more resilient population and reduce the risk of climate-related impacts across the continent.

This will require commitment from all sectors of African society, from governments to businesses to individuals, to ensure that the African continent is able to capitalize on these opportunities and meet the challenges of climate change head-on to develop and secure a brighter future for the continent and its people. In addition to these investments, there is also a need for greater collaboration between governments, businesses, and individuals to ensure that the knowledge and resources that are available to African nations and their people are utilized.

African nations must rise to the challenge and make the necessary investments in their infrastructure, government initiatives, and clean energy sources for the sake of the continent and its people. By doing so, African nations will be able to ensure that they are better equipped to handle the climate-related risks that are now present and guarantee a brighter future for their people and the environment, as well as a healthier and more prosperous economy, in ways that have never been done before. African nations must also commit to investing in their people, including by providing access to education and job training for those who are unable to gain formal employment and encouraging greater collaboration between the private and public sectors to develop innovative solutions that can help improve the well-being of African people and their environment. While all these investments are important, African nations must go beyond simply investing in infrastructure and energy sources and strive to foster a culture of environmental stewardship and sustainability to ensure that the environment is respected and protected to protect the continent.

7.4 Efforts to Mitigate the Impacts of Climate Change in Africa

7.4.1 Government Efforts

Governmental efforts have been increasing in recent years as the continent faces extreme weather conditions such as droughts, floods, and heat waves that threaten lives, livelihoods, and ecosystems, resulting in economic, environmental, and health consequences across the continent. Consequently, African governments have begun to take significant action to adapt to and mitigate climate change by implementing policies such as investing in renewable energy sources, promoting sustainable agriculture, and managing forests in a way that preserves them as carbon sinks. These efforts have started to pay off, with some governments succeeding in their missions to reduce carbon emissions, increase green energy use, and limit their contribution to climate change.

These initiatives are encouraging, but more is needed to protect Africa from extreme weather for long-term climate change adaptation. For these efforts to be successful, African governments must continue to promote sustainable agriculture while also expanding their efforts to promote more efficient and eco-friendly practices, such as the increased use of electric vehicles, efficient waste management systems, and eco-friendly housing initiatives. Additionally, African governments should prioritize education and outreach initiatives to ensure that their citizens adapt to their effects. African governments must also collaborate with other African countries and global partners to promote the sharing of knowledge, resources, and best practices to develop comprehensive climate policies that can be implemented across the continent. In general, increased efficiency and eco-friendly practices, education and outreach initiatives, and collaboration with other African countries and global partners are necessary to develop and implement effective solutions that can protect the continent.

African countries have ratified international agreements to cut emissions and use more renewable energy sources, such as the Paris Climate Agreement and the African Renewable Energy Initiative (AREI). However, there is still much to be done to adequately protect the African continent. The development of comprehensive climate policies and initiatives is key to protecting Africa from both short- and long-term measures to manage and ensure a sustainable and prosperous future. These policies must be implemented through both government and nongovernment actors, such as local communities, businesses, and civil society organizations.

Due to its sparse resources and infrastructure, Africa is suffering disproportionately from the effects of climate change, so government efforts to mitigate these effects have fallen short in many cases. This lack of investment has resulted in a decrease in the continent's capacity to cope with climate change, exacerbating poverty and vulnerability across Africa. Therefore, it is essential that local governments develop mechanisms to ensure that both government and nongovernment actors contribute to the implementation of climate policies and create better incentives for stakeholders to invest in renewable energy sources and other low-carbon technologies. To ensure effective mitigation and adaptation efforts, African governments must prioritize the strengthening of their environmental policies and invest in the necessary technology and infrastructure in their countries.

African governments have made a wide range of extensive efforts, ranging from sustainable agriculture and forestry practices such as agroforestry and conservation tillage, to mitigate climate externalities. However, these efforts have often been hampered by the lack of both financial and human capital investments required to successfully implement climate policies and provide the incentives necessary to encourage stakeholders to invest in low-carbon technologies and renewable energy sources. In addition, there is a need to strengthen the ability of local governments to coordinate and collaborate with other stakeholders, including civil society organizations and businesses, to ensure the effective implementation of climate policies. To address these issues, African governments must focus on creating an enabling environment for investment by removing existing barriers to private-sector financing and encouraging public–private partnerships that bring together the necessary expertise and resources needed from all angles to mitigate climate change. Additionally, governments should seek to strengthen the capacity of local and regional governments by investing in training programs needed to lead climate action and actively engage with local communities, civil society, and other stakeholders. This can be achieved by providing the necessary resources and investment needed to create effective climate policies, strengthening local governments' capacity to coordinate with other stakeholders, incentivizing investments in low-carbon technologies and renewable energy sources, and developing training programs to equip local governments with the knowledge and skills needed to drive climate action at the regional and local levels.

Governmental initiatives have had varying degrees of success, with some nations advancing their environmental laws while others are lagging behind due to a lack of resources, awareness, and expertise needed to bring about meaningful progress on the issue. African governments must create an enabling environment for investment by removing existing barriers to private sector financing, incentivizing investments in low-carbon technologies and renewable energy sources, and strengthening local governments' capacity to coordinate with other stakeholders. As such, it is essential for governments to provide the necessary resources and investment needed to create effective climate policies while also incentivizing investments in low-carbon technologies and renewable energy sources and developing training programs to equip local governments with the knowledge and skills needed to drive climate action at the regional and local levels.

In addition, it is important to support education and public awareness campaigns to improve the understanding of the benefits of low-carbon development and ensure successful implementation of climate change mitigation strategies and broad-based public support for such measures. In conclusion, effective climate change mitigation requires African governments to take meaningful steps to create an enabling environment for investment by removing existing barriers to private sector financing, incentivizing investments in low-carbon technologies and renewable energy sources, and strengthening local governments' capacity to coordinate with other stakeholders, civil society, and other actors to drive the implementation of climate policies at the regional and local levels.

7.4.2 International Aid

International assistance to lessen climate change's effects in Africa has been crucial in recent years. Various organizations and developed countries have come together to provide financial support, technological expertise, and capacity-building initiatives. These efforts aim to enhance Africa's resilience and mitigate the adverse impacts of climate change on its vulnerable populations. One significant area of focus has been renewable energy projects. Through funding and collaboration, Africa has seen a surge in the development of solar, wind, and hydroelectric power systems. These sustainable energy sources not only reduce greenhouse gas emissions but also provide reliable electricity access to remote communities. Additionally, international assistance has supported the implementation of climate-smart agriculture practices across the continent. By promoting sustainable farming techniques, such as agroforestry and conservation agriculture, African farmers can adapt to changing climatic conditions while improving food security and preserving natural resources. Furthermore, capacity-building programs have empowered local communities with the knowledge and skills to tackle climate change challenges effectively. Training sessions on disaster preparedness, early warning systems, and ecosystem management have equipped Africans with the tools needed to respond to emerging threats and mitigate their impact.

In addition, the promotion of renewable energy sources, such as solar and wind power, has reduced dependence on fossil fuels and contributed to the reduction of greenhouse gas emissions. These sustainable practices not only address climate change but also promote economic growth and social development. By diversifying their crops and implementing water-efficient irrigation systems, African farmers have been able to increase productivity while conserving water resources. Moreover, the integration of digital technologies, such as remote sensing and data analytics, has revolutionized agricultural practices by providing real-time information on weather patterns, soil moisture levels, and crop health. This enables farmers to make informed decisions regarding planting schedules, fertilizer application, and pest management. The adoption of climate-smart agriculture has not only improved resilience but also opened up new opportunities for income generation through carbon credits and sustainable agribusiness ventures. With continued investment in research and innovation, Africa is poised to lead the way in sustainable farming practices that can address climate change while ensuring food security for its growing population. By integrating climate data and advanced technology, farmers can optimize their planting schedules based on weather patterns, ensuring that crops are cultivated during favorable conditions. This not only maximizes yield but also minimizes the risk of crop failure due to extreme weather events.

Furthermore, climate-smart agriculture emphasizes the efficient use of fertilizers, reducing both costs and environmental impacts. Through soil testing and precision application techniques, farmers can tailor their fertilizer usage to specific crop needs, avoiding excessive use that can lead to nutrient runoff and water pollution. This sustainable approach not only protects the environment but also improves soil health and long-term productivity. In addition, climate-smart agriculture promotes the adoption of innovative irrigation techniques to optimize water usage. By implementing drip irrigation systems and using moisture sensors, farmers can ensure that crops receive the right amount of water at the right time, reducing water waste and conserving this precious resource. Moreover, climate-smart agriculture encourages the use of cover crops and agroforestry practices to enhance soil quality and biodiversity. Cover crops help prevent soil erosion, retain moisture, and suppress weeds, while agroforestry systems provide shade and windbreaks, creating a favorable microclimate for crops. This holistic approach not only mitigates the effects of extreme weather events but also enhances ecosystem resilience. Additionally, climate-smart agriculture promotes diversification by encouraging farmers to grow a variety of crops instead of relying on monocultures. Diversification not only reduces the risk of crop failure but also improves food security and provides economic opportunities for farmers. Furthermore, climate-smart agricultural practices include integrated pest management strategies that minimize pesticide use while pest management is another crucial aspect of climate-smart agriculture. By monitoring pest populations and using integrated pest management strategies, farmers can minimize the need for chemical pesticides while effectively controlling pests. This reduces chemical exposure for farmers and consumers alike, promoting healthier ecosystems and safer food production. Additionally, climate-smart agriculture emphasizes the importance of water management. By implementing efficient irrigation systems and adopting water-saving techniques, farmers can conserve water resources and reduce their reliance on unsustainable water practices. This not only helps mitigate the impacts of climate change but also ensures a more sustainable and resilient agricultural system. Furthermore, diversification of crops and livestock is a key component of climate-smart agriculture. By growing a variety of crops and raising different types of livestock, farmers can reduce their vulnerability to climate-related risks such as extreme weather events and disease outbreaks. This promotes biodiversity, enhances ecosystem services, and improves overall farm resilience. Finally, climate-smart agriculture encourages the adoption of innovative technologies and practices that promote energy efficiency and reduce greenhouse gas emissions. From renewable energy sources to precision farming techniques, these advancements contribute to a more sustainable agricultural sector while mitigating the effects of climate change. In conclusion, climate-smart agriculture offers a holistic approach that addresses the challenges posed by climate change while promoting sustainable food production and environmental protection.

The adoption of climate-smart agriculture has also opened up new avenues for income generation in Africa. Through carbon credit schemes and sustainable farming practices, farmers can earn additional income while contributing to climate change mitigation. By sequestering carbon in the soil through practices such as agroforestry and conservation agriculture, farmers can sell carbon credits on the global market. This not only provides financial incentives for farmers to adopt climate-smart practices but also helps reduce greenhouse gas emissions. Additionally, climate-smart agriculture promotes the diversification of income sources by encouraging the cultivation of high-value crops and the adoption of innovative farming techniques. This allows farmers to tap into new markets and increase their resilience to climate-related risks. Overall, the integration of pest management strategies and income-generating opportunities makes climate-smart agriculture a holistic approach that benefits both farmers and the environment.

7.4.3 Community-Led Adaptation Strategies

Strategies for reducing emissions in Africa that are driven by the community are a vital tool in addressing the environmental, economic, and social issues posed by climate change. African community-driven strategies for reducing emissions can include promoting renewable energy sources such as solar and wind power. By harnessing the abundant sunlight and wind resources in Africa, communities can decrease their reliance on fossil fuels and reduce greenhouse gas emissions. Additionally, community-led initiatives can focus on improving energy efficiency by implementing energy-saving technologies and practices. This can involve educating community members about the benefits of energy conservation and providing them with tools to make informed decisions about their energy consumption. Furthermore, sustainable agriculture practices can play a significant role in emission reduction efforts. Encouraging organic farming methods, agroforestry, and soil conservation techniques can help sequester carbon dioxide from the atmosphere while ensuring food security for local communities. Another crucial aspect is promoting sustainable transportation options such as public transit systems, cycling infrastructure, and carpooling initiatives. By reducing reliance on individual cars and promoting alternative modes of transportation, communities can decrease carbon emissions from the transportation sector.

Finally, community-driven waste management strategies are essential for reducing emissions in Africa. Implementing recycling programs and composting initiatives can divert organic waste from landfills, reducing methane emissions.

Additionally, educating communities about the importance of waste reduction and proper disposal methods can further contribute to emission reduction efforts. Furthermore, encouraging the use of renewable energy sources such as solar and wind power can significantly reduce greenhouse gas emissions in Africa. Investing in renewable energy infrastructure and providing incentives for individuals and businesses to adopt clean energy technologies will not only help combat climate change but also create job opportunities and stimulate economic growth. Moreover, fostering international collaborations and partnerships can facilitate knowledge sharing and technology transfer, enabling African countries to leapfrog to cleaner and more sustainable development pathways. Ultimately, a comprehensive approach that integrates sustainable agriculture practices, promotes alternative transportation options, implements effective waste management strategies, adopts renewable energy sources, and fosters international cooperation is crucial for Africa to mitigate its carbon emissions and build a resilient future for its communities.

7.4.4 Adoption of Sustainable Agricultural Practices

Strategies to mitigate the use of sustainable agricultural methods in Africa have been developed to improve agricultural production and reduce the reliance on traditional methods that are not sustainable. These strategies include drought-tolerant crops, the adoption of agroforestry practices, and improved soil management techniques such as crop rotation and composting, conservation tillage, and nutrient-rich composting. In addition, the promotion of crop diversification and access to modern technologies is essential to ensure that African farmers are equipped with the right tools to produce more sustainably and maximize their yields.

Consequently, African countries must prioritize the implementation of these strategies to achieve their climate goals in the long term. By emphasizing the use of sustainable agricultural practices, African countries can not only foster economic growth, reduce poverty, and improve overall sectors such as food security and health but also mitigate the effects of climate change. Sustainable agricultural practices, such as agroforestry and organic farming, can help restore degraded lands, enhance soil fertility, and promote biodiversity conservation. Additionally, investing in climate-smart technologies such as precision agriculture and drip irrigation can optimize resource use and reduce greenhouse gas emissions. By adopting these practices, African countries can build resilience against climate shocks and ensure a sustainable future for their people and ecosystems.

These measures can help African farmers become more self-reliant and better able to cope with climate change impacts to achieve food security and improve their agricultural production in the long term. It is clear that African countries must prioritize the implementation of strategies such as crop diversification and access to modern technologies to ensure that they are better equipped to face climate change challenges and ultimately achieve long-term food security and improve their agricultural production. In addition to crop diversification and access to modern technologies, African countries should also focus on promoting sustainable farming practices. This includes adopting agroforestry techniques that combine tree planting with crop cultivation, which can enhance soil fertility, conserve water, and provide additional sources of income through timber and fruit production. Furthermore, investing in irrigation systems and water management infrastructure can help mitigate the effects of droughts and ensure a consistent water supply for crops. Additionally, improving postharvest storage facilities and transportation networks can reduce food losses and increase market access for farmers. Collaborating with international organizations and neighboring countries can facilitate knowledge sharing and promote regional cooperation in tackling climate change challenges. Moreover, providing financial support, training programs, and extension services to smallholder farmers will empower them with the necessary skills and resources to adapt to changing climatic conditions. By implementing these comprehensive strategies, African farmers can become more resilient in the face of climate change impacts while achieving long-term food security and sustainable agricultural production.

With the implementation of sustainable agricultural practices, African countries can achieve a more resilient and self-reliant agricultural sector that is better equipped to tackle climate change while improving economic growth, reducing poverty, and providing long-term food security for the first time. This can be done through the creation of policies that incentivize farmers to adopt and implement conservation agriculture and integrated pest management. These strategies have the potential to not only enable African countries to be more resilient but also help reduce poverty and improve economic growth in the long term.

In general, strategies such as crop diversification, access to modern technologies, and the implementation of sustainable agricultural practices can help African countries become more self-reliant. While these strategies will not be easy feats to achieve, with the correct policies in place and the support of international donors, African countries have a greater potential for economic growth through sustainable agricultural practices, and we can provide technical advice on how to do so. To create a more sustainable future for African countries and their citizens, there must be an effort to adopt sustainable agricultural practices that are tailored to the local environment, provide access to modern technologies and information, diversify crops, increase yields, and create economic opportunities that support small-scale farmers and rural communities.

7.4.5 Reforestation

Reforestation strategies in Africa aim to reduce the impact of climate change as its effects become more apparent. Reforestation-based strategies involve the planting of trees in areas that were once heavily forested but have since become deforested for a variety of reasons, such as unsustainable logging practices or agricultural expansion. Reforestation-based strategies can increase carbon sequestration, increase water retention, prevent soil erosion, and provide other ecosystem services, such as habitat for wildlife and fuel for cooking and heating. Reforestation-based strategies also provide employment opportunities for people in rural areas as well as a source of revenue for local governments through the sale of timber and other forest products.

Reforestation-based strategies play a crucial role in mitigating climate change by absorbing carbon dioxide from the atmosphere and storing it in trees and vegetation. This not only helps reduce greenhouse gas emissions but also contributes to the overall health of the planet. Additionally, reforestation efforts can improve water quality by filtering pollutants and reducing runoff, thus benefiting both human populations and aquatic ecosystems. By preventing soil erosion, reforestation helps maintain fertile soils for agriculture and prevents sedimentation in rivers and streams. Moreover, these restored forests serve as vital habitats for countless species, promoting biodiversity conservation and supporting ecosystem resilience. Last, reforestation initiatives offer employment opportunities for local communities, especially in rural areas where alternative job prospects may be limited. This not only improves livelihoods but also strengthens social cohesion and reduces poverty rates. In conclusion, reforestation-based strategies are multifaceted solutions that address various environmental challenges while simultaneously providing economic benefits and enhancing the overall well-being of both humans and nature.

Furthermore, reforestation-based strategies can also improve soil fertility and productivity, providing local farmers with greater food security and reducing the need to clear forests for agricultural expansion. Reforestation-based strategies have the potential to improve the lives of many African citizens in a variety of ways, including by providing economic opportunities, increasing access to vital resources, and helping to mitigate climate change-related risks that are otherwise beyond their control. In general, reforestation-based strategies in Africa have the potential to bring about positive environmental, economic, and social changes for the region while at the same time creating a more sustainable future for all of its citizens.

Reforestation-based strategies in Africa can have a transformative effect on the environment, economy, and social structure of the region, which could prove essential for the long-term health and prosperity of African communities. Reforestation-based strategies can also provide vital environmental benefits, such as increased carbon sequestration and the protection of habitats for species at risk of extinction. Reforestation-based strategies can also promote economic development, as they can create job opportunities, bring income to local communities, and create a healthier environment for farming and other agricultural activities Reforestation-based strategies can also have a significant impact on the social dynamics of African communities by increasing access to clean drinking water, reducing air pollution, restoring degraded land, and providing resources for fuel, food, and medicine.

Furthermore, reforestation-based strategies help promote biodiversity and provide refuge for endangered species while also protecting against natural disasters such as floods, landslides, and soil erosion. Reforestation-based strategies can provide a number of additional benefits, such as improving air quality, creating new sources of income through sustainable forestry practices, and providing carbon sequestration services that can help reduce global warming, as well as a number of other direct benefits. Additionally, reforestation-based strategies can help preserve biodiversity by providing habitat for plants and animals and helping to protect local ecosystems from the effects of climate change. Reforestation-based strategies can be powerful tools in the effort to promote sustainable development, both economically and environmentally, but it is essential that they are implemented with an understanding of the local context and with consideration for the potential consequences of any action taken.

As such, reforestation-based strategies must be supported by the local community to ensure that they provide long-term benefits and are successful in achieving their goals. Reforestation-based strategies can be effective in the long term when designed with the local community in mind, considering its specific needs and potential consequences, and involving members of the community in the planning and implementation stages of the strategies. In general, reforestation-based strategies have the potential to protect biodiversity and promote sustainable development if implemented thoughtfully with an understanding of the affected members.

All of these positive benefits combined can make reforestation-based strategies an invaluable tool in promoting economic development and ensuring a healthier, more sustainable future for communities across the globe. Reforestation-based strategies are an incredibly powerful tool in the effort to promote economic development and create a more sustainable future for our planet. However, it is important to note that reforestation-based strategies must be implemented carefully and with consideration of their local context to be successful and to ensure that any negative impacts are minimized. To ensure that reforestation-based strategies are implemented in a manner that is both effective and sustainable, careful consideration should be given when designing such plans.

This includes considering the local ecosystem, climate, and economic conditions to create plans that are tailored to the needs of the region being implemented and ensure that any long-term, negative environmental impacts are avoided. As part of this consideration process, it is also important to ensure that the local population is engaged in the development and implementation of reforestation-based strategies to ensure that any cultural and social concerns are taken into account and that reforestation-based strategies become part of a larger conversation around sustainability, climate change, and local economic development. Finally, careful consideration and thoughtful engagement of the local population are essential for any reforestation-based strategies to be effective, sustainable, and beneficial for local communities to have a lasting impact on the region's environmental, social, and economic development goals.

7.5 Conclusion

African nations must implement a comprehensive climate action plan to protect their communities from the impacts of climate change and capitalize on economic opportunities. This plan should include investments in infrastructure, technology, and education to ensure adaptation and capitalize on new economic opportunities. Governments must take responsibility for mitigating climate change's effects on agricultural communities by providing them with resources, knowledge, and technological solutions. This includes access to modern agricultural technologies, supporting local climate adaptation initiatives, and developing strategies that enable them to use traditional knowledge of their land.

Investing in sustainable agriculture and climate-related information and education can help reduce poverty levels, improve public health outcomes, and create a more resilient future for African communities. Investments in renewable energy sources and clean technologies are essential for reducing greenhouse gas emissions and creating employment opportunities for vulnerable groups. African governments must create an enabling environment for investment by removing barriers to private sector financing, incentivizing investments in low-carbon technologies and renewable energy sources, and strengthening local governments' capacity to coordinate with other stakeholders.

International assistance is crucial for ensuring the sustainability of Africa's vulnerable communities and ecosystems due to extreme poverty and reliance on rain-fed agriculture. It can reduce the impact of climate change by providing resources for adaptation and mitigation projects, early warning systems for extreme weather events, financial resources for climate risk insurance and adaptation measures, and strengthening institutions. It can also help African communities build capacity and develop knowledge and skills to protect local livelihoods from climate-related risks and foster an environment of resilience and adaptive capacity.

African-led strategies for climate change mitigation empower African nations to take ownership of the challenge posed by climate change, strengthen relationships between African nations and the global community, and create a more equitable and sustainable global future. External actors can provide support and resources to enable African nations to develop and implement innovative initiatives that foster sustainable and equitable development while preserving their unique cultures, values, and ways of life for future generations.

References

Addaney, M., & Dumenu, W. K. (2023, July 27). *Climate change in Africa: Adaptation, resilience, and policy innovations*. https://doi.org/10.1007/978-3-031-30050-9

Adefisan, E. A. (2022, October). Impacts of climate change on rainfall amount, onset and length of rainy season over West Africa. *Journal of Natural Sciences Research*. https://doi.org/10.7176/jnsr/13-18-05

Adepoju, P. (2021, November 7). Africa worst hit by climate change impacts, COP26 told. *Nature Africa*. https://doi.org/10.1038/d44148-021-00107-z

Ajala, S. B., & Chagwiza, C. (2022). Determinants of smallholder farmers' adaptation strategies in response to a changing climate: A review of sub-Saharan Africa agriculture. *The International Journal of Climate Change: Impacts and Responses, 15*(1), 57–71. https://doi.org/10.18848/1835-7156/cgp/v15i01/57-71

Anekwe, S., Zhou, H., Mkhize, M., & Akpasi, S. (2023). Effects of climate change on agricultural production in South Africa. *The International Journal of Climate Change: Impacts and Responses, 15*(2), 89–114. https://doi.org/10.18848/1835-7156/cgp/v15i02/89-114

Asafu-Adjaye, J. (2010). The economic impacts of climate change in sub-Saharan Africa. *The International Journal of Climate Change: Impacts and Responses, 2*(1), 103–118. https://doi.org/10.18848/1835-7156/cgp/v02i01/37294

Bedford, D., & Cook, J. (2016, July 18). *Climate change: Examining the facts*.

Berck, C. S., Berck, P., & Falco, S. D. (2018, March 8). *Agricultural adaptation to climate change in Africa: Food security in a changing environment*.

DiMento, J. F., & Doughman, P. M. (2007, September 14). *Climate change: What it means for us, our children, and our grandchildren*. https://doi.org/10.1604/9780262042413.

Combating Climate Change in Africa*: Hearing Before the Subcommittee on Africa and Global Health of the Committee on Foreign Affairs, House of Representatives, One Hundred Eleventh Congress, Second Session, April 15, 2010*. (2010, January 1).

Dietz, K., & Engels, B. (2018, June 11). *Climate change in Africa: Social and political impacts, conflicts, and strategies*.

Gbenga, A. (2014, January 1). *Emissions trading? Option for fighting climate change in Africa*.

Guzmán León, A. (2019). Kyoto's governance footprint in the global South: An overview of climate finances and REDD+ in Asia, Latin America, and Africa. *The International Journal of Climate Change: Impacts and Responses, 11*(1), 25–46. https://doi.org/10.18848/1835-7156/cgp/v11i01/25-46

Hassan, M. A., Bailek, N., Bouchouicha, K., Ibrahim, A., Jamil, B., Kuriqi, A., & Nwokolo, S. C. (2022). Evaluation of energy extraction of PV systems affected by environmental factors under real outdoor conditions. *Theoretical and Applied Climatology, 150*, 715. https://doi.org/10.1007/s00704-022-04166-6

Hassan, M. A., Bailek, N., Bouchouicha, K., & Nwokolo, S. C. (2021). Ultrashort-term exogenous forecasting of photovoltaic power production using genetically optimized nonlinear autoregressive recurrent neural networks. *Renewable Energy, 171*(February), 191–209. https://doi.org/10.1016/j.renene.2021.02.103

Igbokwe, E., & Mkpado, M. (2011, August 10). Socio-economic impacts of climate change in Africa. *Agro-Science, 10*(1). https://doi.org/10.4314/as.v10i1.68721

Impacts of Climate Change on Food Security: A Literature Review in Sub Saharan Africa. (2014). *Journal of Earth Science & Climatic Change, 05*(08). https://doi.org/10.4172/2157-7617.1000225.

Impacts of Climate Change on Rainfall Amount, Onset and Length of Rainy Season over West Africa. (2022, October). *Journal of Natural Sciences Research*. https://doi.org/10.7176/jnsr/13-18-05.

Joseph, S., Rubhara, T. T., Antwi, M. A., & Oduniyi, O. S. (2020). Challenges to climate change adaptation practices among citrus farmers in Limpopo Province, South Africa. *The International Journal of Climate Change: Impacts and Responses, 12*(3), 19–32. https://doi.org/10.18848/1835-7156/cgp/v12i03/19-32

Kang, M. (2022). Climate change impacts on agriculture and conflicts in sub-Saharan Africa. *SSRN Electronic Journal*. https://doi.org/10.2139/ssrn.4172520

Letcher, T. (Ed.). (2009, May 8). *Climate change: Observed impacts on planet earth*. Elsevier Science.

Li, X., Stringer, L. C., & Dallimer, M. (2022, October 30). The impacts of urbanization and climate change on the urban thermal environment in Africa. *Climate, 10*(11), 164. https://doi.org/10.3390/cli10110164

Moreland, S., & Smith, E. (2013). Climate change, food security, and population in sub-Saharan Africa: Modelling the linkages. *The International Journal of Climate Change: Impacts and Responses, 4*(2), 29–47. https://doi.org/10.18848/1835-7156/cgp/v04i02/37158

Ndhlovu, G., Woyessa, Y., & Vijayaraghavan, S. (2020). Generating a High Resolution Climate Scenario through Regional Climate Modelling over Part of Southern Africa. *The International*

Journal of Climate Change: Impacts and Responses, 12(4), 29–48. https://doi.org/10.18848/1835-7156/cgp/v12i04/29-48

Njenga, N. W., Muna, M. W., & Muriuki, J. N. (2013). Climate change impacts on small scale farmers in north Kinangop location, Kenya. *The International Journal of Climate Change: Impacts and Responses, 4*(2), 19–28. https://doi.org/10.18848/1835-7156/cgp/v04i02/37157

Nwokolo, S. C., Amadi, S. O., Obiwulu, A. U., Ogbulezie, J. C., & Eyibio, E. E. (2022a). Prediction of global solar radiation potential for sustainable and cleaner energy generation using improved angstrom-Prescott and Gumbel probabilistic models. *Cleaner Engineering and Technology, 6*(January), 100416. https://doi.org/10.1016/j.clet.2022.100416

Nwokolo, S. C., Obiwulu, A. U., & Ogbulezie, J. C. (2022b). Machine learning and analytical model hybridization to assess the impact of climate. *Physics and Chemistry of the Earth, 130*(February), 103389. https://doi.org/10.1016/j.pce.2023.103389

Nwokolo, S. C., & Ogbulezie, J. C. (2018a). A qualitative review of empirical models for estimating diffuse solar radiation from experimental data in Africa. *Renewable and Sustainable Energy Reviews., 92*, 353. https://doi.org/10.1016/j.rser.2018.04.118

Nwokolo, S. C., & Ogbulezie, J. C. (2018b). A qualitative review of empirical models for estimating diffuse solar radiation from experimental data in Africa. *Renewable and Sustainable Energy Reviews, 92*. https://doi.org/10.1016/j.rser.2018.04.118

Nwokolo, S. C., & Ogbulezie, J. C. (2018c). A quantitative review and classification of empirical models for predicting global solar radiation in West Africa. *Beni-Suef University Journal of Basic and Applied Sciences, 7*(4), 367–396. https://doi.org/10.1016/j.bjbas.2017.05.001

Nwokolo, S. C., Ogbulezie, J. C., & Obiwulu, A. U. (2022c). Impacts of climate change and meteo-solar parameters on photosynthetically active radiation prediction using hybrid machine learning with physics-based models. *Advances in Space Research, 70*, 3614. https://doi.org/10.1016/j.asr.2022.08.010

Nwokolo, S., & Ogbulezie, J. (2017a). A critical review of theoretical models for estimating global solar radiation between 2012-2016 in Nigeria. *International Journal of Physical Research, 5*(2), 60. https://doi.org/10.14419/ijpr.v5i2.8160

Nwokolo, S., & Ogbulezie, J. (2017b). Modelling the influence of cloudiness on diffuse horizontal irradiation under various sky conditions in Nigeria. *International Journal of Physical Research, 5*(2), 91. https://doi.org/10.14419/ijpr.v5i2.8312

Nwokolo, S., & Otse, C. (2019). Impact of sunshine duration and clearness index on diffuse solar radiation estimation in mountainous climate. *Trends in Renewable Energy, 5*(3), 307–332. https://doi.org/10.17737/tre.2019.5.3.00107

Nwokolo, S. C., Obiwulu, A. U., Amadi, O., & Ogbulezie, J. C. (2023a). Assessing the impact of soiling, tilt angle, and solar radiation on the performance of solar PV systems. *Trends in Renewable Energy, 9*(2), 121–137. https://doi.org/10.17737/tre.2023.9.2.00156

Nwokolo, S. C., Ogbulezie, J. C., & Ushie, O. J. (2023b). A multimodel ensemble-based CMIP6 assessment of future solar radiation and PV potential under various climate warming scenarios. *Optik, 285*(May), 170956. https://doi.org/10.1016/j.ijleo.2023.170956

Nwokolo, S. C., Proutsos, N., & Meyer, E. L. (2023c). Machine learning and physics-based hybridization models for evaluation of the effects of climate change and urban expansion on photosynthetically active radiation. *Atmosphere, 14*(687), 1–36. https://doi.org/10.3390/atmos14040687

Obiwulu, A. U., Chendo, M. A. C., Erusiafe, N., & Nwokolo, S. C. (2020a). Implicit meteorological parameter-based empirical models for estimating back temperature solar modules under varying tilt-angles in Lagos, Nigeria. *Renewable Energy, 145*. https://doi.org/10.1016/j.renene.2019.05.136

Obiwulu, A. U., Erusiafe, N., Olopade, M. A., & Nwokolo, S. C. (2020b). Modelling and optimization of back temperature models of mono-crystalline silicon modules with special focus on the effect of meteorological and geographical parameters on PV performance. *Renewable Energy, 154*, 404. https://doi.org/10.1016/j.renene.2020.02.103

Obiwulu, A. U., Erusiafe, N., Olopade, M. A., & Nwokolo, S. C. (2022). Modelling and estimation of the optimal tilt angle, maximum incident solar radiation, and global radiation index of the photovoltaic system. *Heliyon, 8*(6), e09598. https://doi.org/10.1016/j.heliyon.2022.e09598

Ogbulezie, J. C., James, U. O., & Chukwujindu, N. S. (2017). A review of regression models employed for predicting diffuse solar radiation in North-Western Africa. *Trends in Renewable Energy*. https://doi.org/10.17737/tre.2017.3.2.0042

Oguntunde, P. G., Abiodun, B. J., & Lischeid, G. (2017, August). Impacts of climate change on hydrometeorological drought over the Volta Basin, West Africa. *Global and Planetary Change, 155*, 121–132. https://doi.org/10.1016/j.gloplacha.2017.07.003

Ojewola, O., Morgan, M., & Wilhelm Stanis, S. (2021). Vulnerability, resilience, and adaptation to climate change impacts: Perceptions of State Park managers. *The International Journal of Climate Change: Impacts and Responses, 14*(1), 91–106. https://doi.org/10.18848/1835-7156/cgp/v14i01/91-106

Pauleit, S., Yeshitela, K., Jørgensen, G., Coly, A., Fohlmeister, S., Gasparini, P., Jørgensen, G., Kabisch, S., Kombe, W. J., Lindley, S., & Simonis, I. (2015, January 1). *Urban Vulnerability and climate change in Africa: A multidisciplinary approach* (Vol. 4).

Perkins, P. E. (2013, April 12). *Water and climate change in Africa: Challenges and community initiatives in Durban*. Routledge.

Pittock, A. B. (2005, December 1). *Climate change: Turning up the heat*. https://doi.org/10.1604/9781844073009.

Romm, J. (2018, May 16). *Climate change: What everyone needs to know®*.

Samuel Chukwujindu, N. (2017). A comprehensive review of empirical models for estimating global solar radiation in Africa. *Renewable and Sustainable Energy Reviews, 78*(October), 955–995. https://doi.org/10.1016/j.rser.2017.04.101

Sokona, Y. (2001, January). Climate change impacts: Can Africa cope with the challenges? *Climate Policy, 1*(1), 117–123. https://doi.org/10.1016/s1469-3062(00)00009-7

Sokona, Y., & Denton, F. (2001, January). Climate change impacts: Can Africa cope with the challenges? *Climate Policy, 1*(1), 117–123. https://doi.org/10.3763/cpol.2001.0110

Strydom, S., & Savage, M. J. (2017, July 5). Potential impacts of climate change on wildfire dynamics in the midlands of KwaZulu-Natal, South Africa. *Climatic Change, 143*(3–4), 385–397. https://doi.org/10.1007/s10584-017-2019-8

Toulmin, C. (2009, October 15). *Climate change in Africa*.

United States Congress, U. S., House of Representatives, U. S., & Foreign Affairs, C. O. (2018, January 1). *Combating climate change in Africa*.

Wright, C. Y., & Kapwata, T. (2023, May 2). An urgent call to address climate change-related human health impacts in southern Africa. *PLOS Climate, 2*(5), e0000204. https://doi.org/10.1371/journal.pclm.0000204

Zewdie, A. (2014). Impacts of climate change on food security: A literature review in Sub Saharan Africa. *Journal of Earth Science & Climatic Change, 05*(08). https://doi.org/10.4172/2157-7617.1000225

Ziervogel, G., New, M., Archer van Garderen, E., Midgley, G., Taylor, A., Hamann, R., Stuart-Hill, S., Myers, J., & Warburton, M. (2014, July 3). Climate change impacts and adaptation in South Africa. *WIREs Climate Change, 5*(5), 605–620. https://doi.org/10.1002/wcc.295

Chapter 8
Scenarios that Could Give Rise to an African Net-Zero Energy Transition

8.1 Introduction

The International Monetary Fund estimates that $130 trillion is needed to achieve global net zero by 2050 (United Nations, 2021). This will involve investments in renewable energy, energy efficiency technologies (Abade, 2020), and resilient infrastructure (Agreement et al., 2021). These investments will unlock new markets (Abdul-Salam et al., 2022), create job opportunities (Ahl et al., 2020), increase food production (Abade, 2020), improve water security (Allen et al., 2020), and reduce environmental impacts (Amadi et al., 2020). However, it is crucial to ensure inclusive and equitable economic growth, especially for vulnerable communities in African countries.

Investing in a sustainable future benefits global economic growth by improving human capital, boosting long-term growth (Amadi et al., 2018), and expanding business opportunities (Ameli et al., 2021). To keep global warming under 2 °C by 2100, the global economy needs $2.4 trillion in investment over the next decade (BP Energy, 2020). Governments must take a coordinated approach, invest in innovation (Bak et al., 2017), education, green jobs (Bataille et al., 2018), renewable energy, transition away from fossil fuels (Bataille, 2020), and implement adaptation measures (Ogbulezie et al., 2017). The economic cost of reducing emissions (Agbor et al., 2023a) and creating a sustainable future is expected to increase (Agbor et al., 2023b).

This means that by 2100, governments around the world must take bold and urgent action to create a low-carbon (IRENA, 2020a), climate-resilient future that can benefit current and future generations by investing in clean energy sources (IRENA, 2020b), building green infrastructure (IRENA, 2021), and transitioning away from high carbon-emitting sources of energy (IRENA, 2012) while also protecting vulnerable populations (IRENA, 2013). With the right investments and initiatives, global warming can be kept below 2 °C by 2100 (IRENA, 2019), and the world can move toward a more sustainable future that is both economically viable

S. C. Nwokolo et al., *Africa's Path to Net-Zero*, CSR, Sustainability, Ethics & Governance, https://doi.org/10.1007/978-3-031-44514-9_8

(Birge & Berger, 2019) and environmentally secure (Bonciu, 2021). This is a momentous task that requires coordinated and unified action from countries all around the world to create an environmentally sustainable future that will ensure economic (Borasio & Moret, 2022) and social progress for current and future generations (Bray et al., 2022). The challenge before us is clear: the world must act with unity (Caetano et al., 2017) and purpose to effectively address climate change and create a more sustainable future that is both economically (Calvillo & Turner, 2020) and environmentally secure and that will ensure that global temperatures stay within 2 °C by the end of this century and that the world can progress and thrive for years to come (United Nations, 2021).

However, it will be a major disruptor for those who cannot transition, with the possibility of significant stranded assets around the world (Celia, 2021). Unresolved is the crucial question of whether Africa will fairly share in this global investment (Chaichan & Kazem, 2018) and growth or, in the absence of adequate economic strength, will continue to lag behind the global net-zero transition plan (Chukwujindu et al., 2022). Consequently, Africa needs to be provided with adequate resources (Cillari et al., 2021) and support to facilitate its transition into a more sustainable (Cossel et al., 2019) and prosperous future (Cotterman et al., 2021). To ensure that Africa and its citizens are not excluded from the global transition (da Silva et al., 2018), governments around the world must commit to providing Africa with financial (Dawson et al., 2022) and technical assistance to allow it to benefit from the opportunities of net-zero transitions (Delafield et al., 2021). Moreover, in addition to global investments in net-zero transitions (Derkenbaeva et al., 2022), governments around the world should also implement policies that prioritize technology transfer (Deutch, 2020) and capacity-building initiatives (Diringer & Perciasepe, 2020), thereby allowing African countries to develop (du Plooy, 2017) and build the necessary infrastructure required for net-zero transitions while also being provided with the resources needed to reduce their greenhouse gas emissions (Dudău & Cătuți, 2019).

Additionally, African nations should be allowed to take full advantage of the various net-zero initiatives (Elsässer et al., 2018) and programs being implemented across the globe by investing in them (Fan & Friedmann, 2021) and providing technical assistance to create more opportunities for African citizens and businesses to participate (Felder et al., 2021). Furthermore, governments around the world must ensure that African countries are adequately represented in all discussions regarding net-zero transitions (Ferreira et al., 2016), as well as in the allocation of resources and investments toward such initiatives (Florent Baarsch & Schaeffer, 2019) so that the continent and its people are not left behind in the global shift toward a more sustainable (Ford & Hardy, 2020), net-zero future.

The global transition toward net-zero emissions requires African countries (Ford et al., 2021) and citizens' inclusion, resources, and opportunities to contribute to the transition and achieve its goals (Freeman, 2021). African countries and partners must work together to fully integrate into the global net-zero transition (French, 2020), combat climate change risks (French, 2020), and contribute to a sustainable future. Therefore, a more concerted effort is required from all stakeholders to ensure that

Africa is adequately represented, provide African countries with the resources, tools, and opportunities necessary for them to successfully participate in and contribute to the global effort to reach net-zero emissions (Furszyfer Del Rio et al., 2022a) and secure a sustainable and prosperous future for their people (Furszyfer Del Rio et al., 2022a), and ensure that the continent's people are not left behind in this global effort (Galgóczi, 2020).

We consider Africa's current energy situation in this context and present potential developments for its fossil fuel (Greig & Uden, 2021) and renewable energy sectors (Godin et al., 2021). The global transition impacts African economic systems, populations, and surroundings (Gozlugol, 2022). We offer perspective on the following as we examine these scenarios: aided and harsher scenarios (scenario 1), collaboration and a calculated transition (scenario 2), unaided and self-sufficient in exports (scenario 3), and a tangled and trapped transition (scenario 4). The chapter also discusses the four theoretical underpinnings of Africa's propensity for a net-zero energy transition as well as the economic and political potential solutions under different scenarios that can catalyze or impede Africa's propensity for a net-zero energy transition (Guan et al., 2021). Through our evaluation of these scenarios, we ultimately hope to propose a roadmap for Africans in their pursuit of net-zero energy and economic prosperity (Gurieff et al., 2021).

The various scenarios outlined above provide us with a glimpse of what the future could hold for African countries in terms of their energy (Hafner et al., 2021) and economic outlook. Each of these scenarios presents unique challenges as well as opportunities for Africa to move toward a net-zero energy transition. In light of this, it is essential that African countries begin to look at energy solutions from an economic and political perspective to ensure their successful implementation (Hassan et al., 2021). A net-zero energy transition offers African countries economic opportunities, but a comprehensive approach is crucial for success (Hassan et al., 2021). The social, economic, and environmental impacts of energy solutions (Hosseini, 2022) should be considered holistically to identify economically viable solutions (John-Jaja et al., 2018) while meeting African citizens' needs and protecting the environment (Hrnčić et al., 2021). A successful net-zero energy transition in African countries requires a comprehensive approach considering economic (Iqbal et al., 2017), social, and environmental impacts, ensuring successful implementation of strategies for a prosperous future (Ituen et al., 2012). African countries must consider economic and political implications for a successful transition to net-zero energy (Jakob et al., 2020), meeting citizens' needs (John-Jaja et al., 2016a), and safeguarding the environment for the first time (John-Jaja et al., 2016b).

8.2 Theoretical Underpinnings of Africa's Propensity for a Net-Zero Energy Transition

Over the ensuing decades, Africa's energy sector will undergo significant structural change. One of the most urgent issues that Africa and the rest of the world must confront is how to finance this change at the required rate. Four potential situations that could result in an African net-zero energy transition have been identified by the authors. These are aided and harsher scenarios (scenario 1), collaboration and a calculated transition (scenario 2), unaided and self-sufficient in exports (scenario 3), and a tangled and trapped transition (scenario 4). In scenario 1, significant foreign funding for renewable energy will be provided to the continent along with strict prohibitions on the revenue from the export of fossil fuels; in scenario 2, significant foreign funding for renewable energy will be provided to Africa along with strict limitations on the revenue from the export of fossil fuels; and in scenario 3, Africa will be self-sufficient and unaided in exports. According to this scenario, financing foreign renewable energy sources without strict limitations on the money earned from the export of fossil fuels will be very detrimental to Africa. In case 4, there would be significant harm to Africa from foreign investments in renewable energy, and exports of fossil fuels would be severely constrained. Africa would be tangled and trapped in a transition, unable to move forward or retreat without significant outside aid.

The overall result of any of these scenarios would be a continent that is not self-sufficient and unable to advance in the transition to renewable energy sources without significant foreign investments in renewable energy as well as stringent regulations on the export of fossil fuels. Not only would this hamper the transition to renewable energy, but it could also lead to immense socioeconomic (Nwokolo & Amadi, 2018) and environmental damage (Nwokolo et al., 2022a) due to the reliance on foreign investments and regulation to make progress. In other words, restrictions on the export of fossil fuels and support from foreign investments in renewable energy are necessary for Africa to switch to renewable energy sources (Nwokolo et al., 2022b). Africa will not be able to make the switch to renewable energy sources necessary to enhance the lives of its citizens (Nwokolo et al., 2023a) and safeguard the environment without foreign investments and regulations (Nwokolo et al., 2023b) .

The situation is dire, but there is hope if foreign investors and governments take appropriate action to provide adequate investments in renewable energy projects (Nwokolo et al., 2023c) and impose regulations restricting fossil fuel exports to successfully navigate the challenges of transitioning to renewable energy and the potential issues associated with it (Nwokolo et al., 2023d). This will ensure that Africa is well equipped to make the transition to renewable energy sources (Nwokolo & Ogbulezie, 2018a) and that it can leverage this as an opportunity to create sustainable (Nwokolo & Ogbulezie, 2018b), environmentally friendly sources of power (Nwokolo & Ogbulezie, 2018c) and ensure a brighter future for its people and its environment (Nwokolo & Ogbulezie, 2017a). Therefore, by switching to renewable energy sources that are good for the environment (Nwokolo & Ogbulezie, 2017b) and profitable for the continent as well as the global economy (Nwokolo & Otse, 2019), Africa has the potential to improve its own future and that of its people.

In general, Africa has enormous potential in regard to switching to renewable energy sources (Obiwulu et al., 2020a), and it is crucial for foreign investors and governments to invest in renewable energy projects (Obiwulu et al., 2020b), and enact regulations on the export of fossil fuels to ensure that Africa can make this transition as efficiently as possible and reap the economic and environmental benefits of this transition while creating a sustainable and environmentally friendly future for its people (Obiwulu et al., 2022).

These four scenarios depict a variety of potential futures for Africa's switch to renewable energy sources as well as the effects of such a switch. The potential advantages and disadvantages of Africa switching to renewable energy sources can be seen in each of these four scenarios in terms of the various financing and export opportunities available to African countries. For instance, Scenario 2 describes how foreign investment in renewable energy projects could create jobs and allow African countries to pursue a more sustainable and financially secure future without restrictions on the export of fossil fuels. However, Scenario 3 describes the risks of a lack of foreign investment in renewable energy projects, including how this could reduce job opportunities and economic growth while increasing the reliance of African countries on exports, fossil fuels.

The increased pollution and health risks brought on by excessive reliance on fossil fuels (Ogbe et al., 2017), as well as the restricted access to renewable energy sources that could be used to power African countries' economic development in the long run (Okem, 2016), are additional potential negative effects of a lack of foreign investment in renewable energy projects (Omar et al., 2022). Therefore, it is important to consider both the potential advantages and disadvantages of Africa's transition to renewable energy sources when assessing the potential for foreign investment in renewable energy projects on the African continent (Omrany et al., 2022). To guarantee a favorable outcome for African nations and their citizens (Ouedraogo, 2019), it is crucial to take both of the aforementioned scenarios into account when assessing the potential effects of foreign investment in renewable energy projects on the African continent (Oyewo et al., 2020).

Long-term investment in renewable energy projects is crucial for African countries if they are to guarantee economic expansion (Oyewo et al., 2019), create jobs (Pablo-Romero et al., 2021), and lower pollution and health risks for their citizens (Perlaviciute et al., 2021). Scenario 3 presents a sobering picture of the dangers of a lack of foreign investment in renewable energy projects in Africa as well as the potential for foreign investors to take advantage of African countries and their people by aiming to maximize their own financial gains without giving much consideration to the effects on the environment (Pineda & Faria, 2019), public health (Piselli et al., 2021), or the economic development of African countries. Without the required foreign investment, African countries are unable to fully benefit from the advantages that renewable energy projects may offer. They also lack the knowledge (Poggi et al., 2015) and expertise required to ensure sustainable economic growth (Poggi et al., 2020), the creation of jobs (Quarton & Samsatli, 2021), and long-term sustainability in the renewable energy sector (Raihan et al., 2022).

African countries must work to create an environment that is appealing to investors while also safeguarding the interests of their citizens by establishing clear environmental (Rajeh & Bensenouci, 2019) and public health regulations (Ramagoma & Adendorff, 2016), putting safety measures in place to stop foreign investors from abusing African countries (Ratlamwala & Dincer, 2018) and their citizens, and laying the groundwork for sustainable economic growth (Raugei et al., 2020) and job creation. Foreign investment in renewable energy projects can provide economic (Razmjoo et al., 2021) and environmental benefits for African countries if managed and regulated effectively (Rennkamp & Boyd, 2015). Countries should focus on maximizing benefits for citizens and the environment while limiting potential exploitation by foreign parties (Rissman et al., 2020).

African nations must have the courage and foresight to recognize the risks of foreign investment in renewable energy projects (Rogelj et al., 2015), but they must also have the strength to create an environment that is attractive to investors while protecting their citizens' interests (Rosa & Mazzotti, 2022). African nations must develop infrastructure (Salleh et al., 2020), knowledge (Scheinin, 2016), and skills to capitalize on the economic (Samuel, 2017) and environmental benefits of foreign renewable energy investments (Scholten et al., 2020), benefiting both investors and the environment (Schreyer et al., 2020).

However, Scenario 2 paints a much brighter picture. Scenario 2 paints a picture of foreign investments in renewable energy projects providing African nations with an opportunity to diversify their energy portfolio (Scott & Gössling, 2021), create jobs, attract foreign capital and technology (Setyowati, 2020), and provide environmental access to electricity (Shamim et al., 2021) and other energy services. All of these benefits are dependent on African nations having the right policy (Shen, 2020) and regulatory frameworks in place (Sokołowski & Heffron, 2022) as well as the capability to manage renewable energy projects effectively (Sokona et al., 2012). Therefore, it is clear that this can be a great boon to African nations, but for this potential to be realized, African nations must focus on creating the necessary infrastructure (Sovacool et al., 2021) and acquiring the knowledge and skills required to make the most of these investments (Stendahl et al., 2022). As such, it is clear that African nations must prioritize the adoption of policies and regulations that can attract foreign investments into renewable energy projects while also building the capacity to ensure that these investments can be managed effectively (Stern, 2021) to maximize their potential benefits (Styring et al., 2021).

African nations can benefit from foreign investments in renewable energy projects to develop economies, create jobs, and enhance public health. They should also mitigate environmental impacts (Sunday et al., 2016a) and improve citizens' quality of life (Sunday et al., 2016b). This will not only bring much-needed economic and social development to African countries but also help them meet their commitments. In general, African countries have an opportunity to reap the rewards of foreign investments in renewable energy projects, but only if they are willing to make the necessary reforms and develop the capacity to ensure that these investments are managed properly (Thompson, 2022) and used in ways that bring tangible economic, social, and environmental benefits to their citizens (Tlili, 2015).

This requires African countries to take a holistic approach and to prioritize the adoption of regulations and policies that both attract foreign investments in renewable energy projects (Tokimatsu et al., 2018) and ensure that these investments are managed effectively and used in ways that maximize the economic (Tokimatsu et al., 2017), social, and environmental benefits for their citizens. African countries need to invest in developing the necessary institutions and regulatory systems for effectively managing foreign investments so that they can take advantage of the opportunities provided by renewable energy investments (Trotter, 2022) while also taking steps to mitigate any potential environmental impacts and improve the quality of life for their citizens (Tsao et al., 2021). This will create an enabling environment for foreign investments in renewable energy projects that is conducive to both environmental protection (Van Den Dobbelsteen et al., 2018) and socioeconomic development on the African continent and enable African countries to reap the full economic (van Soest et al., 2021), social, and environmental benefits of these investments (Veng et al., 2020) while also ensuring that these investments are managed in a sustainable and responsible manner (Venkataraman et al., 2022).

In contrast, Scenario 4 paints a picture of how increased foreign investment in renewable energy projects could create more job opportunities, reduce emissions, and increase economic stability while providing the freedom to pursue both domestic and export markets for fossil fuels. Scenario 4 offers a potential compromise between the two preceding scenarios; it outlines how African nations could benefit from foreign investment in renewable energy projects with strict restrictions on the export of fossil fuels to maintain a balance between economic growth and environmental sustainability. Scenario 1 offers a contrasting view of African nations transitioning to renewable energy sources independently, highlighting the potential benefits of investing in local renewable energy projects without continuing to export fossil fuels to foreign markets. Scenario 1 outlines a potentially beneficial situation in which African nations transition to renewable energy sources without relying on foreign investment and with strict restrictions on exporting fossil fuels to foreign markets.

These four scenarios show a range of possible futures for Africa's transition to renewable energy and the consequences of such a transition. In each of these four scenarios, the potential benefits and harms of Africa's transition to renewable energy sources can be seen in terms of the different financing and export opportunities available to African nations. For example, Scenario 1 shows a scenario in which foreign investments in renewable energy sources are unrestricted while exports of fossil fuels are strictly limited. In contrast, Scenario 4 shows a scenario where foreign investments in renewable energy sources are more restricted and exports of fossil fuels are more severely constrained. This provides an example of how, depending on the policy decisions made by African governments, the transition to renewable energy can have very different outcomes for African nations.

In scenario 1, African countries benefit from increased foreign investment in renewable energy sources, allowing them to take advantage of technology and expertise that they may otherwise not have access to, while in scenario 4, African countries are able to retain more of the profits from their fossil fuel exports as well as limit the potential for foreign exploitation of their resources. In both scenarios, African countries have the opportunity to develop their economies while transitioning to renewable energy sources in an economically (Vieira et al., 2021) and environmentally sustainable way (Viladrich et al., 2016). In conclusion, scenario 4 shows that African countries can take an active role in the transition to renewable energy sources by making strategic policy decisions that are tailored to their own unique needs and goals while still providing the benefits of increased foreign investment (Vogt-Schilb & Hallegatte, 2017) and limiting their potential to be exploited by outside entities (Waugh, 2011). The differences in these two scenarios clearly demonstrate that African countries have the ability to make informed policy decisions to best utilize renewable energy sources for their economic (Weber et al., 2021) and environmental sustainability without sacrificing their autonomy and independence as they develop their crafts (Wiggins, 2014).

These two scenarios demonstrate the flexibility of African countries to transition to renewable energy sources and make strategic policy decisions for their own benefit while still taking advantage of the potential that foreign investment (Williams et al., 2021) and international partnerships have to offer (Winkler & Marquand, 2009). The best scenario for African nations in this situation is Scenario 4, as it enables them to benefit from increased investment without sacrificing their autonomy or their ability to make independent decisions during the switch to renewable energy sources (Winrow, 2004). The most preferable of the two scenarios is Scenario 4, which enables African countries to make the best use of their resources, design policies that are specific to their own needs (Wiseman, 2018) and goals (Wohland et al., 2018), and still reap the rewards of foreign investment while limiting exploitation (Yan et al., 2019). African nations greatly benefit from this autonomy because it enables them to develop a more comprehensive strategy for the transition to renewable energy that takes into account the particular needs and constraints of their own nations while still allowing them to take advantage of the resources (York et al., 2015) and expertise that come from international partnerships (Young et al., 2021).

Scenario 4 is thus the ideal outcome for African countries in this situation, as it allows them to enjoy the best of both worlds: increased investment and foreign experience without sacrificing their autonomy and ability to tailor policy decisions that are best suited to their own needs (Zeweld et al., 2017) while still protecting their sovereignty and decision-making capabilities (Zinola, 2010). In general, scenario 4 is the most advantageous option for African countries in this situation due to its ability to balance increased investment with autonomy and flexibility for African countries in terms of their renewable energy transition plans and decision-making processes and the potential to craft policy decisions that are tailored to their own needs and goals without compromising on the overall renewable energy transition. Scenario 4 is the most advantageous option for African countries, as it allows them to have access to foreign investment and experience while still protecting their autonomy and decision-making capabilities. They also have the freedom to craft renewable energy policies that are tailored to their unique needs and limitations while still protecting their sovereignty and decision-making capabilities. This ultimately provides African countries with the opportunity to move toward a renewable energy transition that is tailored to their specific needs and limitations while still having access to foreign investment, expertise, and experience.

8.3 Analysis of Various Scenarios for Africa's Propensity for a Net-Zero Energy Transition

8.3.1 Scenario 1: Aided and Harsher

Under scenario 1, foreign funding for renewable energy would provide Africa with much-needed capital and help reduce its reliance on fossil fuels, while under scenario 2, the money earned from the export of fossil fuels would be limited and therefore would not be able to sustain Africa's economy as much as it would in scenario 1. This is because, although Africa would benefit from the capital, it would not be able to gain the same degree of economic stability and sustainability as it would if there were no limitations on the money earned from exporting fossil fuels. Therefore, although scenario 2 would provide Africa with some much-needed capital, it would not be nearly as beneficial to the continent in the long run as scenario 1. In conclusion, without strict limitations on the money earned from exporting fossil fuels, financing foreign renewable energy sources in Africa will not be as beneficial to the continent in the long run as it would be if those limitations were present. Scenario 1 should be the priority in Africa, as foreign investment in renewable energy will provide long-term stability and sustainability to the continent in the form of economic growth, employment opportunities, and other means of development that seem to be in direct opposition to Hesiod's conception of justice. Scenario 1 would therefore be more beneficial for Africa in the long run, as it would provide the continent with a more stable and sustainable economic development that could be used to create employment opportunities and promote economic growth in ways that are more consistent with traditional notions of justice than scenario 2, which does not include any such limitations and relies solely on foreign investment for renewable energy sources and limitations that would be difficult to accept as they strived to prove themselves as competent, talented, and well-respected performers.

8.3.2 Scenario 2: Collaborative and Calculated

Scenario 2, while beneficial to Africa in some ways, could end up having a negative impact if the money earned from exporting fossil fuels is not monitored or regulated properly. The positive aspects of scenario 2 are that Africa can benefit from foreign investment in renewable energy sources and have an additional source of income in the form of the export of fossil fuels. However, without proper monitoring and regulation of the money earned from the export of fossil fuels, this could lead to a situation where Africa is using foreign investment to fund unsustainable practices while earning a large amount of money in return. This could lead to a situation in which African countries are using foreign investments to fuel practices that are not in line with their environmental goals while earning large profits from the export of fossil fuels. This could have a long-term negative impact on the environment and could lead to a situation in which Africa is not utilizing its resources sustainably and is not investing in renewable energy sources as it should be. Consequently, it is crucial for African countries to ensure that the money earned from the export of fossil fuels is properly monitored and regulated to prevent this kind of situation from arising and to ensure that the money is being used to fund renewable energy sources and sustainable practices that are in line with the environmental goals of African countries going forward. Scenario 2, while beneficial to Africa in some ways, could end up having a negative impact if the money earned from exporting fossil fuels is not monitored or regulated properly. The positive aspects of scenario 2 are that Africa can benefit from foreign investment in renewable energy sources and have an additional source of income. However, without proper monitoring and regulation of the money earned from the export of fossil fuels, this could lead to a situation where Africa is using foreign investment to fund unsustainable practices while earning a large amount of money in return. This could lead to a situation in which African countries use foreign investments to fuel practices that are not in line with their environmental goals while earning large profits from the export of fossil fuels. This could lead to a situation in which Africa is not utilizing its resources sustainably and is not investing in renewable energy sources as it should. As a result, it is essential for African nations to ensure that the money made from the export of fossil fuels is properly monitored and regulated to avoid situations such as this from occurring and to ensure that the money is used to finance renewable energy sources and sustainable practices that are in line with the continent's long-term environmental goals.

8.3.3 Scenario 3: Unaided and Self-Sufficient in Exports

In scenario 3, Africa will be self-sufficient and export-free. According to this scenario, financing foreign renewable energy sources without strict limitations on the money earned from the export of fossil fuels will be very detrimental to Africa. If Africa were to remain export-free, then countries that rely on the energy exports of

Africa would be unable to make investments. In addition, remaining export-free would mean that African countries would be unable to finance their own renewable energy sources as well as other infrastructure projects necessary for economic growth and development. This would create a vicious cycle of poverty, pollution, and dependency on foreign aid, making it even more difficult for African countries to break out of their current economic situation and move toward a path of sustainable development. In this case, it is clear that the solution to Africa's energy crisis lies in the ability of African countries to finance their own renewable energy sources while still engaging in limited trade of their fossil fuels to generate the additional revenue necessary to invest in these renewable energy sources as well as other critical infrastructure projects necessary for economic growth and development. For these reasons, it is in the best interest of both African countries and countries dependent on Africa's energy exports for Africa to engage in a balanced approach of limited trade while still investing heavily in renewable energy sources and infrastructure projects to move toward a path of sustainable development. This approach promotes economic growth and development in African countries while ensuring a sustainable energy future. It uses fossil fuels as a limited resource for renewable energy investments and infrastructure projects.

Africa's chances of becoming self-sufficient in energy resources, however, may be difficult to achieve without the proper regulations in place if the continent is unable to control foreign investments. Therefore, Africa must be careful to set restrictions and regulations on foreign investments, ensure that those who invest in renewable energy are held accountable for their actions, and ensure that the money earned from investments in renewable energy is spent responsibly and on the development of local resources to achieve self-sufficiency and prevent potential resource exploitation. Africa must take proactive steps to ensure that their renewable energy sources are properly funded and regulated so that they can achieve true self-sufficiency and be free from the exploitation of their resources by foreign investors. In fact, financing renewable energy sources without proper regulations and limitations in place can be detrimental to Africa's ability to attain self-sufficiency and can lead to an increased reliance on foreign investments, which can ultimately undermine the country's economic autonomy and leave it vulnerable to exploitation. This means that Africa must develop an effective regulatory framework to ensure proper accountability and transparency in foreign investments in renewable energy sources and should also invest in domestic renewable energy sources to increase self-sufficiency and energy security. African governments must regulate foreign investments and hold renewable energy investors accountable to protect the continent's resources and maintain economic autonomy. This can be achieved through tax incentives, environmental regulations, and public–private partnerships to ensure proper use of renewable energy and prevent overreliance on foreign investments. This is essential for African countries to gain true energy independence and protect their resources from exploitation by foreign entities while ensuring that the money invested in renewable energy is used properly for the benefit of their citizens.

8.3.4 Scenario 4: Tangled and Trapped

In the fourth and final scenario, it is assumed that Africa will be self-sufficient and unaided in developing renewable energy sources while limiting fossil fuel exports to a degree that would enable it to be self-sustaining. In this fourth and final scenario, Africa will be the driving force behind its own renewable energy revolution and will not rely solely on foreign investments or severe limitations on fossil fuel exports to become self-sufficient in renewable energy sources. In this fourth and final scenario, Africa will be empowered to take its destiny into its own hands and focus on developing renewable energy sources that are appropriate for their unique climate and landscape while simultaneously limiting the emissions of greenhouse gases to protect their citizens, the environment, and their natural resources for the long term. This fourth and final scenario is the most sustainable and beneficial for Africa, as it gives them autonomy over their own future, keeps their emissions low, and ensures that renewable energy investments are beneficial to the continent in both the short-term and long-term time frames. In the fourth and final scenario, Africa will have the chance to positively influence its own future while also contributing to environmental protection and guaranteeing that its citizens can profit from renewable energy investments in a safe and sustainable manner. The fourth and final scenario is that Africa can become a global leader in the production of renewable energy by making its own investments in the industry and sharing its knowledge with other nations that might require assistance in the development of renewable energy resources in their own regions. As a result, Africa can advance toward greater sustainability for the planet, become a global leader in the production of renewable energy, aid in the reduction of global emissions, improve the welfare of its own people, and increase the security of its own energy resources for future generations. All of this will result in a better future for Africa, where they will be able to make their own choices, gain more autonomy, maintain low emissions, and benefit from investments made in renewable energy sources for both their own people and the entire global community.

These are contingent on a variety of variables, such as the rate of global net-zero adoption, the amount of international financial assistance to Africa, and the pace of economic growth in Africa, which will include the revenue from fossil fuel exports. All four situations are possible pathways to a net-zero African energy transition, and each is dependent upon the aforementioned variables. It is important to recognize that, depending on the variables and outcomes, some situations may be more likely than others. The authors emphasize that a net-zero African energy transition is not a one-size-fits-all solution due to its complexity and various variables affecting outcomes.

This complexity is compounded by the fact that African countries will also have to navigate political, economic, and social constraints that could impede or prevent the successful adoption of a net-zero energy transition plan. To achieve a net-zero energy transition, African nations must consider various variables and implement practical solutions based on technical feasibility and social acceptability. As such, an effective net-zero energy transition plan for African countries must consider both the potential advantages and pitfalls of each possible scenario to ensure that their approach is as informed and well-rounded as possible.

In other words, no single situation should be taken as a definitive solution to the African energy transition challenge, but rather a variety of strategies and solutions should be evaluated to best serve the particular needs of each African nation. This can be a daunting task, and it will require a high level of expertise, strategic planning, and coordination across many sectors. Achieving a net-zero energy transition in Africa depends on African countries' commitment to analyzing and selecting feasible options. A comprehensive, informed, and well-coordinated approach can lead to meaningful and lasting change, impacting not only African countries but also the global community for decades to come.

African countries must develop a comprehensive understanding of net-zero energy systems' costs and benefits to create a tailored policy framework. This should include economic costs such as investments, infrastructure maintenance, and worker training, as well as social and environmental benefits such as increased energy access, improved public health, and reduced greenhouse gas emissions. The net-zero energy transition offers African countries a unique opportunity to tackle economic growth and climate change simultaneously. However, proper harnessing requires a policy framework that considers the costs and benefits of net-zero energy systems. This will maximize the potential for African countries and the global community.

8.4 Possible Responses to Different Scenarios

8.4.1 Economic Solutions

8.4.1.1 Scenario 1: Assisted and more Severe (Carrot and Stick)

Aid is accompanied by strict decarbonization targets, and government-to-government climate agreements continue to become harsher, crueler, and more unusual toward countries that are not even rapidly transitioning away from fossil fuels. This "tough love" approach has encountered some opposition from nations that claim that such punitive measures pose a threat to their economies and will cause them to lag further behind other nations in terms of their climate goals. As a result, many proponents of stricter climate regulations now support a strategy that is more cooperative in nature and entails giving these nations financial and technical support to ease the transition to a low-carbon economy while still holding them responsible for meeting their goals. This could include funding for renewable energy initiatives, subsidies for electric vehicles, and technical support in areas such as energy efficiency and sustainable infrastructure development. This could help ease the economic burden on those countries while still providing an incentive for them to move forward with their decarbonization efforts. This "carrot and stick" approach (aided and harsher), as it is often referred to, could be a powerful tool in the fight against climate change and could be the key to getting countries around the world on board by taking meaningful action to reduce their carbon emissions. Wealthy countries can implement a "carrot and stick" approach to assist developing countries in transitioning to renewable energy sources, mitigating climate change, and reducing carbon emissions. This approach provides economic relief and assistance to those in need, making it a powerful tool in the fight against climate change and incentivizing meaningful action to reduce carbon emissions.

8.4.1.2 Scenario 2: Collaborative and Calculated

Financial support for Africa's energy infrastructure is increased in Case 2 by foreign investors. This helps to significantly increase energy production compared to Situation 1 and alleviate Africa's severe energy shortage. In contrast to scenario 1, the funding is more readily available and takes a longer-term approach to the transition. There are also fewer restrictions on a quick exit from its fossil fuel sector. Foreign investors can encourage and promote the development of renewable energy sources that benefit not only Africa but also the entire planet by contributing financially to Africa's energy infrastructure. This helps the African continent lessen its reliance on more polluting energy sources such as coal and oil and gives it access to clean energy, which is desperately needed.

The establishment of programs such as the African Renewable Energy Initiative (AREI) and the Sustainable Energy for All (SE4ALL) program, along with significant technological advancements, have made renewable energy sources such as solar and wind more accessible in Africa. Because of this, African countries now have better access to renewable and sustainable energy sources, allowing them to reduce their reliance on fossil fuels and move toward a more environmentally friendly and sustainable future. By fostering job growth, enhancing regional economies, lowering energy costs, and enhancing rural residents' access to electricity, investments in renewable energy infrastructure can also have a positive economic impact on African nations. This is a very significant development because the use of renewable energy sources can significantly reduce emissions from fossil fuels and aid in the global fight against climate change, which is largely to blame for the environmental harm occurring all over the world.

Additionally, by lowering air pollution caused by the burning of fossil fuels, which can cause a variety of respiratory and cardiovascular diseases in African nations, investing in renewable energy sources can also benefit the general public's health in a number of ways. This makes the switch to renewable energy a crucial step in raising the general standard of living and boosting the economic prosperity of African nations. By generating jobs, boosting local economies, lowering emissions from fossil fuels, battling climate change, and enhancing public health, investments in renewable energy infrastructure can have a positive effect on African nations. This lessens air pollution, safeguards the environment, and gives citizens access to the economy.

8.4.1.3 Scenario 3: Unaided and Self-Sufficient in Exports

Situation 3, where Africa must compete in the open market for financing and export market share, is largely regarded as "business as usual." As investors seek to reduce risk and place their money in "cleaner" and faster-growing markets in the industrialized world, foreign investment declines. Since they only account for approximately 5% of annual global emissions, Africa's carbon emissions are also not seen as a top priority, but the continent lacks the financial capacity to monetize proven reserves. This creates a catch-22 situation where African nations are unable to receive financial assistance or other incentives from the developed world to combat climate change and, at the same time, do not have the means to expand their export market share to compete for funding and export markets in a precarious environment. This has left Africa in a difficult situation where it must develop viable solutions to climate change that do not rely on outside help while at the same time striving to become competitive in an increasingly competitive global marketplace. African countries must develop self-sustaining, climate-resilient economies that prioritize citizens' wellbeing and sustainable growth to combat climate change and protect the environment for future generations.

African countries are in a precarious position: not only must they face the challenges of climate change without external support, but they must also create an economy that prioritizes sustainability and wellbeing to access the funds needed for climate resilience while at the same time striving to become competitive in an increasingly competitive global marketplace. This has created a great challenge for African countries, as they must attempt to balance both the difficult task of combating climate change without external assistance and creating a sustainable and resilient economy that will help them become more competitive in the global market while at the same time maintaining a focus on protecting their citizens' wellbeing and safeguarding their environment for future generations. This is no easy task, and African countries have already been faced with various obstacles while attempting to reach these goals.

However, with the right policies and practices in place, African countries can create a self-sustaining, climate-resilient economy that will support their citizens and promote long-term economic growth while at the same time protecting the environment and safeguarding it for future generations. African countries must find innovative solutions to the difficult challenge of creating a sustainable and resilient economy while simultaneously addressing climate change and protecting citizens' wellbeing and the environment. This can be done through the implementation of innovative climate-smart policies and practices, such as investing in renewable energy sources, protecting and preserving forests and other ecosystems, transitioning to more sustainable agricultural practices, promoting clean air initiatives, increasing access to clean water resources, and improving the efficiency of energy and transportation systems. In addition, African countries must create effective economic incentives for citizens to support these climate-smart initiatives.

8.4.1.4 Scenario 4: Tangled and Trapped

The worst-case scenario for Africa is Situation 4. Because of the continent's inability to comply with emissions standards set by the developed world and/or deteriorating economic conditions, international relations and investment become more fractious. Due to a sizable capital outflow, Africa is left to address its energy shortage without foreign investment and with only a small amount of donor money. The most likely scenario is that African countries, given the current infrastructure and resource constraints, will continue to rely on coal and natural gas for energy production, as they offer reliable and low-cost options in the near term. To balance the need for increased energy production and consumption, as well as developing renewable sources of energy, African countries must look for long-term solutions that involve more efficient use of existing resources while transitioning to cleaner energy sources to meet the energy needs of a growing population while reducing emissions. If the necessary investments are made in efficient and renewable energy technologies, Africa's energy sector could still undergo a transformation by 2050, although 4% of global funding may seem insignificant. Thus, investments in both renewable energy sources and efficiency technologies must be a priority for African governments if they are to ensure the sustainability of the continent's energy sector by 2050 while also lowering emissions.

Some of Africa's quite adaptable economies may be able to survive this transition, which may be advantageous for those who are fortunate enough to have access to the minerals used in the production of the infrastructure for renewable energy sources. However, in this scenario, it is highly likely that the majority of African countries would lack the time and resources necessary to completely phase out fossil fuels. Fossil fuel-dependent economies are likely to experience a serious economic downturn, as well as an increase in unemployment and instability.

The African continent will find it difficult to raise the funds required to address its energy crisis without foreign investment and with a finite amount of donor money, creating a vicious cycle of poverty, debt, and environmental degradation. African societies will struggle to find lasting solutions to their energy crisis as poverty and environmental degradation become more pervasive on the continent. Additionally, the continent's finite energy resources will not be able to keep up with the rising energy demand, worsening the continent's already precarious economic situation and endangering its inhabitants' ecosystem. On the African continent, this vicious cycle of poverty, debt, and environmental destruction has a significant impact on stability, overall economic growth, and energy security.

The African continent faces challenges in developing renewable energy sources such as solar and wind power due to limited foreign investment and donor money. This lack of technology and expertise could alleviate energy crises, promote economic development, and ensure stability for its inhabitants. As a result, it is essential that the African continent receive increased foreign investment and donor aid to generate the funds necessary for addressing its energy crisis, enabling it to move toward a more sustainable future that relies on renewable energy sources, economic stability, and a healthier environment for its inhabitants. This will ensure that the African continent is able to access the necessary resources, technology, and expertise needed to develop renewable energy sources and create an environment of greater economic development, stability, and prosperity for its inhabitants in the future.

8.4.2 Political Support

8.4.2.1 Scenario 1: Supported and Harsher

Strong foreign support for renewable energy without the political backing of the majority of African leaders and combined with income from the export of fossil fuels will almost certainly lead to the undermining of decarbonization goals. African nations are facing worse treatment in climate relations with foreign funding nations due to their reluctance to transition to fossil fuels. The majority of African politicians and political system participants argue that such punitive policies threaten to disrupt their private businesses and investments in nonrenewable industries, causing them to fall further behind other investors in renewable energy and climate-friendly businesses in terms of their climate goals. As a result, this more aggressive strategy has encountered some resistance. Despite this resistance, foreign nations must still actively pursue a policy of engagement with African leaders that emphasizes the importance of transition if they are to make any meaningful progress in the global effort to mitigate climate change. Foreign nations must offer financial and technical support to African countries for a fair, efficient, and beneficial transition.

To achieve this goal, foreign nations should be willing to take on a leading role in helping African countries by providing incentives for businesses to transition to clean energy sources and helping local communities access green technology solutions that can help reduce their dependence on traditional fossil fuel sources of energy. Moreover, foreign nations must also be cognizant of the potential and be prepared to help African nations address any potential disruptions that may result from this transition to ensure that it is a smooth and equitable one that puts the wellbeing of all involved parties at the forefront. Additionally, foreign nations must promote transparency and accountability in their dealings with African governments to ensure that they are held to the same standards of responsibility and fairness as those adopted by developed countries in regard to environmental stewardship and the responsible use of resources. Foreign nations must support African countries in a successful transition, ensuring equitable, efficient, and beneficial support for a sustainable, environmentally friendly future for African nations and their inhabitants.

By taking on such an active role, foreign nations can help African countries reach their goals of transitioning away from traditional fossil fuel sources of energy and toward a more sustainable future while also helping to promote economic growth and stability throughout the continent. Foreign nations must ensure technology and infrastructure for a successful transition from fossil fuels, especially in limited renewable energy access, while African nations must maintain sovereignty and self-determination in international dealings. Actively participating in international agreements, financial support for infrastructure development, and renewable energy access for African countries can lead to mutually beneficial partnerships and improved energy access. This is a critical task for foreign nations to take on if we are to see a successful transition away from traditional fossil fuel sources of energy and toward a more sustainable future for African countries and the people who inhabit them. Foreign nations must promote and support the smooth transition to renewable energy sources for African countries while preserving sovereignty, self-determination, and autonomy. This will benefit their countries and their inhabitants, create economic opportunities, and foster collaboration between African countries and international organizations.

On the other hand, these punitive policies, which are essentially punishing African nations for their perceived failures in renewable energy and decarbonization, can actually serve to further entrench and deepen the existing socioeconomic inequalities between wealthy nations and Africa and can even lead to further political divisions as well as a lack of economic opportunity and access to essential resources such as electricity and clean water in African nations. This has created a moral dilemma in terms of how African nations should be treated in terms of climate policy since there is an inherent tension between providing necessary financial aid and resources while still ensuring that these nations are held to the same standards as other countries in terms of renewable energy and climate mitigation goals. This presents a difficult conundrum, as it is important not only to protect the environment through progressive climate policy but also to be cognizant of the existing disparities between wealthier nations and Africa and to ensure that punitive measures are not taken, which could lead to further socioeconomic and political inequality between countries. The current approach of punishing African nations for their contributions to the climate crisis is unethical and could have dire consequences for their long-term economic and environmental sustainability. An alternative climate policy in Africa should hold African nations accountable for their climate change contributions while recognizing their socioeconomic realities. This can be achieved through financial aid and technological assistance, enabling African countries to transition to cleaner energy sources and protect the environment.

Ultimately, the current approach of relying solely on punitive measures for African countries is not only impractical and ineffective but also morally wrong, as it fails to take into account the unequal power dynamics between wealthy countries and African nations and the resulting disadvantages faced by African countries in transitioning to a more sustainable economic model that would ensure long-term environmental sustainability and socioeconomic equality between countries. By taking an approach to climate policy in Africa that combines punitive measures with financial aid, technological assistance, and other forms of support, African nations can be held accountable for their contributions to climate change while also receiving the resources and support they need to transition to a more sustainable economic model that would enable them to not only protect their environment but also promote socioeconomic equality. A comprehensive approach to climate policy in Africa that balances punitive measures with targeted support for transitioning to cleaner energy sources is the best way to ensure both environmental sustainability and socioeconomic equality between countries.

Resistance to climate change has deepened the divide between nations investing in renewable energy and those attempting to mitigate its effects, despite its understandable nature. Foreign nations must adopt a more conciliatory approach toward African nations resisting fossil fuel transitions to promote renewable energy, climate-friendly business practices, and reduce climate change effects. African nations must embrace a progressive approach to mitigating climate change and consider the long-term implications of their inaction on their environment, communities, and the world. Recognizing the impact of stricter punitive measures on businesses and investments is crucial.

African nations face challenges in regaining lost ground in technological advancements and government-backed investments for climate change mitigation, highlighting the need for global collaboration. Foreign funding nations argue for harsher measures to force African countries to transition from fossil fuels to renewable energy sources, reducing emissions and contributing to global climate change efforts. However, many African nations have argued that while transitioning away from fossil fuels is an important goal, they need more support in terms of financial resources, technological advancements, and investments to make the shift toward renewable energy sources a reality. African countries face unfair pressure from foreign funding nations to transition to fossil fuels without adequate resources and support. As a result, African nations have been placed in an unfair position, as they have been asked to make changes to their energy systems that require significant investments and technological advancements that many of them simply cannot afford. This has left African countries in a difficult situation, as they are required to take more drastic measures to reduce their emissions but do not necessarily have the resources available to implement those measures. African nations face a significant challenge in transitioning from fossil fuels to renewable energy sources, which are better for the environment and more affordable in the long run. This global issue has led to a lack of funds for African countries to make the transition. The imbalance in resources and funds between African and other countries has placed African nations at a disadvantage in achieving sustainable energy transitions for the planet and its inhabitants. African countries face a challenging situation, requiring global environmental protection standards while lacking resources for a transition from fossil fuels to renewable energy. The international community must support and encourage African countries to successfully transition to renewable energy sources and meet global environmental standards.

African governments and political systems must collaborate to transition away from fossil fuels, invest in renewable energy, and create a consensus on strategies for climate-friendly businesses. This will ensure long-term success, mitigate climate change effects, and keep up with peers in technological advancements and economic development. African politicians and political systems must prioritize transitioning from fossil fuels to renewable energy and investing in climate-friendly businesses. This will create a sustainable future, reduce emissions, and stimulate economic growth. Policies should incentivize renewable energy and climate-friendly businesses while creating the right economic and legal framework for long-term success. African politicians and the political system must collaborate to develop a consensus on transitioning from fossil fuels to renewable energy and climate-friendly businesses. This will ensure long-term success and stability for African nations while also promoting sustainable economic growth and development. By investing in these businesses, African governments can create a better world for future generations. African politicians and political system participants must prioritize transitioning from fossil fuels to renewable energy and climate-friendly businesses for a sustainable future. This is crucial for African nations to remain competitive in the global economy and improve living standards for their citizens.

African countries seek wealthy nations to help them transition to renewable energy sources, as developed nations are responsible for most global carbon emissions. They propose providing financial assistance, technology transfer, and grants to reduce greenhouse gas emissions and achieve climate goals. African countries also suggest partnering with solar, wind, and geothermal power sources. African countries seek financial and technical assistance from developed nations to adapt to climate change changes, including rising sea levels and droughts. They call for technical support and training to develop renewable energy infrastructure and reduce emissions. African countries seek international support to reduce fossil fuel reliance and transition to renewable energy sources, protecting the environment, their citizens, and the world. African countries need the support of developed nations to transition to renewable energy sources and manage the effects of climate change. This will safeguard the environment, people, and the world, build resilience, help build infrastructure for renewable energy, and reduce fossil fuel reliance.

8.4.2.2 Scenario 2: Collaborative and Calculated

Due to strong foreign assistance and stringent regulations on income from the export of fossil fuels, African countries have struggled to invest in renewable energy infrastructure over the past 20 years. Financial hardships, rising global temperatures, and extreme weather events have made it difficult to achieve decarbonization goals and have created uncertain futures. To solve these problems, drastic steps must be taken to increase the appeal of renewable energy investments to African nations. Lack of investment in renewable energy is a problem for African nations, which results in lost employment opportunities. Greater political commitment from decision makers is required to remove obstacles, create incentives for the development of renewable energy infrastructure, and spend money on citizen education and training to meet decarbonization goals. Support from abroad and limitations on the export of fossil fuels have contributed to this lack of investment. To build incentive structures and offer direction for the transition to renewable energy infrastructure, African leaders and international support are essential. To achieve African countries' decarbonization goals, greater political commitment and international support are necessary. Access to funding and a stronger regulatory framework are crucial for promoting renewable energy investments. African leaders must also commit to reducing fossil fuel reliance and embracing renewable energy sources such as solar and wind power. This will lead to a prosperous future for citizens, benefiting economies, the environment, and societies.

Due to their limited access to funding, technology, and implementation capacity, African countries have difficulty achieving their decarbonization goals. For the development of a climate for renewable energy investments that is economically, environmentally, and socially equitable, the absence of support from the international community and African leaders is essential. For economies, the environment, and societies to benefit, a successful transition to renewable energy sources must have both financial and technical support. African nations can build a prosperous future for their citizens and realize their full potential in the new era of decarbonization by embracing renewable energy sources such as solar and wind power. The international community must collaborate with African leaders to offer financial support, capacity-building opportunities, and technological solutions to equip African countries with the necessary resources and knowledge for a successful transition to renewable energy infrastructure. This will help achieve decarbonization goals, reduce fossil fuel reliance, and benefit economies, the environment, and societies in the new era of decarbonization. To achieve full potential in decarbonization, nations must create a prosperous future for citizens, benefiting economies, the environment, and societies. Supporting renewable energy and reducing fossil fuel reliance is crucial for achieving decarbonization goals. To create an environment that encourages investments in renewable energy, African nations need the support of the international community and their leaders. For citizens to enjoy a prosperous future and realize their full potential in the new era of decarbonization, a successful transition to solar and wind power is essential. The international community must collaborate with African leaders to offer financial support, capacity-building opportunities, and technological solutions to support African countries in transitioning to renewable energy infrastructure. This will enable them to reduce their reliance on fossil fuels, achieve decarbonization goals, and enjoy a better quality of life. This will ultimately lead to a sustainable future for all. African countries must capitalize on the international community and leaders' support to achieve decarbonization goals, create a prosperous future, and transition to renewable energy sources such as solar and wind power. This will benefit economies, the environment, and societies.

8.4.2.3 Scenario 3: Unaided and Self-Sufficient in Exports

Significant progress in the African energy sector has been hampered by political backing from African leaders, strong foreign support for renewable energy, and the absence of export restrictions on fossil fuels. Despite efforts from domestic and international actors to expand access and lessen reliance on fossil fuels, the development of renewable energy infrastructure in African countries is hampered by significant challenges. Despite the potential economic benefits of renewable energy in African nations, the lack of political support from African leaders impedes the transition from dirty to cleaner energy sources.

Due to income restrictions on fossil fuels and a lack of political will from their leaders, African countries face difficulties developing infrastructure for renewable energy sources, impeding meaningful advancement. The high reliance of African nations on the fossil fuel sector prevents them from investing in renewable energy, which is becoming increasingly desirable for global initiatives to reduce carbon footprints and move away from fossil fuels.

Despite the potential economic advantages of renewable energy, African countries struggle to switch from fossil fuels to clean energy solutions due to income, political will, and dependence on fossil fuels. African nations must overcome obstacles to implementing renewable energy investments and reaping the financial rewards of clean energy, but political and economic factors stand in the way of real progress.

This creates a Catch-22 situation in which African countries cannot benefit from the clean energy revolution taking place around the world due to their own restrictions on investment, leaving them in a difficult position to pursue economic development and reduce emissions simultaneously. Transitioning infrastructure from fossil fuels to renewable energy sources is stagnant in African nations, but political and economic factors impede development because they lack the funds to make investments.

This has created a disincentive for many African governments to invest in renewable energy, as it is viewed as too risky and expensive in comparison with the income from fossil fuel exports and the perceived security of relying on existing infrastructure rather than investing in untested technologies. Given that there is no agreement on renewable energy, there are financial losses, environmental harm, and restricted access to clean energy sources across the continent. This has left African countries in a difficult situation where despite the benefits that renewable energy could bring to their economies, they are unable to make meaningful progress due to the twofold obstacle of income restrictions on fossil fuels and a lack of political support from African governments. The energy sector in Africa has made significant strides, but more needs to be done to lift obstacles and increase political will from leaders to build renewable energy infrastructure for a more just and sustainable future.

This stasis can only be broken if African countries are able to find ways to fund renewable energy investments, either through aid from the international community or by creating an enabling environment within their own borders for investors to come in and take advantage of renewable energy opportunities. This can be done in a number of ways, such as by allowing for more private investment in renewable energy infrastructure, creating favorable regulatory frameworks for investors to take advantage of the opportunities presented by clean energy, and providing financial incentives to drive investments in renewable energy projects. To benefit local communities and advance the development of sustainable energy sources, African governments should place a high priority on encouraging and promoting local ownership of renewable energy projects. Initiatives to inform African populations of the economic, social, and environmental advantages of renewable energy sources should be supplemented by education and awareness-raising activities to promote renewable energy investments in Africa.

To empower stakeholders in the renewable energy transition and ensure efficient and effective progress in Africa's renewable energy initiatives, African nations must cooperate and forge solid partnerships. Governments must foster a favorable environment for private investment and readily available finance, technology, and expertise to support a successful renewable energy transition in Africa. This will make it possible for the nation to use renewable energy sources sustainably. For the continent to transition to clean, renewable energy by 2050 and ensure a sustainable energy future, African governments must set clear goals, policies, and legal frameworks.

To meet rising demand while ensuring cost-effectiveness and environmental sustainability for African citizens, African countries must invest in the development of energy infrastructure. African governments must spread awareness of renewable energy sources and how they benefit both the continent and the environment at large, encouraging a transition. Africa must switch to renewable energy if it wants to ensure its citizens' economic and environmental security.

8.4.2.4 Scenario 4: Tangled and Trapped

Due to a lack of foreign support for renewable energy sources and political support from African leaders, the majority of African countries continue to be highly dependent on fossil fuels as their main source of energy. The reliance on fossil fuels contributes to problems with the environment and human health, such as air pollution in African cities and global warming, which result in thousands of deaths each year from respiratory illnesses.

The reliance of African nations on fossil fuels impedes the development of renewable energy, discourages investment in clean technology, and slows the transition away from fossil fuels. The continued use of fossil fuels threatens public health and the environment and results in environmental degradation. Due to the obvious costs and ongoing effects of inaction, a swift transition away from fossil fuels is needed.

To combat climate change, improve public health, and create economic growth and job opportunities in the renewable energy sector, African countries must transition from fossil fuels to clean energy technologies. For environmental protection, public health, and economic opportunity, African governments must be more aggressive in transitioning to renewable energy. Policy changes, research investments, public awareness campaigns, and scaling up clean energy technologies such as solar, wind, geothermal, hydropower, energy storage, green buildings, and smart grids are all needed.

To lessen the negative effects of fossil fuels on the environment and human health while fostering economic expansion and job creation, African leaders must support renewable energy sources such as wind and solar power. African nations must support renewable energy technologies, increase public awareness, fund research, and provide incentives such as tax breaks and loan guarantees for people and businesses to make the transition successful.

Africa can lessen its reliance on fossil fuels, lessen the effects of climate change, and gain from a growing economy, new job opportunities, and improved public health. African nations can revolutionize energy, foster economic growth, combat climate change, lessen their reliance on fossil fuels, and enhance public health by investing in renewable energy.

Future leaders in African nations must act decisively to make this transition a reality and encourage private sector investment for their nations to become global leaders in the renewable energy revolution. Renewable energy has the potential to help African nations grow economically, create jobs, reduce their reliance on fossil fuels, improve public health, and slow global warming. To reap the many advantages of switching to renewable energy, African nations must join the global movement toward sustainable energy.

8.5 Conclusion

In terms of economic development, job creation, decreased reliance on fossil fuels, enhanced public health, and reduced climate change effects, renewable energy holds great promise for African nations. It is crucial that African leaders take action to make this transition a reality and encourage private sector investment in this sector so that Africa can become a global leader in the transition to renewable energy. African nations should join the global push for sustainable energy and take decisive action to move toward a future fueled by renewable energy so they can benefit from it. The four scenarios that could lead to an African net-zero energy transition are the most crucial information in this text. Unaided and self-sufficient exports, collaboration and a calculated transition, assisted and harsher scenarios, and a tangled and trapped transition are among the possible outcomes. In scenario 1, the continent will receive significant foreign funding for renewable energy along with stringent restrictions on income from the export of fossil fuels. In scenarios 2 and 3, foreign funding will be given with strict restrictions on the revenue and self-sufficiency of exports for Africa, respectively. Case 4: Foreign investments in renewable energy would significantly harm Africa, and exports of fossil fuels would be severely restricted. The key findings from the four scenarios for Africa's propensity for a net-zero energy transition are as follows:

- Legal frameworks, energy access policies, and regional renewable energy markets must be developed, and international organizations such as the UN and the African Development Bank must provide technical assistance and policy guidance. This will make it possible for African countries to utilize their renewable energy resources and guarantee that all citizens have access to reliable, affordable electricity.

- Moreover, African governments must take proactive steps to ensure an equitable energy transition by providing incentives for renewable energy investment and usage, expanding access to finance for clean energy projects, and utilizing the knowledge and resources of their local communities to promote clean energy development.
- By giving African countries and individuals tools and policy advice, they need to harness renewable energy sources and advance a sustainable, secure, and just energy future, and international stakeholders can help Africa make an equitable transition to clean energy.

References

Abade, L. P. (2020). *Climate change awareness: Role of Education*. https://doi.org/10.1007/978-3-319-95885-9_122

Abdul-Salam, Y., Kemp, A., & Phimister, E. (2022). Energy transition in the UKCS–modelling the effects of carbon emission charges on upstream petroleum operations. *Energy Economics*. https://doi.org/10.1016/j.eneco.2022.105898

Ogbulezie, J. C., James, U. O., & Chukwujindu, N. S. (2017). A review of regression models employed for predicting diffuse solar radiation in North-Western Africa. *Trends in Renewable Energy*. https://doi.org/10.17737/tre.2017.3.2.0042

Agbor, M., Udo, S., Ewona, I., Nwokolo, S., & Ogbulezie, J. A. (2023a). Potential impacts of climate change on global solar radiation and PV output using the CMIP6 model in West Africa. *Cleaner Engineering and Technology, 13*, 100630. https://doi.org/10.1016/j.clet.2023.100630

Agbor, M., Udo, S., Ewona, I., Nwokolo, S., Ogbulezie, J., Amadi, S., & Billy, U. (2023b). Effects of angstrom-Prescott and Hargreaves-Samani coefficients on climate forcing and solar PV technology selection in West Africa. *Trends in Renewable Energy, 9*(1), 78–106. https://doi.org/10.17737/tre.2023.9.1.00150

Agency, I. R. E. (2020). Renewable power generation costs in 2019. *International Renewable Energy Agency*.

Agreement, P., Republic, C., Kingdom, U., States, U., Principles, S. K., Zero, I. N., Secretariat, I. E. A., Iea, T., Energy, C., & Programme, T. (2021). Seven key Principles for implementing net Zero. *International Energy Agency*.

Ahl, A., Yarime, M., Goto, M., Chopra, S. S., Kumar, N. M., Tanaka, K., & Sagawa, D. (2020). Exploring blockchain for the energy transition: Opportunities and challenges based on a case study in Japan. *Renewable and Sustainable Energy Reviews., 117*, 109488. https://doi.org/10.1016/j.rser.2019.109488

Allen, S. R., Cohen, B., & Hammond, G. P. (2020). Indicative appraisal of clustered microgenerators for a low-carbon transition in the UK building sector. *Global Transitions., 2*, 83. https://doi.org/10.1016/j.glt.2020.06.001

Amadi, S., Dike, T., & Nwokolo, S. (2020). Global solar radiation characteristics at Calabar and Port Harcourt cities in Nigeria. *Trends in Renewable Energy., 6*, 101. https://doi.org/10.17737/tre.2020.6.2.00114

Amado, M., Poggi, F., Amado, A. R., & Breu, S. (2018). E-city web platform: A tool for energy efficiency at urban level. *Energies, 11*. https://doi.org/10.3390/en11071857

Ameli, N., Dessens, O., Winning, M., Cronin, J., Chenet, H., Drummond, P., Calzadilla, A., Anandarajah, G., & Grubb, M. (2021). *A climate investment trap in developing economies*. Nature Portfolio.

Bak, C., Bhattacharya, A., Edenhofer, O., & Knopf, B. (2017). Towards a comprehensive approach to climate policy, sustainable infrastructure, and finance. *Economics., 11*. https://doi.org/10.5018/economics-ejournal.ja.2017-33

Bataille, C., Åhman, M., Neuhoff, K., Nilsson, L. J., Fischedick, M., Lechtenböhmer, S., Solano-Rodriquez, B., Denis-Ryan, A., Stiebert, S., Waisman, H., Sartor, O., & Rahbar, S. (2018). A review of technology and policy deep decarbonization pathway options for making energy-intensive industry production consistent with the Paris Agreement. *Journal of Cleaner Production., 187*, 960. https://doi.org/10.1016/j.jclepro.2018.03.107

Bataille, C. G. F. (2020). Physical and policy pathways to net-zero emissions industry. *Wiley Interdisciplinary Reviews: Climate Change*. https://doi.org/10.1002/wcc.633

Birge, D., & Berger, A. M. (2019). Transitioning to low-carbon suburbs in hot-arid regions: A case-study of Emirati villas in Abu Dhabi. *Building and Environment*. https://doi.org/10.1016/j.buildenv.2018.09.013

Bonciu, F. (2021). The Dawn of a geopolitics of a hydrogen-based economy. *The Place of European Union. Romanian Journal of European Affairs.*

Borasio, M., & Moret, S. (2022). Deep decarbonization of regional energy systems: A novel modelling approach and its application to the Italian energy transition. *Renewable and Sustainable Energy Reviews., 153*, 111730. https://doi.org/10.1016/j.rser.2021.111730

BP Energy. (2020). *Energy Outlook 2020 edition explores the forces shaping the global energy transition out to 2050 and the surrounding that. BP Energy Outlook 2030, Statistical review.* British Petroleum.

Bray, R., Mejía Montero, A., & Ford, R. (2022). Skills deployment for a 'just' net zero energy transition. *Environmental Innovation and Societal Transitions., 42*, 395. https://doi.org/10.1016/j.eist.2022.02.002

Caetano, T., Merven, B., Hartley, F., & Ahjum, F. (2017). Decarbonization and the transport sector: A socioeconomic analysis of transport sector futures in South Africa. *Journal of Energy in Southern Africa., 28*. https://doi.org/10.17159/2413-3051/2017/v28i4a2945

Calvillo, C. F., & Turner, K. (2020). Analysing the impacts of a large-scale EV rollout in the UK–how can we better inform environmental and climate policy? *Energy Strategy Reviews., 30*, 100497. https://doi.org/10.1016/j.esr.2020.100497

Celia, M. A. (2021). The role of subsurface engineering in the net-zero Energy transition. *IOP Conference Series: Earth and Environmental Science., 861*, 072017. https://doi.org/10.1088/1755-1315/861/7/072017

Chaichan, M. T., & Kazem, H. A. (2018). Photovoltaic experiences in Iraq neighborhood countries. *In Generating Electricity Using Photovoltaic Solar Plants in Iraq*. https://doi.org/10.1007/978-3-319-75031-6_6

Chukwujindu, S., Umunnakwe, A., & Ogbulezie, J. C. (2022). Hybridization of statistical machine learning and numerical models for improving beam, diffuse and global solar radiation prediction. *Cleaner Engineering and Technology, 9*(May), 100529. https://doi.org/10.1016/j.clet.2022.100529

Cillari, G., Franco, A., & Fantozzi, F. (2021). Sizing strategies of photovoltaic systems in nZEB schemes to maximize the self-consumption share. *Energy Reports., 7*, 6769. https://doi.org/10.1016/j.egyr.2021.09.117

Von Cossel, M., Wagner, M., Lask, J., Magenau, E., Bauerle, A., Von Cossel, V., Warrach-Sagi, K., Elbersen, B., Staritsky, I., van Eupen, M., Iqbal, Y., Jablonowski, N. D., Happe, S., Fernando, A. L., Scordia, D., Cosentino, S. L., Wulfmeyer, V., Lewandowski, I., & Winkler, B. (2019). Prospects of bioenergy cropping systems for a more social-ecologically sound bioeconomy. *Agronomy., 9*. https://doi.org/10.3390/agronomy9100605

Cotterman, T., Small, M. J., Wilson, S., Abdulla, A., & Wong-Parodi, G. (2021). Applying risk tolerance and sociotechnical dynamics for more realistic energy transition pathways. *Applied Energy., 291*, 116751. https://doi.org/10.1016/j.apenergy.2021.116751

da Silva, P. P., Cerqueira, P. A., & Ogbe, W. (2018). Determinants of renewable energy growth in sub-Saharan Africa: Evidence from panel ARDL. *Energy, 156*, 45. https://doi.org/10.1016/j.energy.2018.05.068

Dawson, C., Dargusch, P., & Hill, G. (2022). Assessing how big insurance firms report and manage carbon emissions: A case study of Allianz. *Sustainability (Switzerland).* https://doi.org/10.3390/su14042476

Delafield, G., Donnison, C., Roddis, P., Arvanitopoulos, T., Sfyridis, A., Dunnett, S., Ball, T., & Logan, K. G. (2021). Conceptual framework for balancing society and nature in net-zero energy transitions. *Environmental Science and Policy.* https://doi.org/10.1016/j.envsci.2021.08.021

Derkenbaeva, E., Halleck Vega, S., Hofstede, G. J., & van Leeuwen, E. (2022). Positive energy districts: Mainstreaming energy transition in urban areas. *Renewable and Sustainable Energy Reviews., 153*, 111782. https://doi.org/10.1016/j.rser.2021.111782

Deutch, J. (2020). Is net Zero carbon 2050 possible? *In Joule., 4*, 2237. https://doi.org/10.1016/j.joule.2020.09.002

Diringer, E., & Perciasepe, B. (2020). The climate awakening of global capital. *Bulletin of the Atomic Scientists.* https://doi.org/10.1080/00963402.2020.1806574

du Plooy, N. (2017). *Fostering sustainable energy transitions in South Africa: A system dynamics approach to achieving a sustainable electricity sector.* Stellenbosch University.

Dudău, R., & Cătuți, M. (2019). The decarbonization challenge of Southeast Europe: A case study of Romania. *In Intereconomics.* https://doi.org/10.1007/s10272-019-0851-y

Elsässer, J. P., Hickmann, T., & Stehle, F. (2018). The role of cities in South Africa's energy gridlock. *Case Studies in the Environment., 2*, 1. https://doi.org/10.1525/cse.2018.001297

Fan, Z., & Friedmann, S. J. (2021). Low-carbon production of iron and steel: Technology options, economic assessment, and policy. *In Joule., 5*, 829. https://doi.org/10.1016/j.joule.2021.02.018

Felder, F. A., Unel, B., & Dvorkin, Y. (2021). Modelling strategic objectives and behavior in the transition of the energy sector to inform policymaking. *Electricity Journal., 34*, 107009. https://doi.org/10.1016/j.tej.2021.107009

Ferreira, M., Almeida, M., Rodrigues, A., & Silva, S. M. (2016). Comparing cost-optimal and net-zero energy targets in building retrofit. *Building Research and Information., 44*, 188. https://doi.org/10.1080/09613218.2014.975412

Florent, B., & Schaeffer, M. (2019). Climate change impacts on Africa's economic growth. *African Development Bank*, 1–169.

Ford, R., & Hardy, J. (2020). Are we seeing clearly? The need for aligned vision and supporting strategies to deliver net-zero electricity systems. *Energy Policy, 147*, 111902. https://doi.org/10.1016/j.enpol.2020.111902

Ford, R., Maidment, C., Vigurs, C., Fell, M. J., & Morris, M. (2021). Smart local energy systems (SLES): A framework for exploring transition, context, and impacts. *Technological Forecasting and Social Change, 166*, 120612. https://doi.org/10.1016/j.techfore.2021.120612

Freeman, R. (2021). Modelling the sociopolitical feasibility of energy transition with system dynamics. *Environmental Innovation and Societal Transitions., 40*, 486. https://doi.org/10.1016/j.eist.2021.10.005

French, S. (2020). The role of Zero and low carbon hydrogen in enabling the Energy transition and the path to net Zero greenhouse gas emissions. *Johnson Matthey Technology Review.* https://doi.org/10.1595/205651320x15910225395383

Furszyfer Del Rio, D. D., Sovacool, B. K., Foley, A. M., Griffiths, S., Bazilian, M., Kim, J., & Rooney, D. (2022a). Decarbonizing the ceramics industry: A systematic and critical review of policy options, developments and sociotechnical systems. *Renewable and Sustainable Energy Reviews.* https://doi.org/10.1016/j.rser.2022.112081

Furszyfer Del Rio, D. D., Sovacool, B. K., Foley, A. M., Griffiths, S., Bazilian, M., Kim, J., & Rooney, D. (2022b). Decarbonizing the glass industry: A critical and systematic review of developments, sociotechnical systems and policy options. *Renewable and Sustainable Energy Reviews.* https://doi.org/10.1016/j.rser.2021.111885

Galgóczi, B. (2020). Just transition on the ground: Challenges and opportunities for social dialogue. *European Journal of Industrial Relations., 26*, 367. https://doi.org/10.1177/0959680120951704

Godin, K., Sapinski, J. P., & Dupuis, S. (2021). The transition to net zero energy (NZE) housing: An integrated approach to market, state, and other barriers. *Cleaner and Responsible Consumption., 3*, 100043. https://doi.org/10.1016/j.clrc.2021.100043

Gozlugol, A. A. (2022). The clash of "E" and "S" of ESG: Just transition on the path to net zero and the implications for sustainable corporate governance and finance. *Journal of World Energy Law and Business., 15*, 1. https://doi.org/10.1093/jwelb/jwab039

Greig, C., & Uden, S. (2021). The value of CCUS in transitions to net-zero emissions. *Electricity Journal., 34*, 107004. https://doi.org/10.1016/j.tej.2021.107004

Guan, Y., Shan, Y., Huang, Q., Chen, H., Wang, D., & Hubacek, K. (2021). Assessment to China's recent emission pattern shifts. *Earth's Future, 9*. https://doi.org/10.1029/2021EF002241

Gurieff, N., Moghtaderi, B., Daiyan, R., & Amal, R. (2021). Gas transition: Renewable hydrogen's future in eastern Australia's energy networks. *Energies, 14*. https://doi.org/10.3390/en14133968

Hafner, S., Jones, A., & Anger-Kraavi, A. (2021). Economic impacts of achieving a net-zero emissions target in the power sector. *Journal of Cleaner Production., 312*, 127610. https://doi.org/10.1016/j.jclepro.2021.127610

Hassan, M. A., Bailek, N., Bouchouicha, K., Ibrahim, A., Jamil, B., Kuriqi, A., & Nwokolo, S. C. (2022). Evaluation of energy extraction of PV systems affected by environmental factors under real outdoor conditions. *Theoretical and Applied Climatology, August.* https://doi.org/10.1007/s00704-022-04166-6

Hassan, M. A., Bailek, N., Bouchouicha, K., & Nwokolo, S. C. (2021). Ultrashort-term exogenous forecasting of photovoltaic power production using genetically optimized nonlinear autoregressive recurrent neural networks. *Renewable Energy, 171*(February), 191–209. https://doi.org/10.1016/j.renene.2021.02.103

Hosseini, S. E. (2022). Transition away from fossil fuels toward renewables: Lessons from Russia-Ukraine crisis. *Future Energy., 1*, 2. https://doi.org/10.55670/fpll.fuen.1.1.8

Hrnčić, B., Pfeifer, A., Jurić, F., Duić, N., Ivanović, V., & Vušanović, I. (2021). Different investment dynamics in energy transition towards a 100% renewable energy system. *Energy, 237*, 121526. https://doi.org/10.1016/j.energy.2021.121526

International Renewable Energy Agency (IRENA). (2013). Renewable power generation costs in 2012: An overview. *Irena*, 1–8.

International Renewable Energy Agency. (2020a). Renewable Power Generation Costs in 2019. *International Renewable Energy Agency.*

International Renewable Energy Agency (IRENA). (2020b). *Renewable energy and climate pledges: Five years after the Paris Agreement.*

Iqbal, M. I., Himmler, R., & Gheewala, S. H. (2017). Potential life cycle energy savings through a transition from typical to energy plus households: A case study from Thailand. *Energy and Buildings., 134*, 295. https://doi.org/10.1016/j.enbuild.2016.11.002

IRENA. (2019). Renewable {Energy} {Now} {Accounts} for a {Third} of {Global} {Power} {Capacity}. In/newsroom/pressreleases/2019/Apr/Renewable-Energy-Now-Accounts-for-a-Third-of-Global-Power-Capacity.

IRENA. (2021). *World energy transitions outlook: 1.5 degrees pathway.* In International Renewable Energy Agency. https://irena.org/publications/2021/March/World-Energy-Transitions-Outlook

IRENA Working Paper. (2012). *Renewable Energy Techlogies: Cost analysis series, hydropower.* International Renewable Energy Agency.

Ituen, E. E., Esen, N. U., Nwokolo, S. C., & Udo, E. G. (2012). Prediction of global solar radiation using relative humidity, maximum temperature and sunshine hours in Uyo, in The Niger Delta region, Nigeria. *Advances in Applied Science Research, 3*(4), 1923–1937.

Jakob, M., Flachsland, C., Christoph Steckel, J., & Urpelainen, J. (2020). Actors, objectives, context: A framework of the political economy of energy and climate policy applied to India,

Indonesia, and Vietnam. *Energy Research and Social Science., 70*, 101775. https://doi.org/10.1016/j.erss.2020.101775

John-Jaja, S. A., Abdullah, A. R., & Nwokolo, S. C. (2016a). Effects of age variance on repeatability estimates of egg dimensions of Bovan Nera black laying chickens. *Journal of Genetic Engineering and Biotechnology, 14*(1), 219–226. https://doi.org/10.1016/j.jgeb.2016.06.003

John-Jaja, S. A., Abdullah, A. R., & Nwokolo, S. C. (2018). Heritability estimates of external egg quality traits of exotic laying chickens under the influence of age variance in the tropics. *Journal of the Saudi Society of Agricultural Sciences, 17*(4), 359–364. https://doi.org/10.1016/j.jssas.2016.07.007

John-Jaja, S. A., Udoh, U. H., & Nwokolo, S. C. (2016b). Repeatability estimates of egg weight and egg-shell weight under various production periods for Bovan Nera black laying chicken. *Beni-Suef University Journal of Basic and Applied Sciences, 5*(4), 389–394. https://doi.org/10.1016/j.bjbas.2016.11.001

Juma, D., Munda, J., & Kabiri, C. (2020). *Progress in grid interconnection in East Africa: Challenges, Experiences and Opportunities*. 2020 IEEE PES/IAS PowerAfrica, PowerAfrica 2020. https://doi.org/10.1109/PowerAfrica49420.2020.9219924.

Kannan, R., Panos, E., Hirschberg, S., & Kober, T. (2022). A net-zero Swiss energy system by 2050: Technological and policy options for the transition of the transportation sector. *Futures & Foresight Science., 4*. https://doi.org/10.1002/ffo2.126

Karimli, A., & Aghayev, A. (2020). The economic opportunities of renewable energy resources in Azerbaijan: Major challenges and scenarios. *Business & IT., X*, 71. https://doi.org/10.14311/bit.2020.03.06

Kato, E., & Kurosawa, A. (2019). Evaluation of Japanese energy system toward 2050 with TIMES-Japan–deep decarbonization pathways. *Energy Procedia*. https://doi.org/10.1016/j.egypro.2019.01.818

Kato, E., & Kurosawa, A. (2021). Role of negative emissions technologies (NETs) and innovative technologies in transition of Japan's energy systems toward net-zero CO2 emissions. *Sustainability Science., 16*, 463. https://doi.org/10.1007/s11625-021-00908-z

Kayode, R., Akhavan Farshchi, M., & Ford, A. (2019). Analysis of household Energy consumption in Nigeria. Innovative solutions for compliance and research. *Management*.

Khan, M. A. M., Rehman, S., & Al-Sulaiman, F. A. (2018). A hybrid renewable energy system as a potential energy source for water desalination using reverse osmosis: A review. *Renewable and Sustainable Energy Reviews., 97*, 456. https://doi.org/10.1016/j.rser.2018.08.049

Kilkiş, Ş. (2015). Exergy transition planning for net-zero districts. *Energy, 92*, 515. https://doi.org/10.1016/j.energy.2015.02.009

Kincer, J., Moss, T., & Thurber, M. (2022). A coal renaissance is not coming to Africa. *World Development Perspectives., 25*, 100375. https://doi.org/10.1016/j.wdp.2021.100375

Kivimaa, P., Primmer, E., & Lukkarinen, J. (2020). Intermediating policy for transitions towards net-zero energy buildings. *Environmental Innovation and Societal Transitions*. https://doi.org/10.1016/j.eist.2020.01.007

Koengkan, M., Poveda, Y. E., & Fuinhas, J. A. (2020). Globalization as a motor of renewable energy development in Latin American countries. *GeoJournal, 85*, 1591. https://doi.org/10.1007/s10708-019-10042-0

Komninos, N. (2022). Net Zero Energy districts: Connected intelligence for carbon-neutral cities. *Land., 11*. https://doi.org/10.3390/land11020210

Kourgiozou, V., Commin, A., Dowson, M., Rovas, D., & Mumovic, D. (2021). Scalable pathways to net zero carbon in the UK higher education sector: A systematic review of smart energy systems in university campuses. *Renewable and Sustainable Energy Reviews*. https://doi.org/10.1016/j.rser.2021.111234

Kouton, J. (2021). The impact of renewable energy consumption on inclusive growth: Panel data analysis in 44 African countries. *Economic Change and Restructuring., 54*, 145. https://doi.org/10.1007/s10644-020-09270-z

Kuramochi, T., Höhne, N., Schaeffer, M., Cantzler, J., Hare, B., Deng, Y., Sterl, S., Hagemann, M., Rocha, M., Yanguas-Parra, P. A., Mir, G. U. R., Wong, L., El-Laboudy, T., Wouters, K., Deryng, D., & Blok, K. (2018). Ten key short-term sectoral benchmarks to limit warming to 1.5°C. *Climate Policy., 18*, 287. https://doi.org/10.1080/14693062.2017.1397495

LaBelle, M. C., Bucată, R., & Stojilovska, A. (2021). Radical energy justice: A green Deal for Romanian coal miners? *Journal of Environmental Policy and Planning., 25*, 142. https://doi.org/10.1080/1523908X.2021.1992266

Lai, C. S., Strasser, T. I., & Lai, L. L. (2022). Editorial to the special issue on smart cities based on the efforts of the systems, man, and cybernetics society. In *IEEE transactions on systems, man, and cybernetics: Systems*. https://doi.org/10.1109/TSMC.2021.3128990

Leal-Arcas, R. (2019). *Using trade agreements to achieve sustainability: A counter-intuitive conundrum*. https://doi.org/10.1007/978-3-030-23933-6_4.

Leidreiter, A., & Boselli, F. (2015). *100% renewable Energy : Boosting development in Morocco*. World Future Council.

Leidreiter, A., et al. (2015). *100% renewable Energy : Boosting development*. World Future Council.

Liu, L., Workman, M., & Hayes, S. (2022). Net Zero and the potential of consumer data-United Kingdom energy sector case study: The need for cross-sectoral best data practice principles. *Energy Policy*. https://doi.org/10.1016/j.enpol.2022.112803

Lowe, R. J., & Drummond, P. (2022). Solar, wind and logistic substitution in global energy supply to 2050–Barriers and implications. *Renewable and Sustainable Energy Reviews*. https://doi.org/10.1016/j.rser.2021.111720

Lucas, A. R. (2022). Canada's carbon energy overhang. *Journal of Energy and Natural Resources Law., 40*, 17. https://doi.org/10.1080/02646811.2021.2012349

Lynes, M. (2016). *International energy outlook 2016, Transportation Sector*. In International Energy Outlook and Projections.

Mac Dowell, N., Sunny, N., Brandon, N., Herzog, H., Ku, A. Y., Maas, W., Ramirez, A., Reiner, D. M., Sant, G. N., & Shah, N. (2021). The hydrogen economy: A pragmatic path forward. *Joule., 5*, 2524. https://doi.org/10.1016/j.joule.2021.09.014

Manjong, N. B., Oyewo, A. S., & Breyer, C. (2021). Setting the pace for a sustainable Energy transition in Central Africa: The case of Cameroon. *IEEE Access., 9*, 145435. https://doi.org/10.1109/ACCESS.2021.3121000

Martin-Roberts, E., Scott, V., Flude, S., Johnson, G., Haszeldine, R. S., & Gilfillan, S. (2021). Carbon capture and storage at the end of a lost decade. *One Earth*. https://doi.org/10.1016/j.oneear.2021.10.002

Martínez-Gordón, R., Sánchez-Diéguez, M., Fattahi, A., Morales-España, G., Sijm, J., & Faaij, A. (2022). Modelling a highly decarbonized North Sea energy system in 2050: A multinational approach. *Advances in Applied Energy., 5*, 100080. https://doi.org/10.1016/j.adapen.2021.100080

Martins, F., Moura, P., & de Almeida, A. T. (2022). The role of electrification in the Decarbonization of the Energy sector in Portugal. *Energies*. https://doi.org/10.3390/en15051759

Martiskainen, M., & Sovacool, B. K. (2021). Mixed feelings: A review and research agenda for emotions in sustainability transitions. *Environmental Innovation and Societal Transitions., 40*, 609. https://doi.org/10.1016/j.eist.2021.10.023

Maupin, A. (2017). The SDG13 to combat climate change: An opportunity for Africa to become a trailblazer? *African Geographical Review., 36*, 131. https://doi.org/10.1080/19376812.2016.1171156

Mavrigiannaki, A., Pignatta, G., Assimakopoulos, M., Isaac, M., Gupta, R., Kolokotsa, D., Laskari, M., Saliari, M., Meir, I. A., & Isaac, S. (2021). Examining the benefits and barriers for the implementation of net zero energy settlements. *Energy and Buildings*. https://doi.org/10.1016/j.enbuild.2020.110564

Maxim, M. R., & Zander, K. (2020). *Green tax reform and employment double dividend in Australia should Australia follow Europe's footsteps?* Margin. https://doi.org/10.1177/0973801020953310

Medved, S., Domjan, S., & Arkar, C. (2021). Contribution of energy storage to the transition from net zero to zero energy buildings. *Energy and Buildings., 236*, 110751. https://doi.org/10.1016/j.enbuild.2021.110751

Mekonnen, Y., & Sarwat, A. I. (2017). *Renewable energy supported microgrid in rural electrification of Sub-Saharan Africa*. Proceedings–2017 IEEE PES-IAS PowerAfrica Conference: Harnessing Energy, information and communications technology (ICT) for affordable electrification of Africa, PowerAfrica 2017. https://doi.org/10.1109/PowerAfrica.2017.7991293.

Mestemaker, B., Van Den Heuvel, H., & Gonçalves Castro, B. (2020). Designing the zero emission vessels of the future: Technologic, economic and environmental aspects. *International Shipbuilding Progress., 67*, 5. https://doi.org/10.3233/ISP-190276

Miller, W., & Senadeera, M. (2017). Social transition from energy consumers to prosumers: Rethinking the purpose and functionality of eco-feedback technologies. *Sustainable Cities and Society*. https://doi.org/10.1016/j.scs.2017.09.009

Morstyn, T., Savelli, I., & Hepburn, C. (2021). Multiscale design for system-wide peer-to-peer energy trading. *One Earth., 4*, 629. https://doi.org/10.1016/j.oneear.2021.04.018

Mutezo, G., & Mulopo, J. (2021). A review of Africa's transition from fossil fuels to renewable energy using circular economy principles. *Renewable and Sustainable Energy Reviews, 137*-(November 2020), 110609. https://doi.org/10.1016/j.rser.2020.110609

Nelson, S., & Allwood, J. M. (2021). The technological and social timelines of climate mitigation: Lessons from 12 past transitions. *Energy Policy, 152*, 112155. https://doi.org/10.1016/j.enpol.2021.112155

Ngondjeb, D. Y., Atewamba, C., & Macalou, M. (2020). Insights on Africa's future in its transition to the green economy. *Advances in African Economic, Social and Political Development*. https://doi.org/10.1007/978-3-030-44180-7_13

Nhuchhen, D. R., Sit, S. P., & Layzell, D. B. (2022). Towards net-zero emission cement and power production using molten carbonate fuel cells. *Applied Energy*. https://doi.org/10.1016/j.apenergy.2021.118001

Nikas, A., Xexakis, G., Koasidis, K., Acosta-Fernández, J., Arto, I., Calzadilla, A., Domenech, T., Gambhir, A., Giljum, S., Gonzalez-Eguino, M., Herbst, A., Ivanova, O., van Sluisveld, M. A. E., Van De Ven, D. J., Karamaneas, A., & Doukas, H. (2022). Coupling circularity performance and climate action: From disciplinary silos to transdisciplinary modelling science. *Sustainable Production and Consumption*. https://doi.org/10.1016/j.spc.2021.12.011

Nurdiawati, A., & Urban, F. (2021). Towards deep decarbonization of energy-intensive industries: A review of current status, technologies and policies. *Energies., 14*. https://doi.org/10.3390/en14092408

Nwokolo, S. C., & Amadi, S. O. (2018). A global review of empirical models for estimating photosynthetically active radiation. *Trends in Renewable Energy, 4*(2), 236–327. https://doi.org/10.17737/tre.2018.4.2.0079

Nwokolo, S. C., Amadi, S. O., Obiwulu, A. U., Ogbulezie, J. C., & Eyibio, E. E. (2022a). Prediction of global solar radiation potential for sustainable and cleaner energy generation using improved angstrom-Prescott and Gumbel probabilistic models. *Cleaner Engineering and Technology, 6*(January), 100416. https://doi.org/10.1016/j.clet.2022.100416

Nwokolo, S. C., Obiwulu, A. U., & Ogbulezie, J. C. (2022b). Machine learning and analytical model hybridization to assess the impact of climate. *Physics and Chemistry of the Earth, 130*(February), 103389. https://doi.org/10.1016/j.pce.2023.103389

Nwokolo, S. C., & Ogbulezie, J. C. (2018a). A qualitative review of empirical models for estimating diffuse solar radiation from experimental data in Africa. *Renewable and Sustainable Energy Reviews., 92*, 353. https://doi.org/10.1016/j.rser.2018.04.118

Nwokolo, S. C., & Ogbulezie, J. C. (2018b). A qualitative review of empirical models for estimating diffuse solar radiation from experimental data in Africa. *Renewable and Sustainable Energy Reviews, 92*. https://doi.org/10.1016/j.rser.2018.04.118

Nwokolo, S. C., & Ogbulezie, J. C. (2018c). A quantitative review and classification of empirical models for predicting global solar radiation in West Africa. *Beni-Suef University Journal of Basic and Applied Sciences, 7*(4), 367–396. https://doi.org/10.1016/j.bjbas.2017.05.001

Nwokolo, S. C., Ogbulezie, J. C., & Obiwulu, A. U. (2022c). Impacts of climate change and meteo-solar parameters on photosynthetically active radiation prediction using hybrid machine learning with physics-based models. *Advances in Space Research., 70*, 3614. https://doi.org/10.1016/j.asr.2022.08.010

Nwokolo, S., & Ogbulezie, J. (2017a). A critical review of theoretical models for estimating global solar radiation between 2012-2016 in Nigeria. *International Journal of Physical Research, 5*(2), 60. https://doi.org/10.14419/ijpr.v5i2.8160

Nwokolo, S., & Ogbulezie, J. (2017b). Modelling the influence of cloudiness on diffuse horizontal irradiation under various sky conditions in Nigeria. *International Journal of Physical Research, 5*(2), 91. https://doi.org/10.14419/ijpr.v5i2.8312

Nwokolo, S., & Otse, C. (2019). Impact of sunshine duration and clearness index on diffuse solar radiation estimation in mountainous climate. *Trends in Renewable Energy, 5*(3), 307–332. https://doi.org/10.17737/tre.2019.5.3.00107

Nwokolo, S. C., Obiwulu, A. U., Amadi, O., & Ogbulezie, J. C. (2023a). Assessing the impact of soiling, tilt angle, and solar radiation on the performance of solar PV systems. *Trends in Renewable Energy, 9*(2), 121–137. https://doi.org/10.17737/tre.2023.9.2.00156

Nwokolo, S. C., Ogbulezie, J. C., & Ushie, O. J. (2023b). A multimodel ensemble-based CMIP6 assessment of future solar radiation and PV potential under various climate warming scenarios. *Optik, 285*(May), 170956. https://doi.org/10.1016/j.ijleo.2023.170956

Nwokolo, S. C., Proutsos, N., & Meyer, E. L. (2023c). Machine learning and physics-based hybridization models for evaluation of the effects of climate change and Urban expansion on photosynthetically active radiation. *Atmosphere, 14*(687), 1–36. https://doi.org/10.3390/atmos14040687

Nwokolo, S. C., Obiwulu, A. U., & Ogbulezie, J. C. (2023d). Machine learning and analytical model hybridization to assess the impact of climate. *Physics and Chemistry of the Earth, 130* (Feburary 2023), 103389. https://doi.org/10.1016/j.pce.2023.103389

Obiwulu, A. U., Chendo, M. A. C., Erusiafe, N., & Nwokolo, S. C. (2020a). Implicit meteorological parameter-based empirical models for estimating back temperature solar modules under varying tilt-angles in Lagos. *Nigeria. Renewable Energy, 145*. https://doi.org/10.1016/j.renene.2019.05.136

Obiwulu, A. U., Erusiafe, N., Olopade, M. A., & Nwokolo, S. C. (2020b). Modelling and optimization of back temperature models of mono-crystalline silicon modules with special focus on the effect of meteorological and geographical parameters on PV performance. *Renewable Energy, 154*, 404. https://doi.org/10.1016/j.renene.2020.02.103

Obiwulu, A. U., Erusiafe, N., Olopade, M. A., & Nwokolo, S. C. (2022). Modelling and estimation of the optimal tilt angle, maximum incident solar radiation, and global radiation index of the photovoltaic system. *Heliyon, 8*(6), e09598. https://doi.org/10.1016/j.heliyon.2022.e09598

Ogbe, W. M., da Silva, P. P., & Cerqueira, P. (2017). *Determinants of renewable energy growth in africa: Is nigeria different?* In Proceedings of the 3rd International Conference on Energy and Environment (ICEE 2017).

Okem, A. E. (2016). South Africa's transition to a low-carbon economy: The role of cooperatives. *Journal of Social Sciences., 49*, 257. https://doi.org/10.1080/09718923.2016.11893619

Omar, A. I., Khattab, N. M., & Abdel Aleem, S. H. E. (2022). Optimal strategy for transition into net-zero energy in educational buildings: A case study in El-Shorouk City, Egypt. *Sustainable Energy Technologies and Assessments*. https://doi.org/10.1016/j.seta.2021.101701

Omrany, H., Chang, R., Soebarto, V., Zhang, Y., Ghaffarianhoseini, A., & Zuo, J. (2022). A bibliometric review of net zero energy building research 1995–2022. *Energy and Buildings.*, *262*, 111996. https://doi.org/10.1016/j.enbuild.2022.111996

Ouedraogo, N. S. (2019). Opportunities, barriers and issues with renewable energy development in africa: A comprehensible review. *Current Sustainable/Renewable Energy Reports*. https://doi.org/10.1007/s40518-019-00130-7

Oyewo, A. S., Aghahosseini, A., Ram, M., & Breyer, C. (2020). Transition towards decarbonized power systems and its socioeconomic impacts in West Africa. *Renewable Energy, 154*, 1092. https://doi.org/10.1016/j.renene.2020.03.085

Oyewo, A. S., Aghahosseini, A., Ram, M., Lohrmann, A., & Breyer, C. (2019). Pathway towards achieving 100% renewable electricity by 2050 for South Africa. *Solar Energy., 191*, 549. https://doi.org/10.1016/j.solener.2019.09.039

Pablo-Romero, M. P., Sánchez-Braza, A., & Galyan, A. (2021). Renewable energy use for electricity generation in transition economies: Evolution, targets and promotion policies. *Renewable and Sustainable Energy Reviews., 138*, 110481. https://doi.org/10.1016/j.rser.2020.110481

Perlaviciute, G., Steg, L., & Sovacool, B. K. (2021). A perspective on the human dimensions of a transition to net-zero energy systems. *Energy and Climate Change., 2*, 100042. https://doi.org/10.1016/j.egycc.2021.100042

Pineda, A. C., & Faria, P. (2019). Towards a science-based approach to climate neutrality in the corporate sector–discussion paper. *SBTi.*

Piselli, C., Salvadori, G., Diciotti, L., Fantozzi, F., & Pisello, A. L. (2021). Assessing users' willingness-to-engagement towards net Zero Energy communities in Italy. *Renewable and Sustainable Energy Reviews.* https://doi.org/10.1016/j.rser.2021.111627

Poggi, F., Firmino, A., & Amado, M. (2015). Moving forward on sustainable Energy transitions: The smart rural model. *European Journal of Sustainable Development.* https://doi.org/10.14207/ejsd.2015.v4n2p43

Poggi, F., Firmino, A., & Amado, M. (2020). Shaping energy transition at municipal scale: A net-zero energy scenario-based approach. *Land Use Policy, 99*, 104955. https://doi.org/10.1016/j.landusepol.2020.104955

Quarton, C. J., & Samsatli, S. (2021). How to incentivize hydrogen energy technologies for net zero: Whole-system value chain optimization of policy scenarios. *Sustainable Production and Consumption., 27*, 1215. https://doi.org/10.1016/j.spc.2021.02.007

Raihan, A., Begum, R. A., Said, M. N. M., & Pereira, J. J. (2022). Relationship between economic growth, renewable energy use, technological innovation, and carbon emission toward achieving Malaysia's Paris Agreement. *Environment Systems and Decisions.* https://doi.org/10.1007/s10669-022-09848-0

Rajeh, M. A., & Bensenouci, A. (2019). *Design a hybrid wind-solar power system for remote areas of Saudi Arabia.* Proceedings of the International Conference on Industrial Engineering and Operations Management.

Ramagoma, J., & Adendorff, C. (2016). Managing a transition to green energy sources: The perspectives of energy practitioners in the southern African development community region. *Journal of Energy in Southern Africa., 27*, 77. https://doi.org/10.17159/2413-3051/2016/v27i4a1638

Ratlamwala, T. A. H., & Dincer, I. (2018). Sustainable Energy management. *Comprehensive Energy Systems.* https://doi.org/10.1016/B978-0-12-809597-3.00522-8

Raugei, M., Kamran, M., & Hutchinson, A. (2020). A prospective net energy and environmental life-cycle assessment of the UK electricity grid. *Energies*. https://doi.org/10.3390/en13092207

Razmjoo, A., Nezhad, M. M., Kaigutha, L. G., Marzband, M., Mirjalili, S., Pazhoohesh, M., Memon, S., Ehyaei, M. A., & Piras, G. (2021). Investigating smart city development based on green buildings, electrical vehicles and feasible indicators. *Sustainability (Switzerland)., 13*. https://doi.org/10.3390/su13147808

Rennkamp, B., & Boyd, A. (2015). Technological capability and transfer for achieving South Africa's development goals. *Climate Policy.* https://doi.org/10.1080/14693062.2013.831299

Rissman, J., Bataille, C., Masanet, E., Aden, N., Morrow, W. R., Zhou, N., Elliott, N., Dell, R., Heeren, N., Huckestein, B., Cresko, J., Miller, S. A., Roy, J., Fennell, P., Cremmins, B., Koch Blank, T., Hone, D., Williams, E. D., & de la Rue du Can, S., ... Helseth, J. (2020). Technologies and policies to decarbonize global industry: Review and assessment of mitigation drivers through 2070. *Applied Energy.* https://doi.org/10.1016/j.apenergy.2020.114848

Rogelj, J., Luderer, G., Pietzcker, R. C., Kriegler, E., Schaeffer, M., Krey, V., & Riahi, K. (2015). Energy system transformations for limiting end-of-century warming to below 1.5°C. *Nature Climate Change.* https://doi.org/10.1038/nclimate2572

Rosa, L., & Mazzotti, M. (2022). Potential for hydrogen production from sustainable biomass with carbon capture and storage. *Renewable and Sustainable Energy Reviews., 157*, 112123. https://doi.org/10.1016/j.rser.2022.112123

Salleh, S. F., Mohd Roslan, M. E., Abd Rahman, A., Shamsuddin, A. H., Tuan Abdullah, T. A. R., & Sovacool, B. K. (2020). Transitioning to a sustainable development framework for bioenergy in Malaysia: Policy suggestions to catalyse the utilization of palm oil mill residues. *Energy, Sustainability and Society.* https://doi.org/10.1186/s13705-020-00269-y

Samuel Chukwujindu, N. (2017). A comprehensive review of empirical models for estimating global solar radiation in Africa. *Renewable and Sustainable Energy Reviews, 78*(October), 955–995. https://doi.org/10.1016/j.rser.2017.04.101

Scheinin, M. (2016). International human rights law. In *A Companion to European Union Law and International Law.* https://doi.org/10.1002/9781119037712.ch29.

Scholten, D., Bazilian, M., Overland, I., & Westphal, K. (2020). The geopolitics of renewables: New board, new game. *Energy Policy, 138*(August 2018), 111059. https://doi.org/10.1016/j.enpol.2019.111059

Schreyer, F., Luderer, G., Rodrigues, R., Pietzcker, R. C., Baumstark, L., Sugiyama, M., Brecha, R. J., & Ueckerdt, F. (2020). Common but differentiated leadership: Strategies and challenges for carbon neutrality by 2050 across industrialized economies. *Environmental Research Letters.* https://doi.org/10.1088/1748-9326/abb852

Scott, D., & Gössling, S. (2021). Destination net-zero: What does the international energy agency roadmap mean for tourism? *Journal of Sustainable Tourism., 30*, 14. https://doi.org/10.1080/09669582.2021.1962890

Setyowati, A. B. (2020). Mitigating energy poverty: Mobilizing climate finance to manage the energy trilemma in Indonesia. *Sustainability (Switzerland)., 12.* https://doi.org/10.3390/su12041603

Shamim, J. A., Hsu, W. L., Paul, S., Yu, L., & Daiguji, H. (2021). A review of solid desiccant dehumidifiers: Current status and near-term development goals in the context of net zero energy buildings. *In Renewable and Sustainable Energy Reviews.* https://doi.org/10.1016/j.rser.2020.110456

Shen, W. (2020). China's role in Africa's energy transition: A critical review of its intensity, institutions, and impacts. *Energy Research and Social Science., 68*, 101578. https://doi.org/10.1016/j.erss.2020.101578

Sokołowski, M. M., & Heffron, R. J. (2022). Defining and conceptualizing energy policy failure: The when, where why, and how. *Energy Policy, 161*, 112745. https://doi.org/10.1016/j.enpol.2021.112745

Sokona, Y., Mulugetta, Y., & Gujba, H. (2012). Widening energy access in Africa: Towards energy transition. *Energy Policy, 47*, 3. https://doi.org/10.1016/j.enpol.2012.03.040

Sovacool, B. K., Bazilian, M., Griffiths, S., Kim, J., Foley, A., & Rooney, D. (2021). Decarbonizing the food and beverages industry: A critical and systematic review of developments, sociotechnical systems and policy options. *Renewable and Sustainable Energy Reviews., 143*, 110856. https://doi.org/10.1016/j.rser.2021.110856

Stendahl, M.-F., Dubois, M.-C., Forgues, D., & Hjelseth, E. (2022). Building information modelling for environmental impact assessment in early design phases: A literature review. *Open Journal of Applied Sciences., 12*, 59. https://doi.org/10.4236/ojapps.2022.121006

Stern, J. (2021). The role of gases in the European energy transition. *Russian Journal of Economics., 6*, 390. https://doi.org/10.32609/J.RUJE.6.55105

Styring, P., Duckworth, E. L., & Platt, E. G. (2021). Synthetic fuels in a transport transition: Fuels to prevent a transport underclass. *Frontiers in Energy Research., 9*. https://doi.org/10.3389/fenrg.2021.707867

Sunday, E., Agbasi, O., & Samuel, N. (2016a). Modelling and estimating photosynthetically active radiation from measured global solar radiation at Calabar, Nigeria. *Physical Science International Journal*. https://doi.org/10.9734/psij/2016/28446

Sunday, E., Samuel, N., Agbasi, O., & Sylvia, J.-J. (2016b). Analysis of photosynthetically active radiation over six tropical ecological zones in Nigeria. *Journal of Geography, Environment and Earth Science International., 7*, 1. https://doi.org/10.9734/jgeesi/2016/27945

Thompson, S. (2022). Green renewable power and policy in Canada: A just Energy transition to net-Zero. *Journal of Geoscience and Environment Protection., 10*, 1. https://doi.org/10.4236/gep.2022.103001

Tlili, I. (2015). Renewable energy in Saudi Arabia: Current status and future potentials. *Environment, Development and Sustainability., 17*, 859. https://doi.org/10.1007/s10668-014-9579-9

Tokimatsu, K., Höök, M., McLellan, B., Wachtmeister, H., Murakami, S., Yasuoka, R., & Nishio, M. (2018). Energy modelling approach to the global energy-mineral nexus: Exploring metal requirements and the well-below 2 °C target with 100 percent renewable energy. *Applied Energy*. https://doi.org/10.1016/j.apenergy.2018.05.047

Tokimatsu, K., Wachtmeister, H., McLellan, B., Davidsson, S., Murakami, S., Höök, M., Yasuoka, R., & Nishio, M. (2017). Energy modelling approach to the global energy-mineral nexus: A first look at metal requirements and the 2°C target. *Applied Energy*. https://doi.org/10.1016/j.apenergy.2017.05.151

Trotter, P. A. (2022). The slow transition to solar, wind and other nonhydro renewables in Africa–Responding to and building on a critique by Kincer, Moss and Thurber (2021). *World Development Perspectives*. https://doi.org/10.1016/j.wdp.2021.100376

Tsao, Y. C., Vu, T. L., & Lu, J. C. (2021). Pricing, capacity and financing policies for investment of renewable energy generations. *Applied Energy., 303*, 117664. https://doi.org/10.1016/j.apenergy.2021.117664

United Nations. (2021). *The Energy transition, towards the achievement of SDG 7 and net-zero emissions*. In UN Energy.

Van Den Dobbelsteen, A., Martin, C. L., Keeffe, G., Pulselli, R. M., & Vandevyvere, H. (2018). *From problems to potentials-the urban energy transition of Gruž*. Energies. https://doi.org/10.3390/en11040922

van Soest, H., Reis, L. A., Baptista, L. B., Bertram, C., Després, J., Drouet, L., den Elzen, M., Fragkos, P., Fricko, O., Fujimori, S., Grant, N., Harmsen, M., Iyer, G., Keramidas, K., Köberle, A., Kriegler, E., Malik, A., Mittal, S., Oshiro, K., & van Vuuren, D. (2021). A global roll-out of nationally relevant policies bridges the emissions gap. Nature. *Communications*.

Veng, V., Suryadi, B., Pranadi, A. D., & Shani, N. (2020). A review of renewable energy development and its policy under nationally determined contributions in ASEAN. *International Journal of Smart Grid and Clean Energy.*, 149. https://doi.org/10.12720/sgce.9.1.149-161

Venkataraman, M., Csereklyei, Z., Aisbett, E., Rahbari, A., Jotzo, F., Lord, M., & Pye, J. (2022). Zero-carbon steel production: The opportunities and role for Australia. *Energy Policy*. https://doi.org/10.1016/j.enpol.2022.112811

Vieira, L. C., Longo, M., & Mura, M. (2021). Are the European manufacturing and energy sectors on track for achieving net-zero emissions in 2050? An empirical analysis. *Energy Policy*. https://doi.org/10.1016/j.enpol.2021.112464

Viladrich, C., Leveque, F., Togna, C., & Mellouki, L. (2016). *PSPP ' s and their role for the integration of variable renewable energy sources in Morocco*. 28th AHR Symposium on Hydraulic Machinery and Systems, Grenoble, July 4-8.

Vogt-Schilb, A., & Hallegatte, S. (2017). Climate policies and nationally determined contributions: Reconciling the needed ambition with the political economy. *Wiley Interdisciplinary Reviews: Energy and Environment*. https://doi.org/10.1002/wene.256

Waugh, C. (2011). The politics and culture of climate change: US actors and global implications. *In Advances in Global Change Research.* https://doi.org/10.1007/978-94-007-0934-8_6

Weber, R., Mueller, C., & Reinhart, C. (2021). *Building for Zero*. SSRN Electronic Journal. https://doi.org/10.2139/ssrn.3939009

Wiggins, S. (2014). Presidential address: African agricultural development: Lessons and challenges. *Journal of Agricultural Economics.*

Williams, J. H., Jones, R. A., & Torn, M. S. (2021). Observations on the transition to a net-zero energy system in the United States. *Energy and Climate Change., 2*, 100050. https://doi.org/10.1016/j.egycc.2021.100050

Winkler, H., & Marquand, A. (2009). Changing development paths: From an energy-intensive to low-carbon economy in South Africa. *Climate and Development., 1*, 47. https://doi.org/10.3763/cdev.2009.0003

Winrow, G. M. (2004). Turkey and the east-west gas transportation corridor. *Turkish Studies., 5*, 23. https://doi.org/10.1080/1468384042000228585

Wiseman, J. (2018). The great energy transition of the 21st century: The 2050 Zero-carbon world oration. *Energy Research and Social Science.* https://doi.org/10.1016/j.erss.2017.10.011

Wohland, J., Witthaut, D., & Schleussner, C. F. (2018). Negative emission potential of direct air capture powered by renewable excess Electricity in Europe. *Earth's Future., 6*, 1380. https://doi.org/10.1029/2018EF000954

Yan, X., Ge, J., Lei, Y., & Duo, H. (2019). China's low-carbon economic transition: Provincial analysis from 2002 to 2012. *Science of the Total Environment., 650*, 1050. https://doi.org/10.1016/j.scitotenv.2018.09.071

York, T. A., Brent, A. C., & Musango, J. K. (2015). *Infrastructure implications of a green economy transition in the Western cape province of South Africa*. IAMOT 2015–24th International Association for Management of technology Conference: Technology, Innovation and Management for Sustainable Growth, Proceedings.

Young, J., Dawes, M., Smith, A., Lake, K., & Lawton, K. (2021). *Financing net zero: Addressing technology risk for financial investments in the energy transition*. Society of Petroleum Engineers–SPE Offshore Europe Conference and Exhibition 2021, OE 2021. https://doi.org/10.2118/205463-MS.

Zeweld, W., Van Huylenbroeck, G., Tesfay, G., & Speelman, S. (2017). Smallholder farmers' behavioural intentions towards sustainable agricultural practices. *Journal of Environmental Management.* https://doi.org/10.1016/j.jenvman.2016.11.014

Zinola, C. F. (2010). *Production, storage, use, and delivery of hydrogen in the electrochemical conversion of energy*. Computational, Experimental, and Industrial Aspects. https://doi.org/10.1201/9781420045451

Chapter 9
Africa's Awakening to Climate Action

9.1 Africa's Awakening to Climate Action

The transition to a net-zero energy economy in Africa is a choiceless adventure (Chukwujindu et al., 2022) because life and our planet "Earth" have no duplicates (Agbor et al., 2023a), and it is up to us, as the custodians of this planet, to take action for its preservation (Agbor et al., 2023b) and ensure that our future generations can also enjoy the same environment we have (Nwokolo et al., 2023a). Contaminating the atmosphere has a correspondingly negative effect on humans (Nwokolo et al., 2023b) and their eco-friendliness because of the increased temperatures (Nwokolo et al., 2023c) and the increased occurrence of extreme weather conditions that accompany global warming (Ogbulezie et al., 2017), such as floods (Amadi et al., 2020), droughts (Nwokolo et al., 2022a), and storms (Nwokolo et al., 2022b). These events, in turn, affect the economic situation of African nations, as they reduce agricultural production (Hassan et al, 2021) and increase the number of displaced people due to the destruction of their homes and communities (Hassan et al., 2022). Careless management or slowness in addressing climate change problems in African countries (Nwokolo et al., 2023d), as seen in many countries, especially in the global South (Obiwulu et al., 2022), is a sign of irresponsibility in controlling our beloved planet and in raising humanity to a higher level of sustainability (Ituen et al., 2012). As such, it is essential for African countries to move toward a net-zero energy economy as quickly as possible to maintain the delicate balance between nature (Obiwulu et al., 2020a) and humans and protect their citizens from the devastating effects of climate change and global warming (Obiwulu et al., 2020b).

Policies are not enough. However, it is not enough to just implement policies (Abade, 2020) and legislation to mitigate the effects of climate change in African countries (Abdul-Salam et al., 2022); education and a commitment to sustainability must be present at all levels of society to ensure that everyone is aware of the consequences of climate change and the importance of reducing their carbon footprint. In addition to these policies, African countries should prioritize funding for

S. C. Nwokolo et al., *Africa's Path to Net-Zero*, CSR, Sustainability, Ethics & Governance, https://doi.org/10.1007/978-3-031-44514-9_9

research and development of sustainable and renewable energy sources (Samuel et al., 2017), as well as invest in measures that would increase the efficiency of the energy sector and incentivize the adoption of sustainable practices (IRENA, 2020). This includes investing in infrastructure to facilitate the use of renewable energy sources (Agreement et al., 2021), such as solar panels (Ahl et al., 2020) and wind turbines (Allen et al., 2020), and encouraging sustainable farming practices that are better suited to the changing climate (Amado et al., 2018).

The 2016 Paris Agreement is active, but progress is slow due to the lack of resources and infrastructure in many African countries (IRENA, 2020), and much of the continent's population remains unaware or uneducated about the dangers of climate change and the need to address the issue. An effect must be applied to save our beloved continent (Ameli et al., 2021a), and this begins with investing in African countries to promote the adoption of sustainable energy sources, which will serve to reduce emissions (Ameli et al., 2021b), increase access to clean energy (Azevedo et al., 2021), and create jobs that help strengthen the economy (Bak et al., 2017). We believe that a net-zero transition can be achieved through innovative research (Barry et al., 2011) and technological development as well as community-driven and government-backed investments in renewable energy (Bataille et al., 2018) and the development of resilient energy systems that are tailored to the unique needs of African nations. Carbon capture and removal technology is feasible (Bataille, 2020) and can be implemented in Africa to reduce emissions and combat climate change, as well as help facilitate a transition toward a carbon-neutral economy (BP Energy, 2020) while also improving the well-being of African people through improved air quality (Bray et al., 2022) and more accessible energy solutions (Caetano et al., 2017).

The use of renewable energy in all aspects of human activity on the African continent is equally realistic (Cillari et al., 2021) and can be achieved through investment in education and training programs to increase the capacity of African citizens in the field of renewable energy (Coelho et al., 2010), as well as incentivizing private investments in clean energy solutions that leverage the potential of Africa's abundant resources (Cotterman et al., 2021). In addition, regulations should be put in place to phase out fossil fuel consumption (da Silva et al., 2018) and drive investment in low-carbon energy production technologies (Delafield et al., 2021), such as solar and wind. Furthermore, the development of a pan-African renewable energy market will enable African countries to benefit from each other's advancements and share the costs of research and development (Derkenbaeva et al., 2021), leading to faster decarbonization across the continent and reduced reliance on external sources of finance for renewable energy projects (Deutch, 2020). In general, carbon capture and removal technology and renewable energy solutions offer an effective way to reduce emissions (Felder et al., 2021), combat climate change, and improve the well-being of African citizens by providing a reliable (Ferreira et al., 2016), affordable, and sustainable source of energy that can be used to power economic development (Florent Baarsch & Schaeffer, 2019), create jobs, and reduce poverty in the region (Fol & Le, 2012).

It only requires that you and I, inventors, research institutes, academic researchers, scientific and non-scientific organizations, the world community, and leaders reflect over and over again to take meaningful action to reduce emissions and address climate change in Africa and across the world. To this end, global collaboration is essential for a successful transition to a low-carbon future (Ford & Hardy, 2020). The development of carbon capture and removal technology and renewable energy solutions (Ford et al., 2021) can also lead to improved access to electricity for African citizens, contributing significantly to an improvement in quality of life across the continent. Through the deployment of these technologies, African countries can enjoy the economic and social benefits that come with reliable (Freeman, 2021) and affordable access to clean energy and mitigate the risk of further climate-related disasters (French, 2020) such as floods, droughts, and rising sea levels (Garcia, 2008). Additionally, education and awareness campaigns are important tools to promote the adoption of energy-efficient practices in the residential (Godin et al., 2021), commercial, and industrial sectors to transition to a low-carbon economy (Gozlugol, 2022). By leveraging the potential of global collaboration and technology (Greig & Uden, 2021), African countries have the opportunity to transition toward a sustainable (Grobbelaar et al., 2014) and low-carbon future while also providing improved access to electricity (Hafner et al., 2021), a reduction in energy poverty, and greater economic and social benefits (Hrnčić et al., 2021).

As we stand on the precipice of a low-carbon economy (IRENA, 2013), African countries can take advantage of this opportunity and make progress toward energy security through the deployment of cutting-edge technologies such as solar and wind power, electric vehicles (IRENA, 2020), and more efficient appliances (IRENA, 2021), which can help reduce energy demand, increase energy efficiency, and make renewable energy more accessible to those who may not otherwise have access to it (IRENA, 2012). In doing so, African countries will be able to reap the benefits of a low-carbon economy (Kilkiş, 2015) while also taking steps toward energy security (Kincer et al., 2022) and sustainability that can help improve the lives of people living in those countries (Kivimaa et al., 2020). Additionally, the adoption of policies that incentivize and encourage the use of renewable energy (Klunne, 2012) and energy efficiency will be essential to ensuring that the transition to a low-carbon economy is successful (Komninos, 2022) and that the economic, social (Kourgiozou et al., 2021), and environmental benefits are felt across Africa. Not only does this help create jobs, generate economic growth (Kouton, 2021), and increase energy security, but it also helps reduce carbon emissions in the atmosphere (Lowe & Drummond, 2022), which in turn helps mitigate climate change and its effects and preserve the health of our planet for future generations (Lucas, 2022). Moreover, with the adoption of a low-carbon economy and energy security measures (Manjong et al., 2021), African countries can begin to move away from fossil fuels and toward more sustainable energy sources that are not only more reliable but also more affordable in the long term (Maupin, 2017). The implications of this shift to a low-carbon economy in Africa are immense and could be the difference between a future of economic prosperity (Mavrigiannaki et al., 2021) and one of environmental destruction. For these reasons, it is clear that transitioning to a low-carbon economy in Africa is necessary and critical for the future of the continent, its people, and the planet as a whole.

As much as scientists could invest in telescopes (Mekonnen & Sarwat, 2017), the Internet, and communication devices, climate change is not an insurmountable challenge for the world's high-tech (Mutezo & Mulopo, 2021) and scientific communities if the African continent is to make a successful transition to a low-carbon economy (Mestemaker et al., 2020) and its populations become more resilient to the impacts of global climate change to prevent further destruction of the environment (Nelson & Allwood, 2021). It all requires joining hands and lending a hand to the outstretched hands of our toxicating planet, which is nearing eternal destruction through the extremes of climate change (United Nations, 2021).

Over the years, we have seen improvements in the human development index and fitness resulting from advanced and transitional paradigms in the extractive (Ngondjeb et al., 2020), manufacturing, and service sectors across the globe (Nhuchhen et al., 2022), inspired by scientific innovation, inventions, and technological discoveries (Nikas et al., 2022). The discovery of computers, for example, fostered progress in all areas of human endeavor (Okem, 2016), from its developmental stage to the evolutionary stage of microsize, speed, capacity, etc.

Exploration of coal, oil, and natural gas, nonrenewable resources (Omar et al., 2022), in its present stage from time immemorial to its evolutionary stage was equally made possible by numerous technological and managerial innovations (Omer, 2009). To crown it all, the evolving degree of management and marketing of coal (Ouedraogo, 2019), oil, and natural gas energy sources cannot register the global value chains experienced in these industries without the use of high-tech (Rogelj et al., 2015) and the Internet of Things (Oyewo et al., 2020). The time will not fail me to enumerate many products obtained from large-scale concentrated deposits of refined natural gas oil across the various nations of the Earth (Oyewo et al., 2019).

I can boldly point out that various tangible ornaments (Rissman et al., 2020), materials, and splendors that are optically visible on Earth are traceable to fossil products. Economic wealth (Pablo-Romero et al., 2021), high employment (Perlaviciute et al., 2021), the human development index (Ramagoma & Adendorff, 2016), innovations (Rennkamp & Boyd, 2015), and the beauty spectrum of many nations are direct derivatives that are deeply blessed by nature with a high concentration of recently discovered fossil deposits around the globe and powered by the instrumentation of fossil fuel sources of energy.

Therefore, nations with coal, oil, and natural gas deposits export these natural resources for high financial remittances in such a large amount that the weight of wealth accrued from it has stimulated rapid development in some nations, such as Norway, the United Arab Emirates, Saudi Arabia, the Russian Federation, Ecuador, the United States, Iran, Qatar, Oman, Nigeria, South Africa, Botswana, Libya, Algeria, Angola, and Canada, as well as all major OPEC member states. The unprecedented influx of wealth in these countries led to greed among most political leaders. The result of this was to push the wealth of the nation into private pockets. This was made possible by the selfish leaders by weakening institutional quality. Some even made themselves permanent presidents or national leaders in some countries operating a democratic system of government. The main goal of these ambitious and democratic leaders and presidents is to obtain abundant wealth to oppress and silence their citizens because wealth is an added power, especially when they are already in political power. In fact, in most resource-rich countries, the fate of these citizens is in the hands of political leaders or presidents.

This has become so severe that most rich countries have encountered the Dutch Disease phenomenon as a result of mismanagement or misappropriation of public funds due to their high dependence on exports of fossil fuel-generated products. To date, almost every nation in the Global South has fossil fuel deposits, but rapid development has been denied despite a steady influx of finances into the country.

According to development economics, Dutch diseases often damage the economy due to rising exchange rates, investing heavily in thriving tradable sectors while ignoring the thriving commercial sectors of manufacturing, mining, and agriculture. Therefore, compared with countries rich in fossil resources, countries with fewer resources have developed economically. The reason is that these resource-poor countries pay great attention to the manufacturing and agricultural sectors, which use oil, coal, and natural gas resources purchased from resource-rich countries to produce various products they need and are most suitable for the market. Resource-rich countries cannot convert their original fossil deposits into end-user products. This has led to the rapid economic development of these countries without the natural deposits of large fossil deposits, detectable signs of Dutch disease, or excellent industrial and high-tech quality. Most of these resource-poor countries include Singapore, Taiwan, Thailand, Japan, and China.

Today, resource-poor China provides many funds to Nigeria and other resource-rich African countries, which may be difficult to repay in the next 20 years due to their size. Resource-rich countries should realize that the dominance of fossil energy in the energy sector is gradually disappearing. This is because the reservoirs from which it is derived are scarce, because climate change is a factor, and because the decentralized energy system model interferes with how humans distribute their energy.

For example, the United States is about to cut off the world's major oil exporter. By the beginning of World War II, the United States was supplying more than 60% of the world's oil needs. Countries that pioneered World War II against the United States include Germany and Japan. Unfortunately, Germany and Japan relied primarily on oil imports from the United States for economic and military survival. The United States applied the technique of energy dominance, that is, its ability to supply abundant oil to other countries, and the total denial of full access to Germany and Japan was one of the strategies to win World War II more than 70 years ago. During the last world war, the principle of energy dominance saved the United States from the hands of nations with fewer resources than powers such as Germany and Japan.

Judging by the way the pattern of the world system is advancing today, energy dominance may not work as effectively as it did in the days of World War II. What I call the dominance of the high-tech brain through artificial intelligence (AI), high-tech communications, the Internet of Things, biotech arsenals, and ultimately the end-use of space will replace the dominance of fossil energy. Therefore, the world's power balance could change if the United States refuses to reinforce China's ambiguous desire and dominates it.

On a global scale, the clock for mitigating the external effects of climate change is measured in microseconds. Many countries, multinational companies, agricultural production industries, manufacturing industries, extractive industries, and service industries have made a series of commitments to reduce their carbon emissions to net zero by 2050 or later, which has inspired a new kind of optimism. Realizing these ambitions remains a huge challenge. This has had a great impact on the world's energy infrastructure. In other words, how to use steel and fuel for heating buildings and power plants to make cars and airplanes. The consequences of this change will affect the lives of billions of people on the planet today and in the future. Some industries will grow significantly, while others will inevitably decline. When looking for the definition of the global transition policy goals required by the 2016 Paris Agreement, politicians, policymakers, and important ministerial officials scattered in various institutions often delay the implementation of renewable energy policies. This is evidence from global South countries. As such, exposing the human race to unavoidable environmental, ecological, and health dangers and risks means forgetting about people and not even about climate change as a meteorological parameter that requires immediate environmental mitigation. There is a lot we need to do to limit global warming to ensure a better future for everyone on the planet.

Efforts to address climate change through clean energy change must also help citizens seize opportunities to resist rising unemployment and its negative impacts on society and optimally navigate between producer disruptions and distractions from fossil fuels toward continuous transformation and the transition to a cleaner and greener economy. Governments must understand and address the social and economic implications of the clean energy transition for individuals, businesses, and communities. Policies are only successful if they ensure that people can benefit from these changes and protect those who might be harmed by them. Many countries are now focusing on these issues and looking for opportunities to exchange ideas on these issues and learn from each other's experiences of employment, participation, and justice.

References

Abade, L. P. (2020). *Climate change awareness: Role of education* https://doi.org/10.1007/978-3-319-95885-9_122.

Abdul-Salam, Y., Kemp, A., & Phimister, E. (2022). Energy transition in the UKCS–modelling the effects of carbon emission charges on upstream petroleum operations. *Energy Economics., 108*, 105898. https://doi.org/10.1016/j.eneco.2022.105898

Agbor, M., Udo, S., Ewona, I., Nwokolo, S., & Ogbulezie, J. A. (2023a). Potential impacts of climate change on global solar radiation and PV output using the CMIP6 model in West Africa. *Cleaner Engineering and Technology, 13*, 100630. https://doi.org/10.1016/j.clet.2023.100630

Agbor, M., Udo, S., Ewona, I., Nwokolo, S., Ogbulezie, J., Amadi, S., & Billy, U. (2023b). Effects of angstrom-Prescott and Hargreaves-Samani coefficients on climate forcing and solar PV technology selection in West Africa. *Trends in Renewable Energy, 9*(1), 78–106. https://doi.org/10.17737/tre.2023.9.1.00150

Agency, I. R. E. (2020). Renewable power generation costs in 2019. *International Renewable Energy Agency.*

Agreement, P., Republic, C., Kingdom, U., States, U., Principles, S. K., Zero, I. N., Secretariat, I. E. A., Iea, T., Iea, T., Iea, T., Energy, C., & Programme, T. (2021). Seven key Principles for implementing net Zero. *International Energy Agency.*

Ahl, A., Yarime, M., Goto, M., Chopra, S. S., Kumar, N. M., Tanaka, K., & Sagawa, D. (2020). Exploring blockchain for the energy transition: Opportunities and challenges based on a case study in Japan. *Renewable and Sustainable Energy Reviews., 117*, 109488. https://doi.org/10.1016/j.rser.2019.109488

Allen, S. R., Cohen, B., & Hammond, G. P. (2020). Indicative appraisal of clustered microgenerators for a low-carbon transition in the UK building sector. *Global Transitions., 2*, 83. https://doi.org/10.1016/j.glt.2020.06.001

Amadi, S., Dike, T., & Nwokolo, S. (2020). Global solar radiation characteristics at Calabar and Port Harcourt cities in Nigeria. *Trends in Renewable Energy., 6*, 101. https://doi.org/10.17737/tre.2020.6.2.00114

Amado, M., Poggi, F., Amado, A. R., & Breu, S. (2018). E-city web platform: A tool for energy efficiency at urban level. *Energies, 11*. https://doi.org/10.3390/en11071857

Ameli, N., Dessens, O., Winning, M., Cronin, J., Chenet, H., Drummond, P., Calzadilla, A., Anandarajah, G., & Grubb, M. (2021a). *A climate investment trap in developing economies*. Nature Portfolio.

Ameli, N., Dessens, O., Winning, M., Cronin, J., Chenet, H., Drummond, P., Calzadilla, A., Anandarajah, G., & Grubb, M. (2021b). Higher cost of finance exacerbates a climate investment trap in developing economies. *Nature Communications, 12*, 4046. https://doi.org/10.1038/s41467-021-24305-3

Azevedo, I., Bataille, C., Bistline, J., Clarke, L., & Davis, S. (2021). Introduction to the special issue on net-Zero Energy systems. *Energy and Climate Change., 2*, 100066. https://doi.org/10.1016/j.egycc.2021.100066

Bak, C., Bhattacharya, A., Edenhofer, O., & Knopf, B. (2017). Towards a comprehensive approach to climate policy, sustainable infrastructure, and finance. *Economics., 11*. https://doi.org/10.5018/economics-ejournal.ja.2017-33

Barry, M. L., Steyn, H., & Brent, A. (2011). Selection of renewable energy technologies for Africa: Eight case studies in Rwanda, Tanzania and Malawi. *Renewable Energy*. https://doi.org/10.1016/j.renene.2011.04.016

Bataille, C. G. F. (2020). Physical and policy pathways to net-zero emissions industry. *Wiley Interdisciplinary Reviews: Climate Change*. https://doi.org/10.1002/wcc.633

Bataille, C., Åhman, M., Neuhoff, K., Nilsson, L. J., Fischedick, M., Lechtenböhmer, S., Solano-Rodriquez, B., Denis-Ryan, A., Stiebert, S., Waisman, H., Sartor, O., & Rahbar, S. (2018). A review of technology and policy deep decarbonization pathway options for making energy-intensive industry production consistent with the Paris Agreement. *Journal of Cleaner Production., 187*, 960. https://doi.org/10.1016/j.jclepro.2018.03.107

BP Energy. (2020). *Energy Outlook 2020 edition explores the forces shaping the global energy transition out to 2050 and the surrounding that. BP Energy Outlook 2030, Statistical review*. British Petroleum.

Bray, R., Mejía Montero, A., & Ford, R. (2022). Skills deployment for a 'just' net zero energy transition. *Environmental Innovation and Societal Transitions., 42*, 395. https://doi.org/10.1016/j.eist.2022.02.002

Caetano, T., Merven, B., Hartley, F., & Ahjum, F. (2017). Decarbonization and the transport sector: A socioeconomic analysis of transport sector futures in South Africa. *Journal of Energy in Southern Africa., 28*. https://doi.org/10.17159/2413-3051/2017/v28i4a2945

Chukwujindu, S., Umunnakwe, A., & Ogbulezie, J. C. (2022). Hybridization of statistical machine learning and numerical models for improving beam, diffuse and global solar radiation prediction. *Cleaner Engineering and Technology, 9*(May), 100529. https://doi.org/10.1016/j.clet.2022.100529

Cillari, G., Franco, A., & Fantozzi, F. (2021). Sizing strategies of photovoltaic systems in nZEB schemes to maximize the self-consumption share. *Energy Reports., 7*, 6769. https://doi.org/10.1016/j.egyr.2021.09.117

Coelho, B., Oliveira, A. C., & Mendes, A. (2010). Concentrated solar power for renewable electricity and hydrogen production from water–A review. *Energy and Environmental Science., 3*, 1398. https://doi.org/10.1039/b922607a

Cotterman, T., Small, M. J., Wilson, S., Abdulla, A., & Wong-Parodi, G. (2021). Applying risk tolerance and sociotechnical dynamics for more realistic energy transition pathways. *Applied Energy., 291*, 116751. https://doi.org/10.1016/j.apenergy.2021.116751

da Silva, P. P., Cerqueira, P. A., & Ogbe, W. (2018). Determinants of renewable energy growth in sub-Saharan Africa: Evidence from panel ARDL. *Energy, 156*, 45. https://doi.org/10.1016/j.energy.2018.05.068

Delafield, G., Donnison, C., Roddis, P., Arvanitopoulos, T., Sfyridis, A., Dunnett, S., Ball, T., & Logan, K. G. (2021). Conceptual framework for balancing society and nature in net-zero energy transitions. *Environmental Science and Policy., 125*, 189. https://doi.org/10.1016/j.envsci.2021.08.021

Derkenbaeva, E., Halleck Vega, S., Hofstede, G. J., & van Leeuwen, E. (2022). Positive energy districts: Mainstreaming energy transition in urban areas. *Renewable and Sustainable Energy Reviews., 153*, 111782. https://doi.org/10.1016/j.rser.2021.111782

Deutch, J. (2020). Is net Zero carbon 2050 possible? *In Joule., 4*, 2237. https://doi.org/10.1016/j.joule.2020.09.002

Felder, F. A., Unel, B., & Dvorkin, Y. (2021). Modelling strategic objectives and behavior in the transition of the energy sector to inform policymaking. *Electricity Journal., 34*, 107009. https://doi.org/10.1016/j.tej.2021.107009

Ferreira, M., Almeida, M., Rodrigues, A., & Silva, S. M. (2016). Comparing cost-optimal and net-zero energy targets in building retrofit. *Building Research and Information., 44*, 188. https://doi.org/10.1080/09613218.2014.975412

Florent, B., & Schaeffer, M. (2019). Climate change impacts on Africa ' s economic growth. *African Development Bank*, 1–169.

Fol, Y. L. (2012). *Renewable Energy transition for a sustainable future in Namibia*. In Aalborg University.

Ford, R., & Hardy, J. (2020). Are we seeing clearly? The need for aligned vision and supporting strategies to deliver net-zero electricity systems. *Energy Policy, 147*, 111902. https://doi.org/10.1016/j.enpol.2020.111902

Ford, R., Maidment, C., Vigurs, C., Fell, M. J., & Morris, M. (2021). Smart local energy systems (SLES): A framework for exploring transition, context, and impacts. *Technological Forecasting and Social Change, 166*, 120612. https://doi.org/10.1016/j.techfore.2021.120612

Freeman, R. (2021). Modelling the sociopolitical feasibility of energy transition with system dynamics. *Environmental Innovation and Societal Transitions., 40*, 486. https://doi.org/10.1016/j.eist.2021.10.005

French, S. (2020). The role of Zero and low carbon hydrogen in enabling the Energy transition and the path to net Zero greenhouse gas emissions. *Johnson Matthey Technology Review*. https://doi.org/10.1595/205651320x15910225395383

Garcia, D. (2008). The climate security divide: Bridging human and national security in Africa. *African Security Review., 17*, 2. https://doi.org/10.1080/10246029.2008.9627480

Godin, K., Sapinski, J. P., & Dupuis, S. (2021). The transition to net zero energy (NZE) housing: An integrated approach to market, state, and other barriers. *Cleaner and Responsible Consumption., 3*, 100043. https://doi.org/10.1016/j.clrc.2021.100043

Gozlugol, A. A. (2022). The clash of "E" and "S" of ESG: Just transition on the path to net zero and the implications for sustainable corporate governance and finance. *Journal of World Energy Law and Business., 15*, 1. https://doi.org/10.1093/jwelb/jwab039

Greig, C., & Uden, S. (2021). The value of CCUS in transitions to net-zero emissions. *Electricity Journal., 34*, 107004. https://doi.org/10.1016/j.tej.2021.107004

Grobbelaar, S., Gauché, P., & Brent, A. (2014). Developing a competitive concentrating solar power industry in South Africa: Current gaps and recommended next steps. *Development Southern Africa., 31*, 475. https://doi.org/10.1080/0376835X.2014.891971

Hafner, S., Jones, A., & Anger-Kraavi, A. (2021). Economic impacts of achieving a net-zero emissions target in the power sector. *Journal of Cleaner Production., 312*, 127610. https://doi.org/10.1016/j.jclepro.2021.127610

Hassan, M. A., Bailek, N., Bouchouicha, K., Ibrahim, A., Jamil, B., Kuriqi, A., & Nwokolo, S. C. (2022). Evaluation of energy extraction of PV systems affected by environmental factors under real outdoor conditions. *Theoretical and Applied Climatology, 150*, 715. https://doi.org/10.1007/s00704-022-04166-6

Hassan, M. A., Bailek, N., Bouchouicha, K., & Nwokolo, S. C. (2021). Ultrashort-term exogenous forecasting of photovoltaic power production using genetically optimized nonlinear autoregressive recurrent neural networks. *Renewable Energy, 171*(February), 191–209. https://doi.org/10.1016/j.renene.2021.02.103

Hrnčić, B., Pfeifer, A., Jurić, F., Duić, N., Ivanović, V., & Vušanović, I. (2021). Different investment dynamics in energy transition towards a 100% renewable energy system. *Energy, 237*, 121526. https://doi.org/10.1016/j.energy.2021.121526

International Renewable Energy Agency. (2020). *Renewable Power Generation Costs in 2019*. International Renewable Energy Agency.

International Renewable Energy Agency (IRENA). (2013). Renewable power generation costs in 2012: An overview. *Irena*, 1–8.

International Renewable Energy Agency (IRENA). (2020). *Renewable energy and climate pledges: Five years after the Paris Agreement*.

IRENA. (2021). *World energy transitions outlook: 1.5 degrees pathway*. In International Renewable Energy Agency. https://irena.org/publications/2021/March/World-Energy-Transitions-Outlook

Irena Working Paper. (2012). *Renewable Energy Technologies: Cost analysis series, hydropower*. International Renewable Energy Agency.

Ituen, E. E., Esen, N. U., Nwokolo, S. C., & Udo, E. G. (2012). Prediction of global solar radiation using relative humidity, maximum temperature and sunshine hours in Uyo, in The Niger Delta region, Nigeria. *Advances in Applied Science Research, 3*(4), 1923–1937.

Kilkiş, Ş. (2015). Exergy transition planning for net-zero districts. *Energy, 92*, 515. https://doi.org/10.1016/j.energy.2015.02.009

Kincer, J., Moss, T., & Thurber, M. (2022). A coal renaissance is not coming to Africa. *World Development Perspectives., 25*, 100375. https://doi.org/10.1016/j.wdp.2021.100375

Kivimaa, P., Primmer, E., & Lukkarinen, J. (2020). Intermediating policy for transitions towards net-zero energy buildings. *Environmental Innovation and Societal Transitions., 36*, 418. https://doi.org/10.1016/j.eist.2020.01.007

Klunne, W. J. (2012). Current status and future developments of small and micro hydro in southern Africa. *Energy Policy*.

Komninos, N. (2022). Net Zero Energy districts: Connected intelligence for carbon-neutral cities. *Land., 11*. https://doi.org/10.3390/land11020210

Kourgiozou, V., Commin, A., Dowson, M., Rovas, D., & Mumovic, D. (2021). Scalable pathways to net zero carbon in the UK higher education sector: A systematic review of smart energy systems in university campuses. *Renewable and Sustainable Energy Reviews., 147*, 111234. https://doi.org/10.1016/j.rser.2021.111234

Kouton, J. (2021). The impact of renewable energy consumption on inclusive growth: Panel data analysis in 44 African countries. *Economic Change and Restructuring., 54*, 145. https://doi.org/10.1007/s10644-020-09270-z

Lowe, R. J., & Drummond, P. (2022). Solar, wind and logistic substitution in global energy supply to 2050–Barriers and implications. *Renewable and Sustainable Energy Reviews., 153*, 111720. https://doi.org/10.1016/j.rser.2021.111720

Lucas, A. R. (2022). Canada's carbon energy overhang. *Journal of Energy and Natural Resources Law., 40*, 17. https://doi.org/10.1080/02646811.2021.2012349

Manjong, N. B., Oyewo, A. S., & Breyer, C. (2021). Setting the pace for a sustainable Energy transition in Central Africa: The case of Cameroon. *IEEE Access., 9*, 145435. https://doi.org/10.1109/ACCESS.2021.3121000

Maupin, A. (2017). The SDG13 to combat climate change: An opportunity for Africa to become a trailblazer? *African Geographical Review., 36*, 131. https://doi.org/10.1080/19376812.2016.1171156

Mavrigiannaki, A., Pignatta, G., Assimakopoulos, M., Isaac, M., Gupta, R., Kolokotsa, D., Laskari, M., Saliari, M., Meir, I. A., & Isaac, S. (2021). Examining the benefits and barriers for the implementation of net zero energy settlements. *Energy and Buildings., 230*, 110564. https://doi.org/10.1016/j.enbuild.2020.110564

Mekonnen, Y., & Sarwat, A. I. (2017). *Renewable energy supported microgrid in rural electrification of Sub-Saharan Africa*. Proceedings–2017 IEEE PES-IAS PowerAfrica Conference: Harnessing Energy, Information and Communications Technology (ICT) for Affordable Electrification of Africa, PowerAfrica 2017. https://doi.org/10.1109/PowerAfrica.2017.7991293.

Mestemaker, B., Van Den Heuvel, H., & Gonçalves Castro, B. (2020). Designing the zero emission vessels of the future: Technologic, economic and environmental aspects. *International Shipbuilding Progress., 67*, 5. https://doi.org/10.3233/ISP-190276

Mutezo, G., & Mulopo, J. (2021). A review of Africa's transition from fossil fuels to renewable energy using circular economy principles. *Renewable and Sustainable Energy Reviews, 137-*(November 2020), 110609. https://doi.org/10.1016/j.rser.2020.110609

Nelson, S., & Allwood, J. M. (2021). The technological and social timelines of climate mitigation: Lessons from 12 past transitions. *Energy Policy, 152*, 112155. https://doi.org/10.1016/j.enpol.2021.112155

Ngondjeb, D. Y., Atewamba, C., & Macalou, M. (2020). Insights on Africa's future in its transition to the green economy. *Advances in African Economic, Social and Political Development*. doi: https://doi.org/10.1007/978-3-030-44180-7_13.

Nhuchhen, D. R., Sit, S. P., & Layzell, D. B. (2022). Towards net-zero emission cement and power production using molten carbonate fuel cells. *Applied Energy., 306*, 118001. https://doi.org/10.1016/j.apenergy.2021.118001

Nikas, A., Xexakis, G., Koasidis, K., Acosta-Fernández, J., Arto, I., Calzadilla, A., Domenech, T., Gambhir, A., Giljum, S., Gonzalez-Eguino, M., Herbst, A., Ivanova, O., van Sluisveld, M. A. E., Van De Ven, D. J., Karamaneas, A., & Doukas, H. (2022). Coupling circularity performance and climate action: From disciplinary silos to transdisciplinary modelling science. *Sustainable Production and Consumption., 30*, 269. https://doi.org/10.1016/j.spc.2021.12.011

Nwokolo, S. C., Amadi, S. O., Obiwulu, A. U., Ogbulezie, J. C., & Eyibio, E. E. (2022a). Prediction of global solar radiation potential for sustainable and cleaner energy generation using improved angstrom-Prescott and Gumbel probabilistic models. *Cleaner Engineering and Technology, 6*(January), 100416. https://doi.org/10.1016/j.clet.2022.100416

Nwokolo, S. C., Obiwulu, A. U., Amadi, O., & Ogbulezie, J. C. (2023a). Assessing the impact of soiling, tilt angle, and solar radiation on the performance of solar PV systems. *Trends in Renewable Energy, 9*(2), 121–137. https://doi.org/10.17737/tre.2023.9.2.00156

Nwokolo, S. C., Obiwulu, A. U., & Ogbulezie, J. C. (2022b). Machine learning and analytical model hybridization to assess the impact of climate. *Physics and Chemistry of the Earth, 130*(February), 103389. https://doi.org/10.1016/j.pce.2023.103389

Nwokolo, S., & Ogbulezie, J. (2017a). A critical review of theoretical models for estimating global solar radiation between 2012-2016 in Nigeria. *International Journal of Physical Research, 5*(2), 60. https://doi.org/10.14419/ijpr.v5i2.8160

Nwokolo, S., & Ogbulezie, J. (2017b). Modelling the influence of cloudiness on diffuse horizontal irradiation under various sky conditions in Nigeria. *International Journal of Physical Research, 5*(2), 91. https://doi.org/10.14419/ijpr.v5i2.8312

Nwokolo, S. C., & Ogbulezie, J. C. (2018a). A qualitative review of empirical models for estimating diffuse solar radiation from experimental data in Africa. *Renewable and Sustainable Energy Reviews., 92*, 353. https://doi.org/10.1016/j.rser.2018.04.118

Nwokolo, S. C., & Ogbulezie, J. C. (2018c). A quantitative review and classification of empirical models for predicting global solar radiation in West Africa. *Beni-Suef University Journal of Basic and Applied Sciences, 7*(4), 367–396. https://doi.org/10.1016/j.bjbas.2017.05.001

Nwokolo, S. C., Ogbulezie, J. C., & Obiwulu, A. U. (2022c). Impacts of climate change and meteo-solar parameters on photosynthetically active radiation prediction using hybrid machine learning with physics-based models. *Advances in Space Research, 70*, 3614. https://doi.org/10.1016/j.asr.2022.08.010

Nwokolo, S. C., Ogbulezie, J. C., & Ushie, O. J. (2023b). A multimodel ensemble-based CMIP6 assessment of future solar radiation and PV potential under various climate warming scenarios. *Optik, 285*(May), 170956. https://doi.org/10.1016/j.ijleo.2023.170956

Nwokolo, S., & Otse, C. (2019). Impact of sunshine duration and clearness index on diffuse solar radiation estimation in mountainous climate. *Trends in Renewable Energy, 5*(3), 307–332. https://doi.org/10.17737/tre.2019.5.3.00107

Nwokolo, S. C., Proutsos, N., & Meyer, E. L. (2023c). Machine learning and physics-based hybridization models for evaluation of the effects of climate change and urban expansion on photosynthetically active radiation. *Atmosphere, 14*(687), 1–36. https://doi.org/10.3390/atmos14040687

Nwokolo, S. C., Obiwulu, A. U., & Ogbulezie, J. C. (2023d). Machine learning and analytical model hybridization to assess the impact of climate. *Physics and Chemistry of the Earth, 130* (Feburary 2023), 103389. https://doi.org/10.1016/j.pce.2023.103389

Obiwulu, A. U., Chendo, M. A. C., Erusiafe, N., & Nwokolo, S. C. (2020a). Implicit meteorological parameter-based empirical models for estimating back temperature solar modules under varying tilt-angles in Lagos, Nigeria. *Renewable Energy, 145*, 442. https://doi.org/10.1016/j.renene.2019.05.136

Obiwulu, A. U., Erusiafe, N., Olopade, M. A., & Nwokolo, S. C. (2020b). Modelling and optimization of back temperature models of mono-crystalline silicon modules with special focus on the effect of meteorological and geographical parameters on PV performance. *Renewable Energy, 154*, 404. https://doi.org/10.1016/j.renene.2020.02.103

Obiwulu, A. U., Erusiafe, N., Olopade, M. A., & Nwokolo, S. C. (2022). Modelling and estimation of the optimal tilt angle, maximum incident solar radiation, and global radiation index of the photovoltaic system. *Heliyon, 8*(6), e09598. https://doi.org/10.1016/j.heliyon.2022.e09598

Ogbulezie, J. C., James, U. O., & Chukwujindu, N. S. (2017, November). A review of regression models employed for predicting diffuse solar radiation in North-Western Africa. *Trends in Renewable Energy*. https://doi.org/10.17737/tre.2017.3.2.0042

Okem, A. E. (2016). South Africa's transition to a low-carbon economy: The role of cooperatives. *Journal of Social Sciences., 49*, 257. https://doi.org/10.1080/09718923.2016.11893619

Omar, A. I., Khattab, N. M., & Abdel Aleem, S. H. E. (2022). *Optimal strategy for transition into net-zero energy in educational buildings: A case study in El-Shorouk City, Egypt*. Sustainable Energy Technologies and Assessments. https://doi.org/10.1016/j.seta.2021.101701

Omer, A. M. (2009). *Energy and sustainable development in Sudan*. In Energy costs, international developments and new directions.

Ouedraogo, N. S. (2019). Opportunities, barriers and issues with renewable energy development in africa: a comprehensible review. *Current Sustainable/Renewable Energy Reports, 6*, 52. https://doi.org/10.1007/s40518-019-00130-7

Oyewo, A. S., Aghahosseini, A., Ram, M., & Breyer, C. (2020). Transition towards decarbonized power systems and its socioeconomic impacts in West Africa. *Renewable Energy, 154*, 1092. https://doi.org/10.1016/j.renene.2020.03.085

Oyewo, A. S., Aghahosseini, A., Ram, M., Lohrmann, A., & Breyer, C. (2019). Pathway towards achieving 100% renewable electricity by 2050 for South Africa. *Solar Energy., 191*, 549. https://doi.org/10.1016/j.solener.2019.09.039

Pablo-Romero, M. P., Sánchez-Braza, A., & Galyan, A. (2021). Renewable energy use for electricity generation in transition economies: Evolution, targets and promotion policies. *Renewable and Sustainable Energy Reviews., 138*, 110481. https://doi.org/10.1016/j.rser.2020.110481

Perlaviciute, G., Steg, L., & Sovacool, B. K. (2021). A perspective on the human dimensions of a transition to net-zero energy systems. *Energy and Climate Change., 2*, 100042. https://doi.org/10.1016/j.egycc.2021.100042

Ramagoma, J., & Adendorff, C. (2016). Managing a transition to green energy sources: The perspectives of energy practitioners in the southern African development community region. *Journal of Energy in Southern Africa., 27*, 77. https://doi.org/10.17159/2413-3051/2016/v27i4a1638

Rennkamp, B., & Boyd, A. (2015). Technological capability and transfer for achieving South Africa's development goals. *Climate Policy., 15*, 12. https://doi.org/10.1080/14693062.2013.831299

Rissman, J., Bataille, C., Masanet, E., Aden, N., Morrow, W. R., Zhou, N., Elliott, N., Dell, R., Heeren, N., Huckestein, B., Cresko, J., Miller, S. A., Roy, J., Fennell, P., Cremmins, B., Koch Blank, T., Hone, D., Williams, E. D., de la Rue du Can, S., & Helseth, J. (2020). Technologies and policies to decarbonize global industry: Review and assessment of mitigation drivers through 2070. *Applied Energy., 266*, 114848. https://doi.org/10.1016/j.apenergy.2020.114848

Rogelj, J., Luderer, G., Pietzcker, R. C., Kriegler, E., Schaeffer, M., Krey, V., & Riahi, K. (2015). Energy system transformations for limiting end-of-century warming to below 1.5 °C. In Nature Climate Change, *5*, 519. https://doi.org/10.1038/nclimate2572

Samuel Chukwujindu, N. (2017). A comprehensive review of empirical models for estimating global solar radiation in Africa. *Renewable and Sustainable Energy Reviews, 78*(October), 955–995. https://doi.org/10.1016/j.rser.2017.04.101

United Nations. (2021). *The Energy transition, towards the achievement of SDG 7 and net-zero emissions*. UN Energy.

Index

S. C. Nwokolo et al., *Africa's Path to Net-Zero*, CSR, Sustainability, Ethics & Governance, https://doi.org/10.1007/978-3-031-44514-9